S0-CMT-051

PETROLEUM TRANSPORTATION AND PRODUCTION
OIL SPILL AND POLLUTION CONTROL

PETROLEUM TRANSPORTATION AND PRODUCTION Oil Spill and Pollution Control

Marshall Sittig

NOYES DATA CORPORATION
Park Ridge, New Jersey, U.S.A.
1978

Library of Congress Catalog Card Number: 78-54002
ISBN: 0-8155-0701-1
Printed in the United States

Published in the United States of America by
Noyes Data Corporation
Noyes Building, Park Ridge, New Jersey 07656

Foreword

Petroleum in its many forms has been on the move in the United States since 1859 when the first commercially successful oil well was developed in Pennsylvania. First transported by wagon and log raft, petroleum is now en route from far away oil fields or offshore drilling sites via ocean tankers, barges, pipelines, railroad tank cars and tank trucks.

While oil spills onto water and beaches catch the public eye and promote an understanding of what is needed for dealing with such events in the future, there are other wastes that originate in the production and refining processes which must conform to rigorous standards that become more stringent all the time.

The detailed descriptive information in this book is based on authoritative government reports, technical journals and U.S. patents issued since January 1974.

Because the information in this book is taken from many sources, it is possible that certain portions of this book may disagree or conflict with other parts of the book. This is especially true of monetary values and opinions of future potential. We chose to include these different points of view, however, in order to make the book more valuable to the reader. Cost figures provided are those given in the report cited, the date of which is always given. When the date of the cost figures themselves are given, we have included them.

As far as patents are concerned this book serves a double purpose in that it supplies detailed technical information and can be used as a guide to the U.S. patent literature in this field. By indicating all the information that is significant, and eliminating legal jargon and juristic phraseology, this book presents an advanced, commercially oriented review of the subjects indicated on the title page.

The U.S. patent literature is the largest and most comprehensive collection of technical information in the world. There is more practical, commercial, timely process information assembled here than is available from any other source. The technical information obtained from a patent is extremely reliable and comprehensive; sufficient information must be included to avoid rejection for "insufficient disclosure." These patents include practically all of those issued on the subject in the United States during the period under review.

The patent literature covers a substantial amount of information not available in the journal literature. The patent literature is a prime source of basic commercially useful information. This information is overlooked by those who rely primarily on the periodical journal literature. It is realized that there is a lag between a patent application on a new process development and the granting of a patent, but it is felt that this may roughly parallel or even anticipate the lag in putting that development into commercial practice.

Many of these patents are being utilized commercially. Whether used or not, they offer opportunities for technological transfer. Also, a major purpose of this book is to describe the number of technical possibilities available, which may open up profitable areas of research and development. The information contained in this book will allow you to establish a sound background before launching into research in this field.

Advanced composition and production methods developed by Noyes Data are employed to bring our new durably bound books to you in a minimum of time. Special techniques are used to close the gap between "manuscript" and "completed book." Industrial technology is progressing so rapidly that time-honored, conventional typesetting, binding and shipping methods are no longer suitable. We have bypassed the delays in the conventional book publishing cycle and provide the user with an effective and convenient means of reviewing up-to-date information in depth.

The table of contents is organized so it can serve as a subject index. The bibliography at the end of the volume constitutes a list of references giving further details on these timely topics. Other indexes by company, inventor and patent number help in providing easy access to the information contained in this book. There are more than 130 illustrations in this book, mainly of transfer devices and of equipment designed to prevent spilled oil from spreading on watery surfaces.

Some of the illustrations in this book may be less clear than could be desired; however, they are reproduced from the best material available to us.

Contents and Subject Index

INTRODUCTION ... 1
Scope of the Industry ... 1
Trends in the Industry ... 2

PETROLEUM PRODUCTION ... 3
Production Waste Characteristics ... 14
Air Pollution Control ... 20
Hydrogen Sulfide Control ... 20
Water Pollution Control ... 26
Regulatory Constraints ... 26
Leakage Control ... 32
Wastewater Purification ... 39

PETROLEUM STORAGE ... 65
Vapor Emissions ... 65
Floating Roof Tanks ... 66
Fixed Roof Tanks ... 67
Vapor Recovery Systems ... 67
Storage Tank Maintenance ... 68
Wastewater Characteristics ... 72
Design Trends ... 72

PETROLEUM TRANSPORTATION ... 73
Air Pollution Control ... 73
Vapor Control ... 74
Water Pollution Control ... 94
Oil Transfer Operations ... 94
Tank Cleaning ... 100
Ballast Disposal ... 104

Tanker Accidents 116
Dock Operations 132
Land Spill Prevention and Control 136
Pipeline Operations 138
Storage Tanks 138

OIL SPILLS ON WATER 139
Regulatory Constraints 143
Detection 144
Identification 170
Containment 171
Chemical Barriers 175
Floating Booms or Fences 178
Treatment 233
Dispersion 234
Sorption 236
Sinking 247
Gelling/Coagulation 248
Combustion 252
Magnetic Removal 252
Biodegradation 253
Removal from Open Water 257
Skimming Devices 257
Skimming Vehicles 285
Removal from Harbor Areas 302
Removal from Icy Waters 307
Removal from Streams 310
Spill Removal Equipment—Operation and Maintenance 314
Pumps 314
Temporary Oil/Water Storage 315
Shore Support Facilities 319
Cleaning and Restowage of Equipment 320
Manpower Requirements 327
Oil/Water Separators 328
Removal from Coasts and Beaches 340
Disposal of Recovered Spill Material 348

BIBLIOGRAPHY 352

COMPANY INDEX 353
U.S. PATENT NUMBER INDEX 355
INVENTOR INDEX 357

Introduction

This book is designed to give an up-to-date review of an increasingly vital topic—the control of oil spills and other pollution in the production, storage and transportation of petroleum products.

In the ever-widening search for petroleum resources and as it becomes necessary to go to the North Slope of Alaska and to new offshore locations off the heavily settled Eastern U.S., knowledge of pollution sources and their control becomes more and more important.

The general topic of pollution control in the petroleum industry has previously been reviewed by Jones (1). Subsequently, oil spill prevention and removal were treated in a volume by Sittig (2).

SCOPE OF THE INDUSTRY

There are approximately half a million producing oil wells and 126,000 gas and condensate wells in the United States. Of the 30,000 new wells drilled each year, about 55% produce oil or gas.

Oil is presently produced in 32 of the 50 states and from the Outer Continental Shelf (OCS) off Louisiana, Texas, and California. Exploratory drilling is underway on the OCS off Mississippi, Alabama, and Florida. In 1972, the five largest oil-producing states were: Texas, Louisiana, California, Oklahoma, and Wyoming. With development of the North Slope oil fields and construction of the Alaska pipeline, Alaska will become one of the most important oil-producing states.

Offshore oil production is presently concentrated in three areas in the United States: the Gulf of Mexico, the coast of California, and Cook Inlet in Alaska. Offshore oil production in 1973 was approximately 62 million barrels from Cook Inlet, 116 million from California, and 215 million from Louisiana and Texas (3).

TRENDS IN THE INDUSTRY

From 1960 to 1970, the nation's demand for energy increased at an average rate of 4.3%. Table 1 gives the projected national demands for oil and gas through 1985.

U.S. offshore production declined by about 78,500 bpd from 1972 to 1973. Offshore production amounts to approximately 10% of U.S. demand and about 15% of U.S. production.

While offshore production declined slightly from 1972 to 1973, the potential for increasing offshore production is much greater than for increasing onshore production. The Department of the Interior has proposed a schedule of three or four lease sales per year through 1978, mainly on remaining acreage in the Gulf of Mexico and offshore California. Additional areas in which OCS lease sales will very probably be held by 1978 include the Atlantic Coast (George's Bank, Baltimore Canyon, and Georgia Embayment) and the Gulf of Mexico.

Not only will new areas be opened to exploration and ultimate development, but production will move farther offshore and into deeper waters in areas of present development.

Movement into more distant and isolated environments will mean even more self-sufficiency of platform operations, with all production, processing, treatment, and disposal being performed on the platforms. Movement into deeper waters will necessitate multiple-well structures, with a maximum number of wells drilled from a minimum number of platforms.

Offshore leasing, exploration, and development will rapidly expand over the next 10 years, and offshore production will make up an increasing proportion of our domestically produced supplies of gas and oil.

TABLE 1: U.S. SUPPLY AND DEMAND OF PETROLEUM AND NATURAL GAS

	1971	1980	1985
Petroleum (10^6 bpd)			
Projected demand	15.1	20.8	25.0
Percent of total U.S. energy demand	44.1	43.9	43.5
Projected domestic supply	11.3	11.7	11.7
Percent petroleum demand fulfilled by domestic supply	74.9	56.3	46.7
Natural gas (10^{12} cfy)			
Projected demand	22.0	26.2	27.5
Percent of total U.S. energy demand	33.0	28.1	24.3
Projected domestic supply	21.1	23.0	23.8
Percent gas demand fulfilled by domestic supply	96.0	87.8	86.6

Source: Reference (3)

Petroleum Production

The majority of wells drilled by the petroleum industry are drilled to obtain access to reservoirs of oil or gas. A significant number, however, are drilled to gain knowledge of geologic formation. This latter class of wells may be shallow and drilled in the initial exploratory phase of operations, or may be deep exploration seeking to discover oil- or gas-bearing reservoirs.

Most wells are drilled today by rotary drilling methods. Basically the methods consist of:

1) Machinery to turn the bit, to add sections on the drill pipe as the hole deepens, and to remove the drill pipe and the bit from the hole.
2) A system for circulating a fluid down through the drill pipe and back up to the surface.

This fluid removes the particles cut by the bit, cools and lubricates the bit as it cuts, and, as the well deepens, controls any pressures that the bit may encounter in its passage through various formations. The fluid also stabilizes the walls of the well bore.

The drilling fluid system consists of tanks to formulate, store, and treat the fluids; pumps to force them through the drill pipe and back to the surface; and machinery to remove cuttings, fines, and gas from fluids returning to the surface (see Figure 1). A system of valves controls the flow of drilling fluids from the well when pressures are so great that they cannot be controlled by weight of the fluid column. A situation where drilling fluids are ejected from the well by subsurface pressures and the well flows uncontrolled is called a blowout, and the controlling valve system is called the blowout preventer (see Figure 2).

For offshore operations, drilling rigs may be mobile or stationary. Mobile rigs are used for both exploratory and development drilling, while stationary rigs are used for development drilling in a proven field. Some mobile rigs are mounted on barges and rest on the bottom for drilling in shallow waters.

Others, also mounted on barges are jacked up above the water on legs for drilling in deeper water (up to 300 feet). A third class of mobile rigs are on floating units for even deeper operations. A floating rig may be a vessel, with a typical ship's hull, or it may be semisubmersible—essentially a floating platform with special submerged hulls and supporting a rig well above the water. Stationary rigs are mounted on pile-supported platforms.

FIGURE 1: ROTARY DRILLING RIG

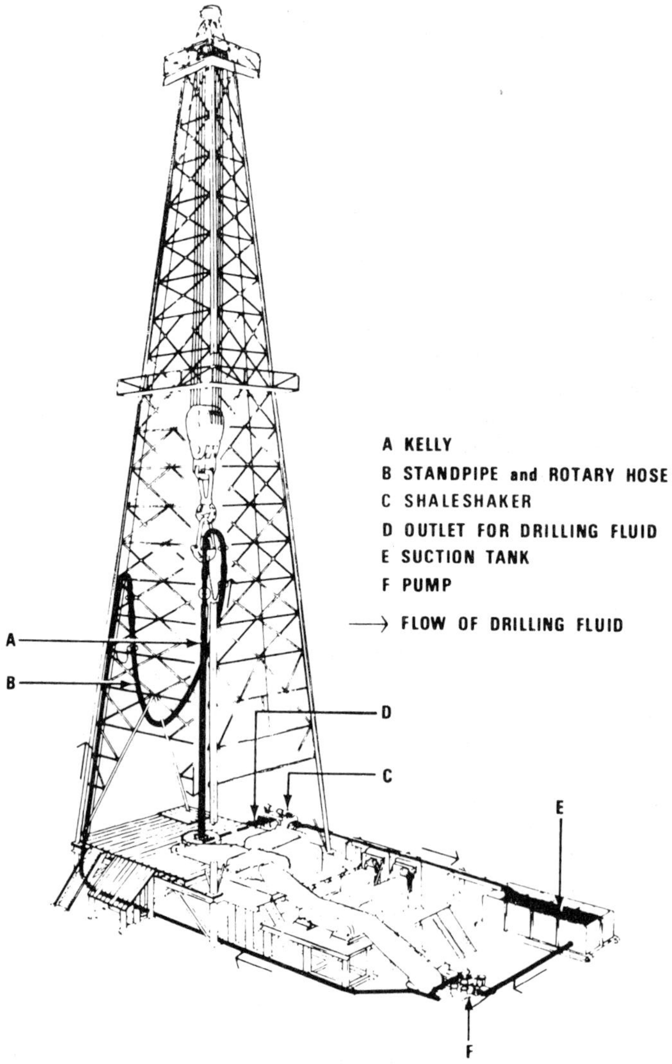

A KELLY
B STANDPIPE and ROTARY HOSE
C SHALESHAKER
D OUTLET FOR DRILLING FLUID
E SUCTION TANK
F PUMP
→ FLOW OF DRILLING FLUID

Source: Reference (3)

FIGURE 2: SHALESHAKER AND BLOWOUT PREVENTER

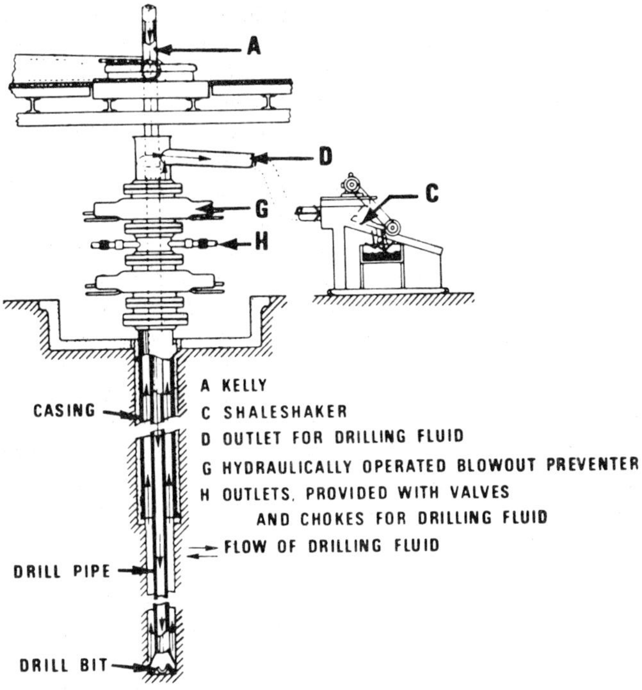

A KELLY
C SHALESHAKER
D OUTLET FOR DRILLING FLUID
G HYDRAULICALLY OPERATED BLOWOUT PREVENTER
H OUTLETS, PROVIDED WITH VALVES AND CHOKES FOR DRILLING FLUID
FLOW OF DRILLING FLUID
CASING
DRILL PIPE
DRILL BIT

Source: Reference (3)

Onshore drilling rigs used today are almost completely mobile. The derrick or mast and all drilling machinery are removed when the well is completed and used again in a new location.

Rigs used in marsh areas are usually barge mounted, and canals are dredged to the drill sites so that the rigs can be floated in.

The major source of pollution in the drilling system is the drilling fluid or "mud" and the cuttings from the bit. In early wells drilled by the rotary method, water was the drilling fluid. The water mixed with the naturally occurring soils and clays and made up the mud. The different characteristics and superior performance of some of these natural muds were evident to drillers, which led to deliberately formulated muds. The composition of modern drilling muds is quite complex and can vary widely, not only from one geographical area to another, but also in different portions of the same well.

The drilling of a well from top to bottom is not a continuous process. A well is drilled in sections, and as each section is completed it is lined with a section of pipe or casing (see Figure 2). The different sections may require different types of mud. The mud from the previous section must either be disposed of or converted for the next section. Some mud is left in the completed well.

Basic mud components include: bentonite or attapulgite clays to increase viscosity and create a gel; barium sulfate (barite), a weighting agent; and lime and caustic soda to increase the pH and control viscosity. (Additional conditioning constituents may consist of polymers, starches, lignitic material, and various other chemicals.) Most muds have a water base, but some have an oil base. Oil-based muds are used in special situations and present a much higher potential for pollution. They are generally used where bottom hole temperatures are very high or where water-based muds would hydrate water-sensitive clays or shales. They may also be used to free stuck drill pipes, to drill in permafrost areas, and to kill producing wells.

As the drilling mud is circulated down the drill pipe, around the bit, and back up in annulus between the bore hole and the drill pipe, it brings with it the material cut and loosened by the bit, plus fluids which may enter the hole from the formation (water, oil, or gas). When the mud arrives at the surface, cuttings, silt, and sand are removed by shaleshakers, desilters, and desanders. Oil or gas from the formation is also removed, and the cleansed mud is cycled through the drilling system again. With offshore wells, the cuttings, silt and sand are discharged overboard if they do not contain oil. Some drilling mud clings to the sand and cuttings, and when this material reaches the water the heavier particles (cuttings and sand) sink to the bottom while the mud and fines are swept down current away from the platform.

Onshore, discharges from the shaleshakers and cyclone separators (desanders or desilters) usually go to an earthen (slush) pit adjacent to the rig. To dispose of this material the pit is backfilled at the end of the drilling operations.

The removal of fines and cuttings is one of a number of steps in a continuing process of mud treatment and conditioning. This processing may be done to keep the mud characteristics constant or to change them as required by the drilling conditions. Many constituents of the drilling mud can be salvaged when the drilling is completed, and salvage plants may exist either at the rig or at another location, normally at the industrial facility that supplies mud or mud components.

Where drilling is more or less continuous, such as on a multiple-well offshore platform, the disposal of mud should not be a frequent occurrence since it can be conditioned and recycled from one well to another.

The drilling of deeper, hotter holes may increase use of oil-based mud. However, new mud additives may permit use of water-based muds where only oil muds would have served before. Oil muds always present disposal problems.

Crude oil, natural gas, and gas liquids are normally produced from geological reservoirs through a deep bore well into the surface of the earth. The fluid produced from oil reservoirs normally consists of oil, natural gas, and salt water or brine containing both dissolved and suspended solids. Gas wells may produce dry gas but usually also produce varying quantities of light hydrocarbon liquids (known as gas liquids or condensate) and salt water.

As in the case of oil field brines, the water contains dissolved and suspended solids and hydrocarbon contaminants. The suspended solids are normally sands, clays, or other fines from the reservoir. The oil can vary widely in its physical and chemical properties. The most important properties are its density and viscosity. Density is usually measured by the "API Gravity" method which assigns a number to the oil based on its specific gravity. The oil can range from very light gasoline-like materials (called natural gasolines) to heavy, viscous asphalt-like materials.

The fluids are normally moved through tubing contained within the larger cased bore hole. For oil wells, the energy required to lift the fluids up the well can be supplied by the natural pressures in the formation, or it can be provided or assisted by various man-made operations at the surface. The most common methods of supplying man-made energy to extract the oil are: to inject fluids (normally water or gas) into the reservor to maintain pressure, which would otherwise drop during withdrawal; to force gas into the well stream in order to lighten the column of fluid in the bore and assist in lifting as the gas expands up the well; and to employ various types of pumps in the well itself.

As the fluids rise in the well to the surface, they flow through various valves and flow control devices which make up the wellhead. One of these is an orifice (choke) which maintains required back pressure on the well and controls, by throttling the fluids, the rate at which the well can flow. In some cases, the choke is placed in the bottom of the well rather than at the wellhead.

Once at the surface, the various constituents in the fluids produced by oil and gas wells are separated: gas from the liquids, oil from water, and solids from liquids (see Figure 3). The marketable constituents, normally the gas and oil, are then removed from the production area, and the wastes, normally the brine and solids, are disposed of after further treatment. At this stage, the gas may still contain significant amounts of hydrocarbon liquids and may be further processed to separate the two.

The gas, oil, and water may be separated in a single vessel or, more commonly, in several stages. Some gas is dissolved in the oil and comes out of solution as the pressure on the fluid drops. Fluids from high-pressure reservoirs may have to be passed through a number of separating stages at successively lower pressures before the oil is free of gas. The oil and brine do not separate as readily as the gas does. Usually, a quantity of oil and water is present as an emulsion. This emulsion can occur naturally in the reservoir or can be caused by various processes which tend to mix the oil and water vigorously together and cause droplets to form.

Passage of the fluids into and up the well tends to mix them. Passage through wellhead chokes, through various pipes, headers, and control valves into separation chambers, and through any centrifugal pumps in the system, tends to increase emulsification. Moderate heat, chemical action, and/or electrical charges tend to cause the emulsified liquids to separate or coalesce, as does the passage of time in a quiet environment. Other types of chemicals and fine suspended solids tend to retard coalescence. The characteristics of the crude oil also affect the ease or difficulty of achieving process separation.

Fluids produced by oil and gas wells are usually introduced into a series of vessels for a two-stage separation process. Figure 4 shows a gas separator for separating gas from the well stream. Liquids (oil or oil and water) along with particulate matter leave the separator through the dump valve and go on to

FIGURE 3: CENTRAL TREATMENT FACILITY IN ESTUARINE AREA

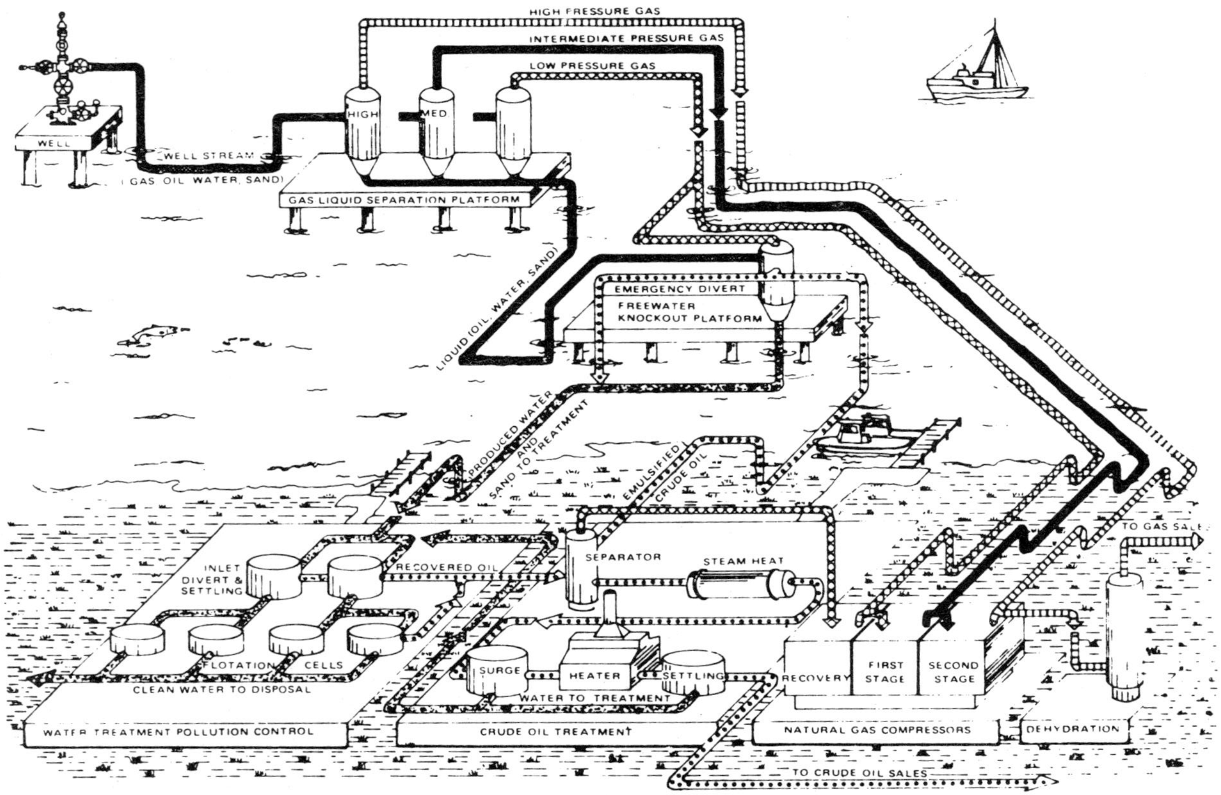

Source: Reference (3)

the next stage: oil-water separation. Because gas comes out of solution as pressure drops, gas-oil separators are often arranged in series. High-pressure, intermediate, and low-pressure separators are the most common arrangement, with the high-pressure liquids passing through each stage in series and gas being taken off at each stage.

Fluids from lower-pressure wells would go directly to the most appropriate separator. The liquids are then piped to vessels for separating the oil from the produced water. Water which is not emulsified and separates easily may be removed in a simple separation vessel called a free water knockout.

FIGURE 4: HORIZONTAL GAS SEPARATOR

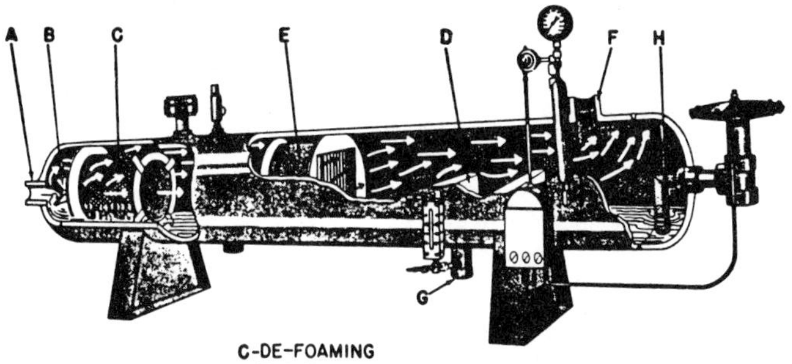

A-OIL AND GAS INLET
B-IMPACT ANGLE
C-DE-FOAMING ELEMENT
D-WAVE BREAKER AND SELECTOR PLATE
E-MIST EXTRACTOR
F-GAS OUTLET
G-DRAIN
H-OIL OUTLET (DUMP VALVE)

Source: Reference (3)

The remaining oil-water mixture will continue to another vessel for more elaborate treatment (see Figure 5). In this vessel (which may be called a heater-treater, electric dehydrator, gun barrel, or wash tank, depending on configuration and the separation method employed), there is a relatively pure layer of oil on the top, relatively pure brine on the bottom, and a layer of emulsified oil and brine in the middle. There is usually a sensing unit to detect the oil-water interface in the vessel and regulate the discharge of the fluids. Emulsion breaking chemicals are often added before the liquid enters this vessel, the vessel itself is often heated to facilitate breaking the emulsion, and some units employ an electrical grid to charge the liquid and to help break the emulsion. A combination of treatment methods is often employed in a single vessel. In three-phase separation, gas, oil, and water are all separated in one unit. The gas-oil and oil-water interfaces are detected and used to control rates of influent and discharge.

Oil from the oil-water separators is usually sufficiently free of water and sediment (less than 2%) so as to be marketable. The produced water or produced water/solids mixtures discharged at this point contain too much oil to be disposed of into a water body. The object of processing through this point is to produce marketable products (clean oil and dry gas). In contrast, the next stages of treatment are necessary to remove sufficient oil from the produced water so that it may be discharged. These treatment operations do not significantly increase the quality or quantity of the saleable product. They do decrease the impact of these wastes on the environment.

Typical produced water from the last stage of processing would contain several hundred to perhaps a thousand or more parts per million of oil. There are two methods of disposal: treatment and discharge to surface (salt) waters or injection into a suitable subsurface formation in the earth. Surface discharge is normally used offshore or near shore where bodies of salt or brackish water are available for disposal. Injection is widely used onshore where bodies of salt water are not available for surface disposal. (Produced water to be disposed of by injection may still require some treatment.)

FIGURE 5: VERTICAL HEATER-TREATER

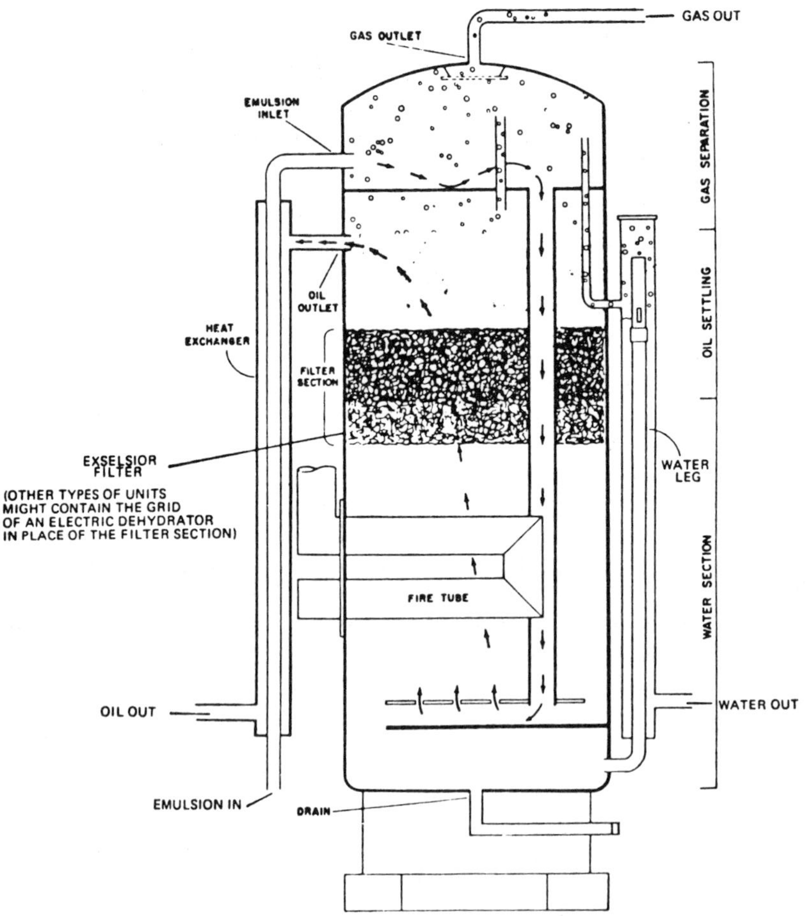

Source: Reference (3)

Some of the same operations used to facilitate separation in the last stage of processing (chemical addition and retention tanks) may be used in wastewater treatment, and other methods such as filtering, and separation by gas flotation are also used. In addition, combinations of these operations can be ued to advantage to treat the wastewater. The vast majority of present offshore and near shore (marsh) facilities in the Gulf of Mexico and most facilities in Cook Inlet, Alaska, treat and dispose of their produced water to surface salt or brackish water bodies.

The sophistication of the treatment employed by dischargers of produced water is dependent upon the regulation governing such discharges. For instance in the Appalachian states most produced water is discharged to local streams after only treatment in ponds; while in California dischargers utilize a high degree of treatment. The state of Wyoming allows discharge for beneficial use if the produced water meets oil and grease and total dissolved solids (TDS) requirements.

Several options are available in injection systems. Often water will be injected into a producing oil reservoir to maintain reservoir pressure, and stabilize reservoir conditions. In a similar operation called waterflooding, water is injected into the reservoir in such a way as to move oil to the producing wells and increase ultimate recovery. This process is one of several known as secondary recovery since it produces oil beyond that available by primary production methods.

A successful waterflooding project will increase the amount of oil being produced at a field. It will also increase produced water volume and thus affect the amount of water that must be treated. Pressure maintenance of water injection may also increase the amount of water produced and treated. Injection is also feasible solely as a disposal method. It (injection) is extensively used in onshore production areas except in the Appalachian states of Pennsylvania, West Virginia, New York and Kentucky, where useable shallow horizons do not exist. In California, produced water from offshore facilities is transported to shore for disposal by reinjection.

The treatment associated with produced water disposal by injection is dependent upon the permeability of the receiving formation. In most all cases corrosion-inhibiting chemicals are necessary, but the treatment can range from simple skim tanks to gas flotation followed by mixed-media filtration.

Early offshore development tended to place wells on individual structures, bringing the fluids ashore for separation and treatment (see Figure 3). As the industry moved farther offshore, the wells still tended to be located on individual platforms with the output to a central platform for separation, treatment, and discharge to a pipeline or barge transportation system.

With increasing water depth, multiple-well platforms were developed with 20 or more wells drilled directionally from a single platform. Thus an entire field or a large portion of a field could be developed from one structure. Offshore Louisiana multiple-well platforms include all processing and treatment; in offshore California and in Cook Inlet facilities, gas separation takes place on the platforms, with the liquids usually sent ashore for separation and treatment.

All forms of primary and secondary recovery as well as separation and treatment are performed on platforms, which may include compressor stations for gas lift wells and sophisticated water treatment facilities for water flood projects. Platforms far removed from shore are practically independent production units.

Platform design reflects the operating environment. Cook Inlet platforms are enclosed for protection from the elements and have a structural support system designed to withstand ice flows and earthquakes. Gulf Coast platforms are usually open, reflecting a mild climate. Support systems are designed to withstand hurricane-generated waves.

A typical onshore production facility would consist of wells and flowlines, gas-liquid and oil-water production separators, a wastewater treatment unit (the level of treatment being dependent on the quality of the wastewater and the demands of the injection system and receiving reservoir), surge tank, and injection well. Injection might either be for pressure maintenance and secondary recovery or solely for disposal. In the latter case, the well would probably be shallow and operate at lower pressure. The system might include a pit to hold wastewater should the injection system shut down.

A more recent production technique and one which may become a significant source of waste in the future is called "tertiary recovery." The process usually involves injecting some substance into the oil reservoir to release or carry out additional oil not recovered by primary recovery (flowing wells by natural reservoir pressure, pumping, or gas lift) or by secondary recovery.

Tertiary recovery is usually classified by the substance injected into the reservoir and includes:

1) Thermal recovery
2) Miscible hydrocarbon
3) Carbon dioxide
4) Alcohols, soluble oil, micellar solutions
5) Chemical floods, surfactants
6) Gas, gas/water, inert gas
7) Gas repressuring, depletion
8) Polymers
9) Foams, emulsions, precipitates

The material is injected into the reservoir and moves through the reservoir to the producing wells. During this passage, it removes and carries with it oil remaining in pores in the reservoir rocks or sands. Oil, the injected fluid, and water may all be moved up the well and through the normal production and treatment system.

At this time very little is known about the wastes that will be produced by these production processes. They will obviously depend on the type of tertiary recovery used.

A number of satellite industries specialize in providing certain services to the production side of the oil industry. Some of these service industries produce a particular class of waste that can be identified with the service they provide. Of the waste-producing service industries, drilling (which is usually done by a contractor) is the largest. Drilling fluids and their disposal have already been discussed. Other services include completions, workovers, well acidizing, and well fracturing.

When a company decides that an oil or gas well is a commercial producer, certain equipment will be installed in the well and on the wellhead to bring the well into production. The equipment from this process—called "completion"—normally consists of various valves and sealing devices installed on one or more strings of tubing in the well. If the well will not produce sufficient fluid by natural flow, various types of pumps or gas lift systems may be installed in the well. Since heavy weights and high lifts are normally involved, a rig is usually used. The rig may be the same one that drilled the well, or it may be a special (normally smaller) workover rig installed over the well after the drilling rig has been moved.

After a well has been in service for a while it may need remedial work to keep it producing at an acceptable rate. For example, equipment in the well may malfunction, different equipment may be required, or the tubing may become plugged up by deposits of paraffin. If it is necessary to remove and reinstall the tubing in the well, a workover rig will be used. It may be possible to accomplish the necessary work with tools mounted on a wire and lowered into the well through the tubing. This is called a wire line operation. In another system, tools may be forced into the well by pumping them down with fluid. Where possible, the use of a rig is avoided, since it is expensive.

In many wells, the potential for production is limited by impermeability in the producing geological formation. This condition may exist when the well is first drilled, it may worsen with the passage of time, or both situations may occur. Several methods may be used, singly or in combination, to increase the well flow by altering the physical nature of the reservoir rock or sand in the immediate vicinity of the well.

The two most common methods to increase well flow are acidizing and fracturing. Acidizing consists of introducing acid under pressure through the well and into the producing formation. The acid reacts with the reservoir material, producing flow channels which allow a larger volume of fluids to enter the well. In addition to the acid, corrosion inhibitors are usually added to protect the metal in the well system. Wetting agents, solvents, and other chemicals may also be used in the treatment.

In fracturing, hydraulic pressure forces a fluid into the reservoir, producing fractures, cracks, and channels. Fracturing fluids may contain acids so that chemical disintegration, as well as fracturing takes place. The fluids also contain sand or some similar material that keeps the fracture propped open once the pressure is released.

When a new well is being completed or when it is necessary to pull tubing to work over a well, the well is normally "killed"—that is, a column of drilling mud, oil, water, or other liquid of sufficient weight is introduced into the well to control the down hole pressures.

When the work is completed, the liquid used to kill the well must be removed so that the well will flow again. If mud is used, the initial flow of oil from the well will be contaminated with the mud and must be disposed of. Offshore, it may be disposed of into the sea if it is not oil contaminated, or it may be salvaged. Onshore, the mud may be disposed of in pits or may be salvaged. Contaminated oil is usually disposed of by burning at the site.

In acidizing and fracturing, the spent fluids used are wastes. They are moved through the production, process, and treatment systems after the well begins to flow again. Therefore, initial production from the well will contain some of these fluids. Offshore, contaminated oil and other liquids are barged ashore for treatment and disposal; contaminated solids are buried.

The fines and chemicals contained in oil from wells put on stream after acidizing or fracturing have seriously upset the wastewater treatment units of production facilities. When the sources of these upsets have been identified, corrective measures can prevent or mitigate the effects.

PRODUCTION WASTE CHARACTERISTICS

Wastes generated by the oil and gas industry are produced by drilling exploratory or development wells, by the production or extraction phase of the industry, and, in the case of offshore facilities, sanitary wastes generated by personnel occupying the platforms. Drilling wastes are generally in the form of drill cuttings and mud, and production wastes are generally produced water. Additionally, well workover and completion operations can produce wastes, but they are generally similar to those from drilling or production operations (3).

Approximately half a million producing oil wells onshore generate produced water in excess of 20 million bpd of which it is estimated 50% is reinjected for recovery purposes. Approximately 17,000 wells have been drilled offshore in U.S. waters, and approximately 11,000 are producing oil or gas. The offshore Louisiana OCS alone produces approximately 410,000 barrels of water per day; by 1983, coastal Louisiana production will generate an estimated 1.54 million barrels of water per day.

This section characterizes the types of wastes that are produced at offshore and onshore wells and structures. The discussion of drilling wastes can be applied to any area of the United States since these wastes do not change significantly with locality.

Other than oils, the primary waste constituents considered are oxygen-demanding pollutants, heavy metals, toxicants, and dissolved solids contained in drilling muds or produced water.

Sanitary wastes are also produced during both drilling and production operations both onshore and offshore, but they are discussed only for offshore situations where sanitary wastes are produced from fixed platforms or structures. Drilling or exploratory rigs that are vessels are not part of this discussion.

Production wastes include produced waters associated with the extracted oil, sand and other solids removed from the produced waters, deck drainage from the platform surfaces, sanitary wastes, and domestic wastes.

The produced waters from production platforms generate the greatest concern. The wastes can contain oils, toxic metals, and a variety of salts, solids and organic chemicals. The concentrations of the constituents vary somewhat from one geographical area to another, with the most pronounced variance in chloride levels. Table 2 shows the waste constituents in offshore Louisiana production facilities in the Gulf of Mexico. The data were obtained during the verification survey conducted by EPA in 1974. The only influent data obtained in the survey were on oil and grease.

In planning the verification survey, it was decided that offshore produced water treatment facilities would have virtually no effect on metals and salinity levels in the influent, and that these constituents could be satisfactorily characterized by analyzing only the effluent.

Total organic carbon (TOC) is also tabulated under effluent in Table 2, but it is reasonable to assume that actual analysis of the influent would be higher. Since TOC is a measurement of all organic carbon in the sample and oil is a major source of organic carbon, it is logical to assume removal of some organic carbon when oil is removed in the treatment process. Suspended solids are also expressed as effluent data, and this parameter would be expected to be reduced by the treatment process.

TABLE 2: POLLUTANTS IN PRODUCED WATER LOUISIANA COASTAL*

Pollutant Parameter	Range, mg/l	Average, mg/l
Oil and grease	7-1,300	202
Cadmium	<0.005-0.675	<0.068
Cyanide	<0.01-0.01	<0.01
Mercury	–	<0.0005
Total organic carbon	30-1,580	413
Total suspended solids	22-390	73
Total dissolved solids	32,000-202,000	110,000
Chlorides	10,000-115,000	61,000
Flow, bpd	250-200,000	15,000

*Results of 1974 EPA survey of 25 discharges.

Source: Reference (3)

Industry data for offshore California describes a broader range of parameters (see Table 3). Similar data were provided for offshore Texas (see Table 4). Except as noted in the tables, all data are from effluents.

Sand and other solids are produced along with the produced water. Observations made by EPA personnel during field surveys indicated that drums of these sands stored on the platform had a high oil content. Sand has been reported to be produced at approximately 1 barrel sand per 2,000 barrels oil.

As part of a recent EPA study (1976) to collect information on treatment technologies and costs, surveys were made of onshore production facilities in California, Wyoming, Texas, Louisiana and Pennsylvania. The data represented in Tables 5 through 9 are from the effluent of the treatment facilities

TABLE 3: POLLUTANTS CONTAINED IN PRODUCED WATER COASTAL CALIFORNIA*

Pollutant Parameter	Range, mg/l
Arsenic	0.001-0.08
Cadmium	0.02-0.18
Total chromium	0.02-0.04
Copper	0.05-0.116
Lead	0.0-0.28
Mercury	0.0005-0.002
Nickel	0.100-0.29
Silver	0.03
Zinc	0.05-3.2
Cyanide	0.0-0.004
Phenolic compounds	0.35-2.10
BOD	370-1,920
COD	400-3,000
Chlorides	17,230-21,000
TDS	21,700-40,400
Suspended solids	
Effluent	1-60
Influent	30-75
Oil and grease	56-359

*Some data reflect treated waters for reinjection.

Source: Reference (3)

TABLE 4: RANGE OF CONSTITUENTS IN PRODUCED FORMATION WATER—OFFSHORE TEXAS

Pollutant Parameter	Range, mg/l
Arsenic	<0.01-<0.02
Cadmium	<0.02-0.193
Total chromium	<0.10-0.23
Copper	<0.10-0.38
Lead	<0.01-0.22
Mercury	<0.001-0.13
Nickel	<0.10-0.44
Silver	<0.01-0.10
Zinc	0.10-0.27
Phenolic compounds	53
BOD	126-342
COD	182-582
Chlorides	42,000-62,000
TDS	806-169,000
Suspended solids	12-656

Source: Reference (3)

prior to reinjection for secondary recovery or disposal. It could be expected that the quality of the untreated produced water from the production separator would range from 200–1,000 mg/l oil and grease and 100–400 mg/l suspended solids. The remainder of the analyzed constituents such as TDS, phenols and heavy metals would be unaffected by treatment.

Prior to the utilization of the Freon extraction method for oil and grease, the samples were screened for organic acids and if they were present in quantities greater than 100 mg/l the sample was not acidified. Therefore, the results for oil and grease as reported in Tables 5 through 9, particularly in California where organic acids are known to be a part of the crude oil, are not comparable to data in other parts of this report and are shown only for information.

TABLE 5: RANGE OF CONSTITUENTS IN PRODUCED FORMATION WATER—ONSHORE CALIFORNIA

Pollutant Parameter	Range, mg/l	Median, mg/l
Oil and grease	16-191	75
Suspended solids	3-51	31
Total dissolved solids	580-27,300	6,300
Phenol	0.07-0.15	0.11
Arsenic	<0.01-0.03	0.11
Chromium	<0.01	<0.01
Cadmium	<0.005-0.02	<0.005
Lead	<0.05	<0.05
Barium	<0.2-0.4	0.3

Source: Reference (3)

TABLE 6: RANGE OF CONSTITUENTS IN PRODUCED FORMATION WATER—WYOMING

Pollutant Parameter	Range, mg/l	Median, mg/l
Oil and grease	1.5-205	67
Suspended solids	<1-64	12.8
Total dissolved solids	345-90,400	13,800
Phenol	0.07-0.33	0.16
Arsenic	<0.01-0.06	0.01
Chromium	<0.01	<0.01
Cadmium	<0.005-0.023	<0.005
Lead	<0.05-0.08	<0.05
Barium	<0.2-9.7	0.9

Source: Reference (3)

TABLE 7: RANGE OF CONSTITUENTS IN PRODUCED FORMATION WATER—PENNSYLVANIA

Pollutant Parameter	Range, mg/l	Median, mg/l
Oil and grease	<0.2-114	25
Suspended solids	1.4-666	107
Total dissolved solids	1,500-109,400	29,000
Phenol	0.06-0.35	0.19
Arsenic	<0.01	<0.01
Chromium	<0.01-0.025	<0.01
Cadmium	<0.005-0.013	<0.005
Lead	<0.05-0.50	<0.05
Barium	0.1-36	8.6

Source: Reference (3)

TABLE 8: RANGE OF CONSTITUENTS IN PRODUCED FORMATION WATER—ONSHORE LOUISIANA

Pollutant Parameter	Range, mg/l	Median, mg/l
Oil and grease	16-441	165
Suspended solids	20.8-155	82
Total dissolved solids	42,600-132,000	73,900

Source: Reference (3)

TABLE 9: RANGE OF CONSTITUENTS IN PRODUCED FORMATION WATER—ONSHORE TEXAS

Pollutant Parameter	Range, mg/l	Median, mg/l
Oil and grease	57-1,200	460
Suspended solids	30-473	143
Total dissolved solids	42,600-132,000	94,000

Source: Reference (3)

Drill cuttings are composed of the rock, fines, and liquids contained in the geologic formations that have been drilled through. The exact makeup of the cuttings varies from one drilling location to another, and no attempt has been made to qualitatively identify cuttings.

The two basic classes of drilling muds used today are water-based muds and oil-muds. In general, much of the mud introduced into the well hole is eventually displaced out of the hole and requires disposal or recovery.

Water-based muds are formulated using naturally occurring clays such as bentonite and attapulgite and a variety of organic and inorganic additives to achieve the desired consistency, lubricity, or density. Fresh or salt water is the liquid phase for these muds. The additives are used for such functions as pH control, corrosion inhibition, lubrication, weighting, and emulsification.

The additives that should be scrutinized for pollution control are ferrochrome lignosulfonate and lead compounds. Ferrochrome lignosulfonate contains 2.6% iron, 5.5% sulfur, and 3.0% chromium. In an example presented by the Bureau of Land Management in an Environmental Impact Statement for offshore development, the drilling operation of a typical 10,000-foot development well (not exploratory) used 32,900 pounds of ferrochrome lignosulfonate mud which contained 987 pounds of chromium. Table 10 presents the volumes of cuttings and muds used in the Bureau's example of a "typical" 10,000-foot drilling operation. The amount of lead additives used in mud composition varies from well to well, and no examples are available.

TABLE 10: VOLUME OF CUTTINGS AND MUDS IN TYPICAL 10,000-FOOT DRILLING OPERATION

Interval (ft)	Hole Size (in)	Volume of Cuttings (bbl)	Weight of Cuttings (lb)	Drilling Mud	Volume of Mud Components (bbl)	Weight of Mud Components (lb)
0-1,000	24	562	505,000	seawater and natural mud	variable	—
1,000-3,500	16	623	545,000	gelled seawater	700	81,500
3,500-10,000	12	915	790,000	lime base	950	424,000

Source: Reference (3)

Drilling constituents for onshore operations will parallel those for offshore, except for the water used in the typical mud formulation. Onshore drilling operations normally use a fresh water-based mud, except where drilling operations encounter large salt domes. Then the mud system would be converted either to a salt clay mud system with salt added to the water phase, or to an oil-based mud system. This change in the liquid phase is intended to prevent dissolving salt in the dome, enlarging the hole, and causing solution cavities in the formation.

In offshore operations, the direct discharge of cuttings and water-based muds creates turbidity. Limited information is available to accurately define the degree of turbidity, or the area or volume of water affected by such turbid discharges, but experienced observers have described the existence of substantial plumes of turbidity when muds and cuttings are discharged.

Oil-based muds contain carefully formulated mixtures of oxidized asphalt, organic acids, alkali, stabilizing agents and high-flash diesel oil. The oils are the principal ingredients and so are the liquid phase. Muds displaced from the well hole also contain solids from the hole. There are two types of emulsified oil muds: 1) oil emulsion muds, which are oil-in-water emulsions;

and 2) inverted emulsion muds, which are water-in-oil emulsions. The principal differences between these two muds and oil-based muds is the addition of fresh or salt water into the mud mixture to provide some of the volume for the liquid phase. Newer formulations can contain from 20 to 70% water by volume. The water is added by adding emulsifying and stabilizing agents. Clay solids and weighting agents can also be added.

The sanitary wastes from offshore oil and gas facilities are composed of human body waste and domestic waste such as kitchen and general housekeeping wastes. The volume and concentration of these wastes vary widely with time, occupancy, platform characteristics, and operational situation. Usually the toilets are flushed with brackish water or seawater. Due to the compact nature of the facilities the wastes have less dilution water than common municipal wastes. This results in greater waste concentrations. Table 11 indicates typical waste flow for offshore facilities and vessels.

TABLE 11: TYPICAL RAW COMBINED SANITARY AND DOMESTIC WASTES FROM OFFSHORE FACILITIES

No. of Men	Flow (gpd)	. . BOD_5, mg/l. . .		Suspended Solids, ... mg/l.		Total Coliform (x10)
		Average	Range	Average	Range	
76	5,500	460	270-770	195	14-543	10-180
66	1,060	875	–	1,025	–	–
67	1,875	460	–	620	–	–
42	2,155	225	–	220	–	–
10-40	2,900	920	–	–	–	–

Source: Reference (3)

AIR POLLUTION CONTROL

Air pollution is not generally considered as much of a problem for production operations, many of which occur at remote locations.

Hydrogen Sulfide Control

During the drilling of an oil well, the drilling mud that is circulated through the borehole frequently becomes contaminated with gases encountered in the formations. Since it is economically unfeasible to discard the contaminated drilling mud and because of the danger of the gases in the drilling mud escaping into the atmosphere creating dangerous conditions at the drill site, it is necessary to process the drilling mud to remove the gases and recirculate the degassed drilling mud through the borehole. The contaminating gases may be poisonous or highly explosive. The release of such gases into the atmosphere would present a substantial risk to personnel in the drilling area. The presence of gases in the drilling mud decreases its weight and viscosity and renders it unsuitable for recirculation through the borehole. When gases are contained in the drilling mud being circulated through the borehole, it increases the danger of a blowout in the well.

A "Notice to Lessees and Operators of Federal Oil and Gas Leases in the Outer Continental Shelf, Gulf of Mexico Area" was released May 7, 1974 by the United States Department of the Interior Geological Survey Gulf of Mexico area, relating to hydrogen sulfide in drilling operations. The notice outlines requirements for drilling operations when there is a possibility or probability of penetrating reservoirs known or expected to contain hydrogen sulfide. Section 3, f. provides that "drilling mud containing H_2 gas shall be degassed at the optimum location for the particular rig configuration employed. The gases so removed shall be piped into a closed flare system and burned at a suitable remote stack."

It is desirable that any system for removing gas from drilling mud without allowing the gas to escape to the atmosphere contain alarms that will be actuated should conditions exceed a predetermined level. The actuation of the alarms will allow proper procedures to be taken to correct the problem. The system must be totally enclosed and of a fail-safe design. The system must start at the wellhead and contain the gas within the system until it is treated at a flare or disposed of by other means. In order for the degassing system to operate effectively, it is essential that a partial vacuum be maintained in the degassing vessel at all times. In addition, controlled flow rates should be provided to the degassing vessel to insure efficient operation of the degassing system.

A process developed by *P.H. Griffin, III, W.A. Rehm, M.L. Talley, M.J. Sharki and W.E. Renfro; U.S. Patent 4,010,012; March 1, 1977; assigned to Dresser Industries, Inc.* is one for removing gas from drilling mud without allowing the gas to escape to the atmosphere before the gas is treated to prevent contamination. The system includes a mud-gas separator vessel with a drilling mud inlet, means for maintaining a minimum mud level in the mud-gas separator vessel, a separator vessel gas outlet, and a separator vessel drilling mud outlet. A degassing means includes a drilling mud inlet located in a position higher than the minimum mud level. Enclosed conduit means connects the separator vessel drilling mud outlet to the degassing means drilling mud inlet.

The apparatus for the conduct of such a process is shown in Figure 6. Drilling mud is directed to a mud-gas separator vessel **3** through line **1**. The drilling mud enters the mud-gas separator vessel at drilling mud inlet **2**. The mud-gas separator vessel **3** acts as a surge tank and gas knockout. Free gas escaping from the drilling mud is exhausted from the mud-gas separator vessel **3** to a flare line **5** through a gas outlet **4**. The flare line carries the gas to a remote point where it is burned or treated in some fashion to remove any objectionable components.

A minimum mud level **6** is maintained in the mud-gas separator vessel. The minimum mud level **6** is maintained by a valve **13** and connecting linkage from valve **13** that extends through a stuffing box **9** to a float rod **8** and float **7**. The float **7** is connected to the stuffing box by the float rod **8**. The mud-gas separator vessel **3** has a mud outlet **11** and a mud outlet line **12** extending from mud outlet **11**. The valve **13** is located in the mud outlet line **12**. The connecting linkage connects valve **13** to rod **8** through the stuffing box. Changes in the mud level **6** cause the float **7** and rod **8** to move, thereby moving the connecting linkage and adjusting valve **13** to either increase or decrease drilling mud flow as necessary to maintain the minimum mud level **6**.

FIGURE 6: DRILLING MUD DEGASSING AND GAS CONTAINMENT SYSTEM

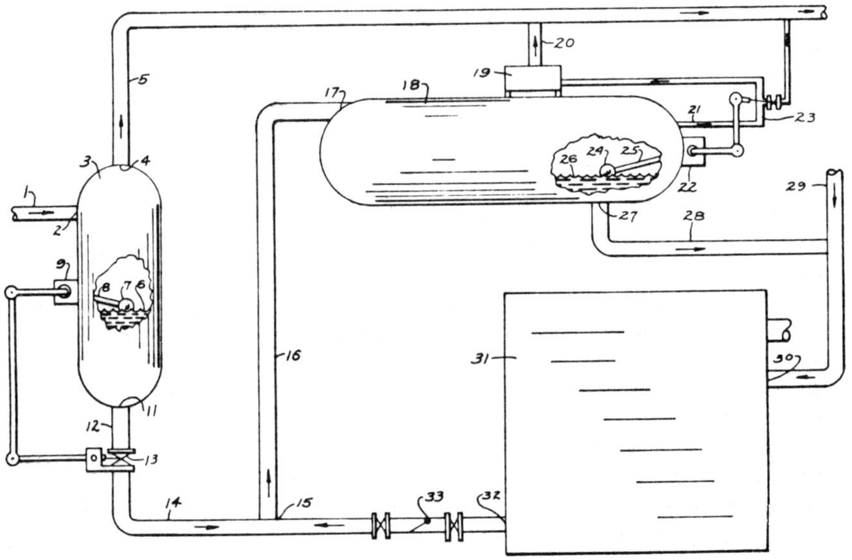

Source: U.S. Patent 4,010,012

Drilling mud passing through the valve enters line **14** and is transmitted to a T-junction **15** where it enters suction line **16** leading to a degassing vessel **18**. The drilling mud enters the degassing vessel through a drilling mud inlet **17**. The drilling mud inlet is located at a higher level than the minimum level in the mud-gas separator vessel. This insures that a partial vacuum in the degassing vessel suction line will be required to create flow of drilling mud from the mud-gas separator vessel through drilling mud inlet into the degassing vessel.

A vacuum pump unit **19** is positioned on or near the upper surface of the degassing vessel for maintaining a partial vacuum in the degassing vessel. Gas is exhausted from the degassing vessel by pump unit **19** through exhaust line **21** from the degassing vessel to the pump unit and through exhaust line **20** extending from the pump unit to the flare line **5**. A near maximum mud level **26** is maintained in the degassing vessel by a valve **23** that controls the vacuum applied to the degassing vessel. A mechanical linkage extends from valve **23** to a float rod **25** and float **24** through a stuffing box **22**.

Changes in the mud level cause the float and rod to move, thereby moving the mechanical linkage and adjusting valve **23** to either increase or decrease the vacuum applied to the degassing vessel as necessary to maintain the near maximum mud level **26**. A line from the flare line to valve **23** allows exhaust gas to be admitted into vacuum unit **19** and into the degassing vessel when the vacuum applied to the degassing vessel is to be decreased. This also will insure proper operation of vacuum pump unit **19**.

Degassed mud is drawn from the degassing vessel through a drilling mud outlet **27** below the operating mud level. The mud is carried through a line **28** and passes into a mud tank **31** through an inlet **30**. A jet line **29** connected to line **28** assists in withdrawing the degassed mud from the degassing vessel.

An outlet **32** in the mud tank allows degassed drilling mud from the mud tank to be transmitted through line **14** to T-junction **15** and drawn into degassing vessel **18** through suction line **16**. A check valve **33** prevents gas-contaminated mud from entering mud tank **31** through the mud outlet and contaminating the degassed mud in the mud tank. The provision of a source of gas-free mud to the suction line **16** and drilling mud inlet **17** of the degassing vessel insures continuity of efficient operating conditions in the degassing vessel. The minimum mud level of the mud-gas separator vessel is maintained at a higher level than the maximum level of mud in the mud tank to assure a preferential flow of mud from the mud-gas separator vessel to the degassing vessel. This allows flow from the mud tank **31** through outlet **32**, check valve **33**, T-junction **15** and suction line **16** to the degassing vessel only to supplement the preferential flow from the mud-gas separator vessel to maintain continuity of total flow volume rate to the degassing vessel.

The structural details of the degassing system having been described, the operation of the degassing system will be considered. The process provides an effective, entirely enclosed system for safely and effectively removing dangerous gases from drilling mud from a source such as the mud return system of a well drilling operation and venting of the gases to a flare line burn at a safe distance from the drill site. The mud-gas separator vessel vents free gases entering the system through line **1** to a flare line **5**. The mud-gas separator vessel discharges gas-contaminated drilling mud through outlet **11**. The outlet prevents the accumulation of solid debris in the mud-gas separator vessel. A minimum mud level **6** is maintained in the mud-gas separator vessel at all times by a float **7** connected to a valve **13** in the outlet line **12** by a mechanical linkage.

The contaminated drilling mud is transmitted through line **14** to suction line **16** of the degassing vessel. The inlet **17** of the degassing vessel is located at a height sufficiently above the normal operating levels of drilling mud in the mud-gas separator vessel **3** and the mud in mud tank **31** in order that the desired subatmospheric pressure in the degassing vessel will be required to cause the desired rate of flow of mud into the degassing vessel.

Degassed mud from mud tank **31** is provided to the suction line **16** of degassing vessel **18** through line **14**. A check valve **33** in line **14** prevents contaminated mud from mud-gas separator vessel **3** from entering the mud tank **31** and contaminating the degassed mud therein. The mud-gas separator vessel is positioned so that the minimum mud level in the mud-gas separator vessel is a planned height above the maximum level of drilling mud in the mud tank **31**, assuring a preferential flow of the contaminated mud from the mud-gas separator vessel to the degassing vessel. This allows flow from the mud tank to the degassing vessel only to supplement the preferential flow from the mud-gas separator vessel to maintain approximate continuity of total flow volume rate into the degassing vessel.

A vacuum pump **19** provides a subatmospheric pressure in the degassing vessel by drawing gas through line **21** and discharging the gas through line **20** into flare line **5**. A near maximum mud level **26** is maintained in the degassing vessel. Changes in mud level **26** cause the float **24** and rod **25** to move, thereby moving the mechanical linkage and adjusting valve **23** to either increase or decrease the vacuum applied to the degassing vessel as necessary to maintain the near maximum mud level **26**. A line from the flare line **5** to valve **23** allows exhaust gases to be admitted through the valve when open to vacuum unit **19** and into the degassing vessel. Degassed mud is withdrawn from the degassing vessel through line **28** and discharged into mud tank **31**. A source of pressure **29** is provided through a jet nozzle to assist in withdrawing the degassed mud from the degassing vessel.

The production of hydrocarbons from subterranean hydrocarbon-bearing formations by means of a fireflood is well known to those skilled in the art. Concisely, a fireflood operation is effected by igniting hydrocarbons in a subterranean formation, injecting air through an injection well to sustain the combustion, and producing hydrocarbons freed by the heat of combustion from a production well. Combustion product gases are also produced from a production well. When the subterranean hydrocarbon-bearing formation also contains sulfur compounds, H_2S is sometimes formed and produced as part of the combustion product gases from the production well.

H_2S is known to be a highly toxic gas. Safety, environmental concern and governmental regulations prohibit the release of this hydrogen sulfide into the atmosphere. Thus, the combustion product gases from the subterranean formation must often be treated in some manner to remove the H_2S therefrom prior to exhausting into the atmosphere.

The Claus reaction is well known to those skilled in the art whereby H_2S is reacted with oxygen to form elemental sulfur. This process is widely employed to remove H_2S from gaseous streams bearing the H_2S with the concurrent recovery of elemental sulfur.

There is a distinct need in the oil industry for means to recover additional hydrocarbons after primary recovery operations are no longer feasible. There is also a considerable need for processes which mitigate environmental pollution when fireflood operations are employed to recover hydrocarbons from subsurface formations which contain hydrocarbons and sulfur compounds.

A process developed by *V.W. Rhoades, U.S. Patent 3,845,196; October 29, 1974; assigned to Cities Service Oil Company* is one in which an oxygen-containing gas is heated by a combustion engine employed to drive a compressor employed to inject air into a fireflood operation. The heated oxygen-containing gas is contacted with H_2S-containing combustion product gas from the underground combustion of the fireflood operation. The oxygen and the H_2S react to form elemental sulfur, which can be recovered. Pollution of the atmosphere with H_2S is mitigated.

Figure 7 illustrates the essential details of such a process. In the figure, an injection well **105** is cased and completed through cement completion **104** from the earth's surface **101** through overburden **102** into subterranean reservoir **103** which contains a viscous crude oil. Perforations **106**, for example, are provided in order to permit air **131** injected from compressor **130** connected through wellhead tubing **129** into reservoir **103**. The injected air forms an air bank **107** within the formation supplying oxygen for sustaining the combustion front **108** burning the residual oil **109** remaining after primary and/or secondary oil recovery. Residual oil **109** and noxious production gases resulting from the in situ combustion fireflooding of the reservoir **103** are produced through production well **110**, cased and completed by cement **111**, from wellhead **112** and are passed to and separated in a gas-liquid separator **113** to yield recovered liquid **115** from the lower section of the gas-liquid separator **113** and H_2S-containing combustion product gases from the gas outlet **114** located in the upper portion of the gas-liquid separator.

The H_2S-containing combustion product gases are introduced into a catalytic reactor **116** through inlet **150**. The catalytic reactor contains a catalyst to promote the reaction of H_2S with oxygen to produce sulfur and water vapor. The catalyst can also be effective to promote the conversion of CO and oxygen to form CO_2. A stack gas of water vapor, nitrogen, and carbon dioxide **118** is produced from the exhaust stack **117** of the catalytic reactor **116**. The catalyst of the catalytic reactor is continuously recirculated and regenerated for the removal of sulfur. This is accomplished by removing the catalyst from the lower portion of the catalytic reactor **116** through exit **119**, passing the catalytic material through the pump **120** and introducing it into a sulfur recovery unit **121** into which steam **122** is introduced. Water and elemental sulfur **124** are produced through exit line **125** controlled by valve **123**. Regenerated catalyst is then reintroduced into the upper portion of the catalytic reactor **116** through entry **154**.

Also charged to the catalytic reactor through entry **155** via line **126** entirely insulated by insulating covering **156** is an oxygen-containing gas. The oxygen-containing gas which is charged to catalytic reactor **116** through entry **155** is taken from high-pressure air line **129**. The air from high-pressure air line **129** is passed via line **153** through control valve **157** in heat exchange relationship with combustion engine **151** via line **126** to catalytic reactor **116**. Combustion engine **151** drives compressor **130** by means of drive shaft **152**. Exhaust gases **128** are exhausted from combustion engine **151** by way of exhaust pipe **160**. Valve **158** can be opened if desired to pass exhaust gases through line **159** into line **126** and then to reactor **116**. The system including line **160** and **126** are enclosed by insulative jacket **156** to conserve heat.

A control system provides for optimum heat control in the reactor **116** as well as optimum oxygen content for most efficient reaction of oxygen with H_2S and CO in the H_2S-containing combustion product gases from the production well. Thus, temperature and H_2S plus CO content sensors **161** are connected to controller **162** which automatically activates valve **157** controlling the amount of air fed to the reactor, and valves **163** and **166** which control the temperature of air passed to the reactor **116** through entry **155**. Valves **163** and **166** control the amount of oxygen-containing gas passing through shunt **164** and heat radiator **165** thus controlling the temperature of oxygen-containing gas passing to the reactor **116** through entry **155**, and thus the reaction temperature in the catalytic reactor **116**.

FIGURE 7: APPARATUS FOR PREVENTION OF AIR POLLUTION BY H_2S IN FIREFLOOD OPERATIONS

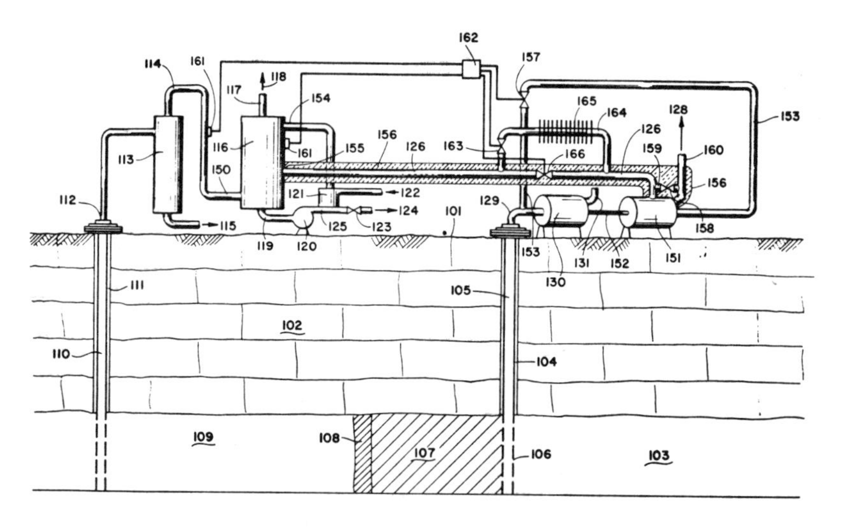

Source: U.S. Patent 3,845,196

WATER POLLUTION CONTROL

The wastes associated with this area of the industry result from the discharge of produced water, drilling muds, drill cuttings, well treatment, and produced sands and additionally, deck drainage, sanitary and domestic wastes for offshore and coastal operations.

Regulatory Constraints

The significant or potentially significant wastewater constituents are oil and grease, fecal coliform, oxygen-demanding parameters, heavy metals and toxic materials. These wastewater constituents have been selected to be the subject of the effluent limitations by EPA (3).

Effluent limitations commensurate with the best practical control technology currently available are proposed for each subcategory. These limitations, listed in Table 12, are explicit numerical values (whenever possible) or some other criteria.

BPCTCA process control measures include the following:

1) Elimination of raw wastewater discharged from free water knockouts or other process equipment.

TABLE 12: OIL AND GAS EXTRACTION INDUSTRY EFFLUENT LIMITATIONS—BPCTCA

Subcategory	Water Source	Oil and Grease, mg/l: Maximum for Any One Day	Oil and Grease, mg/l: Average of Daily Values for 30 Consecutive Days Shall Not Exceed	Residual Chlorine, mg/l
(A) Near offshore	Produced water	72	48*	NA
(B) Far offshore	Deck drainage	72	48*	NA
(D) Coastal	Drilling muds	**	**	NA
	Drill cuttings	**	**	NA
	Well treatment	**	**	NA
	Sanitary M10	NA	NA	>1***
	Sanitary M91M†	NA	NA	NA
	Domestic†	NA	NA	NA
	Produced sand	**	**	NA
(C) Onshore	Produced water	††	NA	NA
(E) Beneficial use	Drilling muds	no discharge	no discharge	–
	Drill cuttings	no discharge	no discharge	–
	Well treatment	no discharge	no discharge	–
	Produced sand	no discharge	no discharge	–

*Not applicable to the coastal subcategory.
**No discharge of free oil to the surface waters.
***Minimum of 1 mg/l and maintained as close to this concentration as possible.
†There shall be no floating solids as a result of the discharge of these materials.
††For the onshore subcategory, no discharge; for the beneficial use subcategory, 45 mg/l.

Source: Reference (3)

2) Supervised operations and maintenance on oil/water level controls, including sensors and dump valves.
3) Redirection or treatment of wastewater or oil discharges from safety valve and treatment unit by-pass lines.

BPCTCA end-of-pipe treatment can consist of some, or all of the following: ing:

1) Equalization (surge tanks, skimmer tanks).
2) Solids removal desanders.
3) Chemical addition (feed pumps).
4) Oil and/or solids removal.
   a. Flotation.
   b. Filters.
   c. Plate coalescers.
   d. Ponds.
   e. Gravity tanks.
5) Subsuface disposal.

Specific treatability studies are required prior to application of a specific treatment sytem to an individual facility.

BPCT for deck drainage is based on control practices used within the oil producing industry and include the following:

1) Installation of oil separator tanks for collection of deck washings.
2) Minimizing of dumping of lubricating oils and oily wastes from leaks, drips and minor spillages to deck drainage collection systems.
3) Segregation of deck washings from drilling and workover operations.
4) O&M practices to remove all of the wastes possible prior to deck washings.

BPCT end-of-pipe treatment technology for deck drainage consists of treating this water with wastewaters associated with oil and gas production. The combined systems may include pretreatment (solids removal and gravity separation) and further oil removal (chemical feed, surge tanks, gas flotation). The system should be used only to treat polluted waters. All storm water and deck washings from platform members containing no oily waste should be segregated as it increases the hydraulic loading on the treatment unit.

The limits for deck drainage are the same as for produced waters offshore.

BPCT for drilling muds includes control practices widely used in both offshore and onshore drilling operations:

1) Accessory circulating equipment such as shaleshakers, agitators, desanders, desilters, mud centrifuges, degassers, and mud handling equipment.
2) Mud saving and housekeeping equipment such as pipe and kelly wipers, mud saver sub, drill pipe pan, rotary table catch pan, and mud saver box.
3) Recycling of oil-based muds.

BPCT end-of-pipe treatment technology is based on existing waste treatment processes currently used by the oil industry in drilling operations.

The limitations for offshore and coastal drilling muds are as follows:

1) Water-based and natural muds shall contain no free oil when discharged.
2) Oil-based and emulsion muds shall not be discharged to surface waters. These muds are to be transported to shore for reuse or disposal in an approved disposal site.

The limitations for onshore drilling muds are as follows:

1) The muds shall not be discharged to surface waters. These muds are to be transported to and disposed of in an approved disposal site.

BPCT for drill cuttings is based on existing treatment and disposal methods used by the oil industry.

The limitations for offshore drill cuttings are as follows:

1) Cuttings in natural or water-based muds shall contain no free oil when discharged.
2) Cuttings in oil-based or emulsion muds shall not be discharged to surface waters. Cuttings should be collected and transported to shore for disposal in an approved disposal site.

The limitation for onshore drill cuttings areas follows:

1) No drill cuttings shall be discharged to surface waters. These drill cuttings are to be transported to and disposed of in an approved disposal site.

Workover fluids other than water used in well treatment or water-based muds are to be recovered and reused. Materials not consumed during workovers and completions are to be transported to and disposed of in an approved site.

The effluent limitations were determined using data supplied by industry and service companies serving the oil producing industry. The limitation for wastes from well treatment offshore is: well treatment wastes shall contain no free oil when discharged.

Alternate handling of wastewater may be necessary when equipment becomes inoperative or requires maintenance. Waste fluids must be controlled during these conditions to prevent discharges of raw wastes into surface waters. Control practices currently used in offshore and coastal operations are:

1) Waste fluids are temporarily stored on board until the waste treatment unit returns to operation.
2) Waste fluids are directed to onshore treatment facilities through a pipeline.
3) Placing waste fluids in a barge for transfer to shore treatment.
4) Waste fluids are piped to a primary treatment unit (gravity separation) to remove free oil and discharged to surface waters.

Effluent limitations commensurate with the best available technology economically achievable are proposed for each subcategory. These effluent limitations are listed in Table 13. The primary end-of-pipe treatment for the near offshore subcategory is the subsurface disposal of production water and for the far offshore subcategory it is similar to that for BPCTCA.

The application of best available technology economically achievable is defined as improved O&M practices and tighter control of the treatment process for the far offshore subcategory. BATEA for the near offshore and coastal subcategories are defined as subsurface disposal for produced waters. BATEA for the onshore, beneficial use, and stripper subcategories are the same as BPTCA. These effluent limitations are to go into effect no later than July 1, 1983. (Now deferred to 1984 by Congressional action, except for toxic pollutants as of November 1977.)

The limitations for all subcategories are the same as BPTCA for drilling muds, drill cuttings, sanitary and domestic wastes, well treatment, and produced sands. Additionally the BATEA limitation for deck drainage in the near offshore subcategory is the same as for BPTCA.

The BATEA limitations for produced water in the coastal and near offshore subcategories is no discharge to surface waters. This can be accomplished by reinjection or by end-of-pipe technologies such as, evaporation ponds and holding pits (when wastes are transferred to shore) or injection to disposal wells. About 40% of those producing facilities with no discharge use one of these end-of-pipe technologies.

Existing no discharge systems were reviewed to select the best technology for the purpose of establishing effluent limitations. Holding pits were found to be the least desirable because of frequent overflow, dike failure, and infiltration of salt water into fresh water aquifers. If properly constructed and lined, evaporation lagoons may result in no discharge in arid and semiarid regions. However, erosion, flooding, and overflow may still occur during wet weather. Disposal well systems which may consist of skim tanks, aeration facilities, filtering systems, backwash holding facilities, clear water accumulators, pumps, and wells provide the best method for disposal of produced water. These systems are equally applicable to onshore and offshore operations and are the primary method used to dispose of produced water on the California coast and in the inland areas.

The BATEA limitations for produced water and deck drainage in the far offshore subcategory are based on the same end-of-pipe technology as used for BPTCA. It is expected that the industry will have gained sufficient experience in the reduction of raw waste loads and operation of end-of-pipe technologies to improve their operation by 1983. In order to define this level of discharge a statistical analysis was carried out on the data from the 27 flotation units used to define BPTCA, to determine if any units were significantly better in effluent quality than the rest. A group of 10 flotation units were separated on that basis and their data analyzed. The resulting BATEA limitations for oil and grease are 52 mg/l daily maximum (composited) and 30 mg/l maximum monthly average.

When the BPTCA limitations were derived, it was concluded that they should be based on what was being achieved by all facilities using the BPTCA.

TABLE 13: OIL AND GAS EXTRACTION INDUSTRY EFFLUENT LIMITATIONS—BATEA AND NEW SOURCE

		Pollutant Parameter–Effluent Limitations		
		Oil and Grease, mg/l		
Subcategory	Water Source	Maximum for Any One Day	Average of Daily Values for 30 Consecutive Days Shall Not Exceed	Residual Chlorine, mg/l
(A) Near offshore	Produced water	no discharge	no discharge	—
(D) Coastal	Deck drainage	72	48	NA
(B) Far offshore	Produced water	52	30	NA
	Deck drainage	52	30	NA
(A) Near offshore	Drilling muds	*	*	NA
(B) Far offshore	Drill cuttings	*	*	NA
(C) Onshore	Well treatment	*	*	NA
(D) Coastal	Sanitary M10	NA	NA	>1
(E) Beneficial use	Sanitary M91M	NA	NA	NA
	Domestic	NA	NA	NA
	Produced sand	*	*	NA

*These BAT and new source limitations are identical to those applicable for each subcategory as for BPCTCA listed in Table 12.

Source: Reference (3)

This conclusion was reached on the basis of industry experience. Since the industry will have, by 1983, 8 additional years of experience in waste abatement, there should be no significant problems in attaining effluent qualities being met by many facilities.

New source performance standards commensurate with the best available demonstrated technology are the same as the BATEA limitations. These effluent limitations are listed in Table 13 also.

Leakage Control

A method developed by *R.J. Chiasson, R.G. Bourg and T.J. Arceneaux; U.S. Patent 3,815,682; June 11, 1974* involves a combination of fire smothering and oil collection in connection with an offshore drilling rig.

Figure 8 shows a protective apparatus being lowered over an offshore rig. The apparatus comprises a crane barge **10** which mounts a crane **12** which suspends telescoping parts **14** by a cable **15**, attached to the lowest of the parts, over a burning offshore oil rig **16**.

FIGURE 8: FIRE SMOTHERING AND OIL CATCHING APPARATUS FOR USE WITH OFFSHORE OIL RIGS

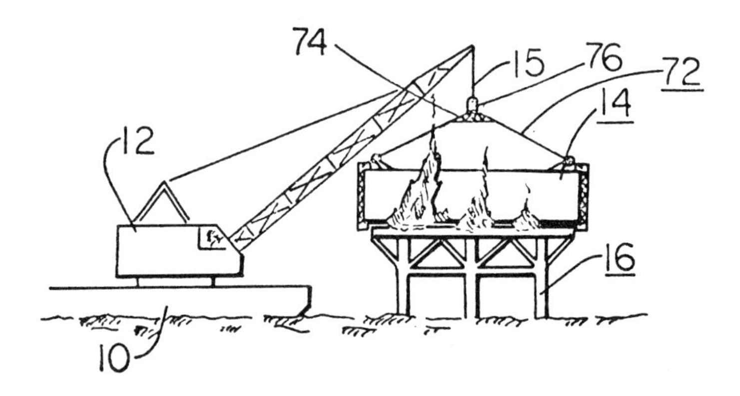

Source: U.S. Patent 3,815,682

The apparatus is suspended over a burning oil rig by a bridle **72** comprising a center framework **74** which is attached by its center **76** to the crane.

The essential part of the apparatus is a cover with depending nesting sides, the upper parts of which telescope into a lower part. The cover and sides adjoining are lined with refractory material. Intermediate the lining and the cover a sprinkler system is mounted to provide a descending spray of cooling and fire extinguishing material. The apparatus is adapted to be lifted by its lower part with its upper parts telescoped therein by a floating crane and lowered by cables over a burning offshore oil rig to enclose it from top to below sea level.

Vertical adjustable jacks are mounted to depend from the cover and engage the top of the working platform of an oil rig to space it from the platform and support the apparatus when it is fully extended downward. The sides, extending below sea level, define an enclosed reservoir area for catching liquid pollutants. A conduit is mounted in the lower part of the sides, an end of which is supported at sea level in the reservoir area, and another end of which extends outwardly of the sides for discharging liquid pollutants in the reservoir area to salvage barges brought alongside. The telescoping sides are roller bearinged between sides and at corners to provide nonfrictional and nonjamming operation. A hatch and depending ladder are mounted in the cover for access to the work platform after the fire smothered for repairing the rig and stopping the flow of oil to the reservoir area.

An apparatus developed by *M.M. Ryan; U.S. Patent 3,839,870; Oct. 8, 1974* is an oil-confining enclosure for an offshore oil well to provide a barrier against the escape of oil leaking from the drilling operation. It includes a floating, generally circular, confining wall encircling an offshore well site and which is supported on a series of interconnected pontoons having operative connection with a floating platform, exteriorly of the confining wall, providing control valves for the connections to the pontoons. Thereby a vessel may approach the floating platform and by means of the valved connections to the pontoons cause the pontoons to be filled with seawater to sink the confining wall below the level of the sea for access by the vessel to the drilling rig and on the way out again connect with the floating platform valves to blow the water out of the pontoons for the flotation of the circular confining wall.

The apparatus is shown in some detail in Figure 9 which contains both a perspective view and an elevation of the apparatus.

In the drawings **10** represents an offshore oil well rig installation including a derrick **11**, one or more platforms **12** as well as all of the usual elements found in wells of this type such as the well house, the tackle for boring, hoisting or lowering and the equipment for handling the well casing sections.

As shown, the well casings **13** are depicted as extending through the ocean floor **14** in drilling for underlying oil and which subsequently to the drilling operation would represent the means by which the oil is withdrawn by pumping means in the rig **10**. As shown also in this figure, it will be seen that oil sometimes leaks from around the wellhead **15** and escapes upwardly through the water to the surface and unless this seepage is confined and recovered an oil slick results which may deleteriously affect the shoreline or wildlife of other things with which it may come in contact or be enveloped thereby.

In order to confine the escaping oil this device provides a generally circular wall structure **16** surrounding and enclosing the rig **10**. This wall is comprised of a series of metal plates **17** arranged in edge to edge relationship and integrally secured together by welding as at **18** to provide a continuous wall of leakproof construction extending entirely around the rig **10** at a distance therefrom such as to catch any and all escaping oil floating upwardly from the wellhead **15** within the area enclosed by the wall. The wall **16** is of a vertical depth such as to extend above the surface of the water to provide a basin for the collection of the escaping oil and extends below the surface as well so that all of the oil that escapes from the wellhead is positively confined within the enclosure formed by the circular wall.

FIGURE 9: OIL WELL LEAKAGE CONTAINER FOR USE IN OFFSHORE OPERATIONS

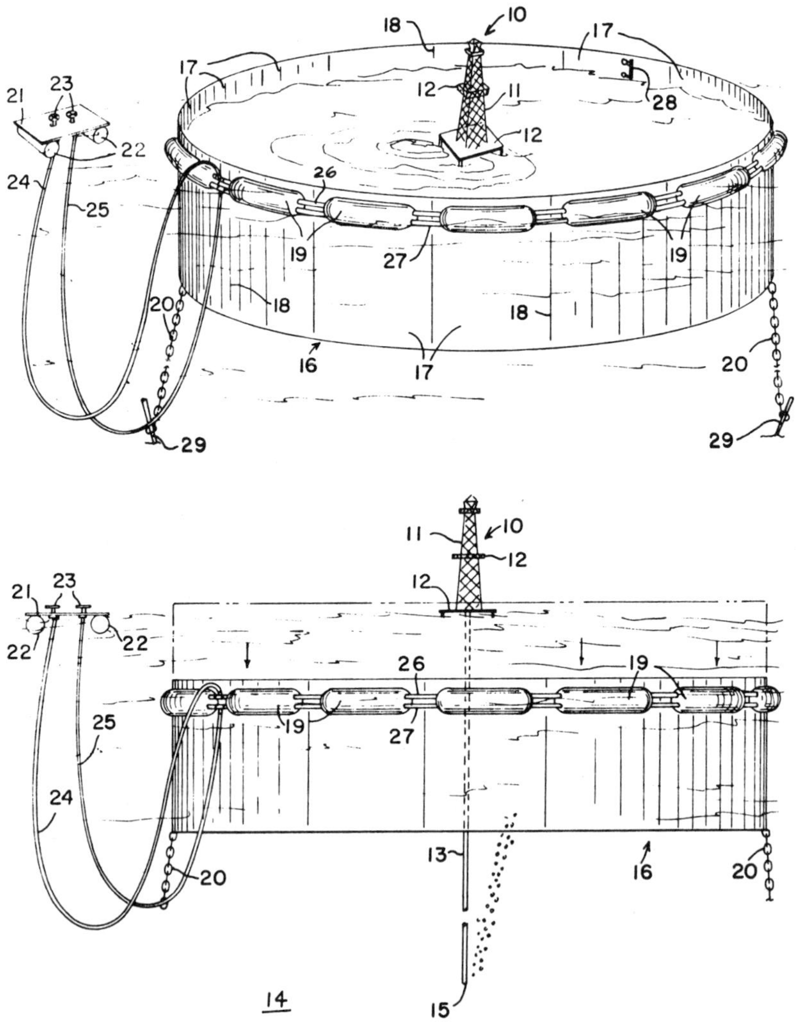

Source: U.S. Patent 3,839,870

The wall structure **16** is supported by pontoons **19** which are disposed around the perimeter of the wall structure and secured to the outside face of the wall at substantially equally spaced intervals so that the entire wall structure floats in the water supported by the buoyancy of the pontoons. The wall structure is disposed in concentric relationship with respect to the rig **10** and this relationship is maintained by anchoring devices arranged at equally spaced intervals around the wall. The anchoring devices include chains **20** which are secured to the wall structure **16** adjacent to the lower area thereof and extend downwardly to the anchors **29** fixed in the ocean floor **14** and to which the chains are also secured.

Thus, the circular wall **16** is confined to a generally concentric position with respect to the rig **10**, but because of the highly flexible nature of the restraining chains **20**, may be lowered into the water for the passage of ships approaching the central rig, or raised to operative position concentrically around the rig, without any adverse effect on the positioning or operation of the confining wall.

A floating platform **21** is provided in the sea outside the combined area encircled by the wall **16** so that the platform is always accessible by boat from outside the wall. One or more pontoons **22** may be utilized to maintain the buoyancy of the platform on the surface of the sea. Valve mechanism **23** is provided on the platform with the controls on the upper side thereof whereby connection can be made to lines **24** and **25** by a ship such as a tanker, or the like, from the sea area surrounding the wall **16**. The lines **24** and **25** have operative connection with the series of pontoons **19** supporting the wall **16** and are shown as being connected respectively with lines **26** and **27** extending between the various adjoining pontoons, but it is contemplated that the lines **24** and **25** might be connected to one or more of the pontoons, if desired, and achieve their purpose. The lines **26** and **27** connect all of the pontoons **19** in a continuous series so that all of the pontoons are fully in communication one with the other entirely around the wall **16**.

When the wall **16** is in its operative raised position around the rig **10** and an approaching ship requires access to the rig, the ship, such as a tanker, first goes to the floating platform **21** and makes operative connection with the lines **24** and **25** through the control valve mechanism **23**. If the basin formed by the wall structure around the central rig does not contain any oil the ship pumps seawater into the pontoons through the lines **24** and **25** with the connecting lines **26** and **27** providing "in-and-out" communication between all of the pontoons.

When sufficient air has been displaced from the pontoons by seawater to reduce their buoyancy the level of the wall structure **16** will be lowered in the water and thereby enable the ship to pass over the wall and go to the rig **10**. When the ship has completed its business at the central rig it again passes over the wall on its way out and again makes connection with the lines **24** and **25** through the control mechanism **23** and blows out the pontoons **19**. Filling the pontoons with air causes the wall **16** to rise again above the level of the sea and thereby form a basin around the rig to collect and confine the oil which leaks from the wellhead **15**.

The wall structure is illustrated in the lower sectional view in a lowered position, as shown in full lines, with the normal operative position raised above the level of the sea, being indicated in broken lines. When the basin formed by the confining wall contains a substantial volume of oil the level of the wall should not be lowered until the confined oil has been removed from the basin. When the basin contains oil this can be pumped out of the basin into a tanker and hauled away for processing.

This operation can be performed at any time that sufficient oil has accumulated in the basin to warrant its removal but the operation can also be completed at the time an approach to the rig **10** must be made. In this latter event a tanker would first pump the oil out of the basin into its hold and then go to the platform **21** to make the connection to the pontoons **19** through the lines **24** and **25** and pump seawater into the pontoons to lower the wall. The tanker would pass over the wall on its way to the central rig, complete whatever business was necessary at the rig and on the way out again pass over the lowered wall and connect up with the lines **24** and **25** to blow out the pontoons with air and thereby cause the confining wall to be restored to operative position.

The pontoons **19** are located on the confining wall structure **16** at a position to float the wall in the sea, with the top edge thereof at a level sufficiently above the surface to form the basin for collecting the escaping oil, and this positioning of the pontoons on the wall might be varied in accordance with particular conditions at a well site and preferences as to the amount of oil to be collected in the basin before signalling the necessity for removing the collected oil from the basin. A transmitting type signal **28** is mounted on the inside face of the wall structure **16** within the enclosure to be responsive to the level of the oil in this basin. When a predetermined level has been reached, the signal **28** will beam a radio wave to a selected location where this information will bring forth the equipment for removing the accumulated oil from the basin.

The wall structure **16**, being constructed from a plurality of adjoining plates welded edge to edge, can be fabricated at the site of the offshore oil well but it is contemplated also that the complete wall structure might be fabricated at some distant location on shore and the completed structure, including the pontoons assembled thereon, brought to the offshore well location by helicopters and lowered into the water around the rig **10** and anchored.

An apparatus developed by *P. Charpentier; U.S. Patent 3,875,998; April 8, 1975; assigned to Entreprise de Recherches et D'Activities Petrolieres (ELF), France* provides for separation on the seabed of the effluents from underwater oil wells.

More specifically, the apparatus consists of a base with negative buoyancy anchored on the seabed, a hermetic caisson attached to this base and preferably articulated on it, equipped with at least one gas/liquid separator. The separator is connected by pipes to each underwater well and from it one pipe leads to a burner, with another pipe to draw off the liquid phase. There is a buoyant tubular column surmounting the caisson and communicating with it. The upper end of this column, which is above water, supports a platform. There are means inside the caisson for controlling the operation of the separator.

Figure 10 shows the apparatus in some detail. The figures which comprise a vertical section and two cross sections show a base **1** with strongly negative buoyancy, preferably a concrete slab, resting on the seabed **2** and anchored to it by being attached to a drilled pile **3**. Naturally, any other method of anchoring the base to the seabed can be used. An articulated joint **4**, allowing movement in two directions, attaches the base to a caisson **5**, at the bottom of which is a cement slab **6**, providing ballast for the caisson and reducing the strain on the articulated joint. The caisson is surrounded by an annular atmospheric separator **7** divided into a large separation compartment **25** and a small monitoring compartment **26**, and fed by a manifold **8** inside the caisson, linked by pipes **9** to the different oil wells (not shown here).

FIGURE 10: INSTALLATION FOR SEPARATION ON THE SEABED OF THE EFFLUENTS FROM UNDERWATER OIL WELLS

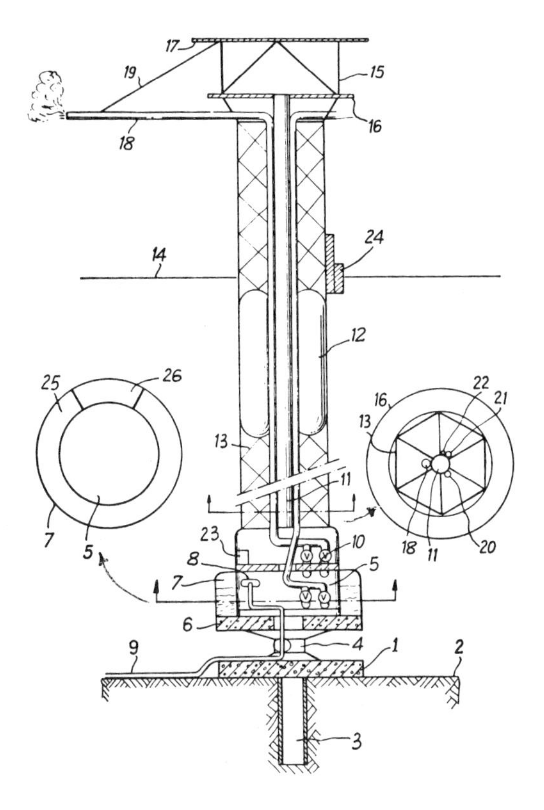

Source: U.S. Patent 3,875,998

These pipes rest on the base **1** and enter the caisson through a central passage in the articulated joint and the base of the caisson. The caisson is divided into two levels by a floor **10**, containing a central passage. A tubular column **11**, the upper end of which is above water, is fixed rigidly and hermetically to the caisson. It is equipped with floats **12**, held in a steel trellis **13** which surrounds the column. The part of the column above the surface of the sea **14** is surmounted by a platform **15**, containing a handling deck **16** and a helicopter pad **17**. The inside of the column **11** communicates with the inside of the caisson, allowing access to caisson from platform, or to platform from caisson. A burner **18**, connected with each of the separator compartments **25** and **26**, places these compartments in communication with the atmosphere, to remove gas released during separation. The burner pipe ascends along the column and is fixed to the platform **15** by a rigid steel unit **19**. Two ventilation shafts **20** and **21** descend from the platform to the caisson, supplying fresh air to the caisson and removing the stale air. Electricity is brought from the platform to a transformer **23** inside the caisson, along an electric cable **22** which also descends along the column. This column may be fitted with mooring equipment for a boat **24**.

A device developed by *C.M. Mason; U.S. Patent 3,879,951; April 29, 1975; assigned to Sun Oil Company* is a pollution control curtain attached to the sea floor on one end and to a collar floating on the surface of the sea on the other end for positioning around a drilling platform to entrap the petroleum fluids seeping from the wellhead or sea floor being serviced by the platform and to convey the fluids to the surface within a predefined area from which they can be collected.

The device is illustrated in Figure 11 which shows a drilling platform, **3**, positioned on a truss support **4** which extends to the floor of the body of water in which the drilling is taking place, **15**. The riser pipe **5** through which the drill bit and conduit pass is shown reaching from the platform to the wellhead **6**. Surrounding the entire truss is the pollution control curtain **7**, which is attached to the truss at point **12** on its lower section and at point **10** by a support member below the water surface. An extension flap, **13**, attached to the bottom of the curtain, extends over an area of the floor **15** which is greater than the cross-sectional area of the curtain. This flap is attached to the floor by cables **14** with a space between the bottom of the flap and the floor itself. The curtain contains numerous slots, **16**, which open inwardly to allow equalization of pressure on both sides of the curtain. The spaces left between the flap and the floor also serve this function.

Attached to the upper end of the curtain at point **10** is a flexible barrier **17** which allows for communication of fluid to the surface within an area defined by a rigid collar **8**, attached to floats **9** and to the upper end of the flexible barrier. As the water level changes due to tidal activity, etc., the collar rides upon the surface and the flexible barrier maintains the connection between the collar and the curtain. The collar is operable whether it is larger or smaller in the surface area than the curtain. It also envisioned (not shown) that a second collar, surrounding the first collar, can be utilized to provide a lock for the ingress and egress of boats. Also, the collar itself can be constructed from a floating material.

Fluids seeping from the wellhead, or from a fault in the sea floor in another embodiment, rise to the surface within the area defined by the control curtain. At the surface, the resulting oil slick is confined to the area defined by the floating collar from which it can be collected. As the water level changes, due to tidal activity, winds, or storms, the collar rides on the surface and the flexible section of the barrier maintains communication between the collar and the control curtain.

Water currents and pressure necessitate inwardly opening flaps **18** and a space between the lower end of the curtain and the sea floor so as to equalize these effects on both sides of the curtain and prevent collapse. However, no flaps should be located in the uppermost section of the curtain which comprises the oil zone.

FIGURE 11: UNDERWATER DRILLING POLLUTION CONTROL CURTAIN

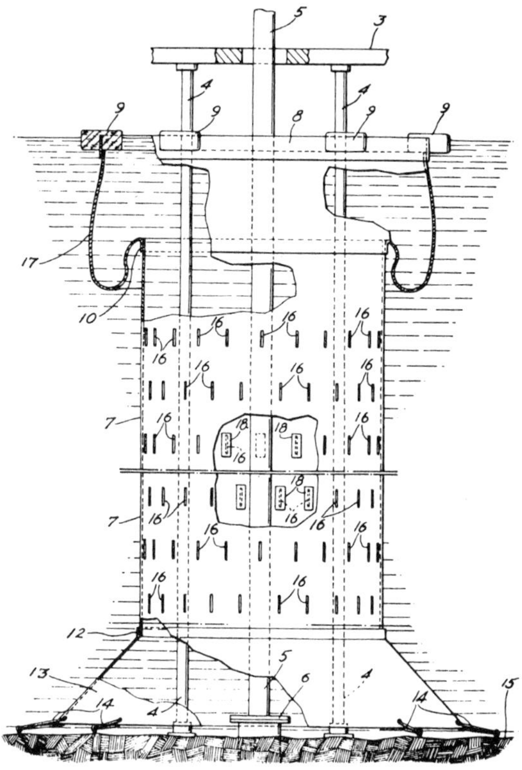

Source: U.S. Patent 3,879,951

Wastewater Purification

Petroleum production, drilling, and exploration wastes vary in quantity and quality from facility to facility. A wide range of control and treatment technologies has been developed to treat these wastes. The results of industry surveys indicate that techniques for in-process controls and end-of-pipe treatment are generally similar for each of the industry subcategories; however, local factors, discharge criteria, availability of space, and other factors influence the method of treatment (3).

In-plant control or treatment techniques are those practices which result in: 1) reduction or elimination of a waste stream; or 2) a change in the character of the constituents which allow the end-of-pipe processes to be more efficient and cost effective.

The two types of in-plant techniques that reduce the waste load to the treatment system or to the environment are reuse and recycle of waste products. Examples of reuse are: 1) reinjection of produced water to increase reservoir pressures; and 2) utilization of treated production water (softened, if necessary) for steam generation. An example of a recycle system is the conservation and reuse of drilling muds.

Examples of character change in waste stream would be: 1) the substitution of a positive displacement pump for a high-speed centrifugal pump; and 2) substitution of a downhole choke for a wellhead choke, thereby reducing the amount of emulsion created.

Proper pretreatment and maintenance practices are also effective in reducing waste flows and improving treatment efficiencies. Return of deck drainage to the process units and elimination of waste crankcase oil from the deck drainage or produced water treatment systems are examples of good offshore pretreatment and maintenance practices.

The single most significant change in process technology is reinjection to the reservoir formation for secondary recovery and pressure maintenance. This is distinguished from injection for disposal purposes only, which is considered as end-of-pipe treatment. Waters used for secondary recovery and pressure maintenance should be free of suspended solids, bacterial slimes, oxygen, sludges, and precipitates.

In some cases the quantity of produced water is insufficient to provide the needed water for a secondary recovery and pressure maintenance system. In this case, additional makeup water must be found, and wells or surface water (including seawater) may be used as a source of makeup water. There may be problems of compatibility between produced water and makeup water. A typical reinjection water treatment facility consists of a surge tank, flotation cell, filters, retention tank, and injection pumps.

Reinjection of produced water for secondary recovery and pressure maintenance is a very common practice onshore. It has been estimated that 60% of all onshore produced water is reinjected for secondary recovery.

Produced water treatment for reinjection is similar, both offshore and onshore. Existing reinjection systems vary from small units which treat less than 100 bpd of brine waste to large complexes which handle over 170,000 bpd. Produced water reinjection systems for pressure maintenance and waterflooding are less common in the Gulf Coast, and none are in use in Cook Inlet, Alaska. (Cook Inlet water is treated and injected for waterflooding, because of compatibility problems with the produced water.)

Produced water treatment and reinjection systems are not generally limited by space availability but must be specifically designed to fit offshore platforms. Two limiting factors which affect produced water reinjection are insufficient quantities of produced water to meet the requirement for reservoir pressure maintenance and incompatibility between makeup seawater and produced water.

With the increasing oil demand, new (tertiary) methods are being developed to recover greater amounts of oil from producing formations. The addition of steam or other fluids into the formation can improve ultimate recovery. A system which reuses produced water for steam generation is operating on the West Coast. The system consists of a typical reinjection treatment unit with water softeners added to the system.

Changes in process technology have also occurred in drilling operations. Environmental considerations and high cost of drilling muds have led to the development of special equipment and procedures to recycle and recondition both water-based and oil-based muds. With the system operating properly, mud losses are limited to deck splatter and the mud clinging to drill cuttings.

The main pretreatment process which is applicable to offshore production systems is the return of deck drainage to the production process units to remove free oil prior to end-of-pipe treatment. This method of pretreatment is not applicable to facilities that flush drilling muds into the deck drainage system during rig washdown or to facilities that pipe all produced crude oil and water to shore for processing and brine treatment.

A key in-plant control is good operation and maintenance practices. Not only do they reduce waste flows and improve treatment efficiencies, but they also reduce the frequency and magnitude of systems upsets. Some examples of good offshore operations are:

1) Separation of waste crankcase oils from deck drainage collection system.
2) Reduction of wastewater treatment system upset from deck washdown by discriminant use of detergents.
3) Reduction of oil spillage through good prevention techniques such as drip pans and other collection methods.
4) Elimination of oil drainage from transfer pump bearings or seals by pumping into the crude oil processing system.
5) Reduction of oil gathered in the pig (pipeline scraper) traps by channeling oil back into the gathering line system instead of the sump system.
6) Elimination of extreme loading of the produced water treatment system, when the process system malfunctions, by redirecting all production to shore for treatment.

Good maintenance practice includes: 1) inspection of dump valves for sand cutting as a preventive measure; 2) use of dual sump pumps for pumping drainage into surge tanks; 3) use of reliable chemical injection pumps for produced water treatment; 4) selection of the best combination of oil-and water-treating chemicals; and 5) use of level alarms for initiating shutdown during major system upsets. Operation and maintenance of a produced water treatment system during startup presents special problems.

As an example, an offshore facility had two problems with the heater-treaters that caused problems with the water treatment system: 1) insufficient heat in the treaters; and 2) malfunctioning level controls which caused excessive oil loading. A change in the type of level controls and reduced production which lowered the heating requirements helped alleviate the problem during startup of the produced water treatment unit. Further improvements were achieved by careful selection of chemicals for treating oil and produced water, and the chemical injection and recycling pumps were replaced.

The preceding paragraph describes an actual case where detailed failure analysis and corrective action ended an upset in the waste treatment system. Evaluation of operational practices, process and treatment equipment and correct chemical use is imperative for proper operation and in the prevention and detection of failures and upsets. The description of these operation and maintenance practices is not intended to advocate their universal application. Nevertheless, good operations and maintenance on an oil/gas production facility can have a substantial impact on the loads discharged to the waste treatment system and the efficiency of the system. Careful planning, good engineering, and a commitment on the part of operating and management personnel are needed to ensure that the full benefits of good operation and maintenance are realized.

End-of-pipe control technology for offshore treatment of produced water from oil and gas production primarily consists of physical/chemical methods. The type of treatment system selected for a particular facility is dependent upon availability of space, waste characteristics, volumes of waste produced, existing discharge limitations, and other local factors. Simple treatment systems may consist of only gravity separation pits without the addition of chemicals, while more complex systems may include surge tanks, clarifiers, coalescers, flotation units, chemical treatment, or reinjection.

Gas Flotation: In a gas flotation unit gas bubbles are released into the body of wastewater to be treated. As the bubbles rise through the liquid, they attach themselves to any oil droplet in their path, and the gas and oil rise to the surface where they may be skimmed off as a froth. Two types of gas flotation systems are used in oil production: dispersed gas flotation and dissolved gas flotation.

Dispersed gas flotation units use specially shaped rotating mines or dispersers to form small gas bubbles which float to the surface with the contacted oil. The gas is drawn down into the water phase through the vortex created by the rotors, from a gas blanket maintained above the surface. The rising bubbles contact the oil droplets and come to the surface as a froth, which is then skimmed off. These units are normally arranged as a series of cells, each one operating as outlined above. The wastewater flows from one cell to the next, with a net oil removal in each cell (some oil is recycled back into the water phase by the rotor action).

Dissolved gas flotation units differ from the dispersed gas flotation because the gas bubbles are created by a change in pressure which lowers the dissolved gas solubility, releasing tiny bubbles. A portion of the wastewater stream is recycled back to the bottom of the cell after wastewater has been gasified. This gasification is accomplished by passing the wastewater through a pump to raise the pressure and then through a contact tank filled with gas. The wastewater leaves the contact tank with a concentration of gas equivalent to the gas solubility at the elevated pressure. When the recycled (gasified) water is released in the bottom of the cell (at atmospheric pressure) the solubility of the gas decreases and the excess gas is released as microscopic bubbles. These bubbles then rise to the surface contacting the oil and bringing it to the surface where it is skimmed off. Dissolved gas flotation units are usually a single cell only.

On production facilities it is usual practice to recycle the skimmed oily froth back through the production oil-water separating units. A flow diagram of the two typical flotation units is shown in Figure 12.

The addition of chemicals can increase the effectiveness of either type of gas flotation unit. Some chemicals increase the forces of attraction between the oil droplets and the gas bubbles. Others develop a floc which eases the capture of oil droplets, gas bubbles, and fine suspended solids, making treatment more effective.

In addition to the use of chemicals to increase the effectiveness of gas flotation systems, surge tanks upstream of the treatment unit also increase its effectiveness. The period of quiescence provided by the surge tank allows some gravity separation and coalescence to take place, and dampens out surges in flow from the process units. This provides a more constant hydraulic loading to the treatment unit, which, in turn, aids in the oil removal process.

The verification survey conducted on coastal Louisiana facilities included 10 flotation systems which varied in design capacities from 5,000 to 290,000 bpd and included both rotor/disperser and dissolved gas units. The designs of waste treatment systems are basically the same for both offshore platform installations and onshore treatment complexes; however, parallel units are provided at two of the onshore installations, permitting greater flexibility in operations.

Information obtained during the field survey of onshore treatment systems for Cook Inlet indicated that one of the four onshore systems utilized a dissolved gas flotation system comparable to those used in the Gulf Coast. This system provides physical/chemical treatment and consists of a surge tank, chemical injection, and a dissolved air flotation unit. In addition, two of the Cook Inlet platforms use flotation cells for treatment of deck drain wastes.

Field surveys on the West Coast found that physical/chemical treatment is the primary method of treating produced water for either discharge to coastal waters or for reinjection and that flotation is the most widely used of the physical/chemical methods. On the West Coast, all treatment systems except one are located onshore and produced fluids are piped to these complexes. The majority of the wastewater treatment systems have been converted to reinjection systems. However, some of those that still discharge are somewhat different from the systems in the Gulf Coast and Cook Inlet. One of the more complex onshore systems consists of pretreatment and grit settling, primary clarification, chemical addition (coagulating agent), chemical mixing, final clarification, aeration, chlorination, and air flotation. This system handles 50,000 bpd.

Surveys of onshore production facilities in California revealed induced

FIGURE 12: ROTOR-DISPERSER AND DISSOLVED GAS FLOTATION PROCESSES FOR TREATMENT OF PRODUCED WATER

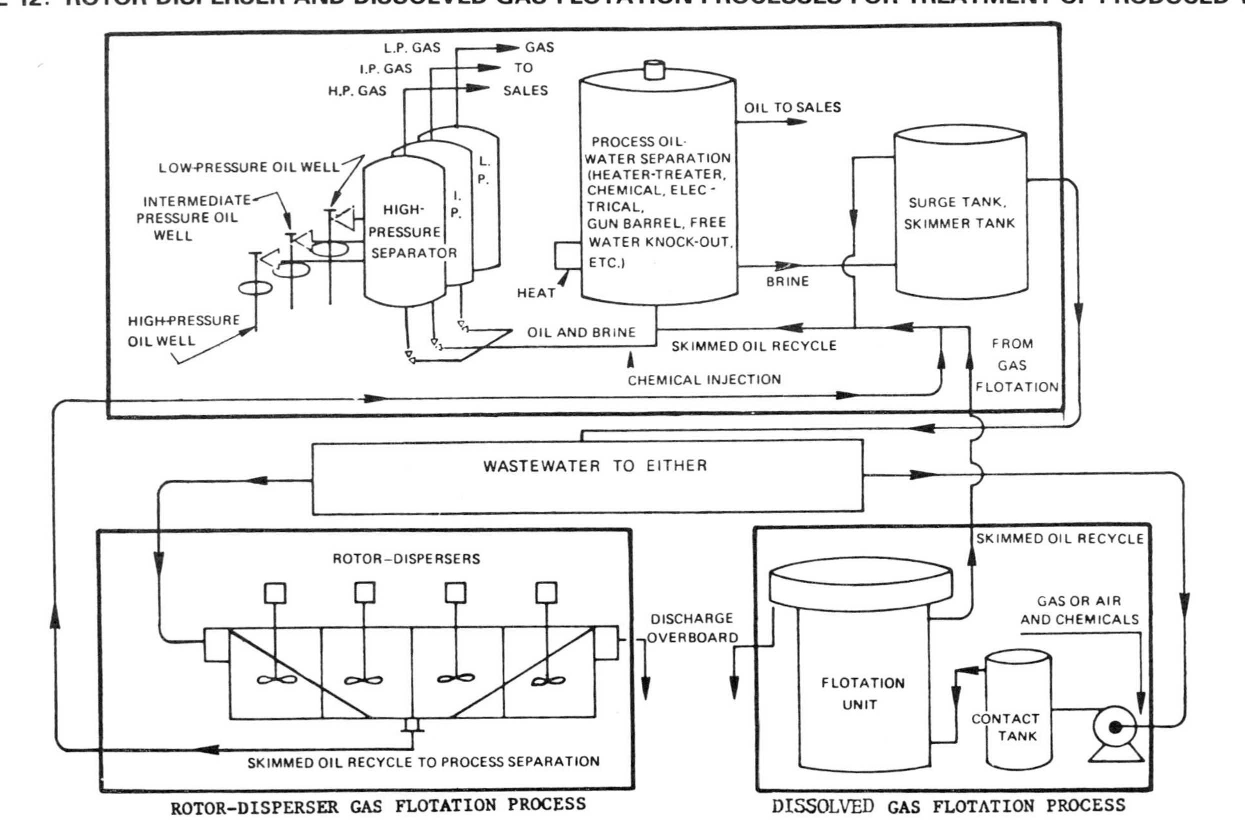

Source: Reference (3)

gas flotation being used for treatment of produced water for recovery, disposal by reinjection and discharge. A total of seven units were observed, three of which were utilized ahead of sand filters and one unit was followed by a pond. The size range of the entire group was from 10,500 to 350,000 bpd. Surge tanks normally preceded the flotation units with the floc going to a sump or being recycled.

In Wyoming two dispersed air flotation systems were observed, both of which discharged and reinjected for recovery the treated produced water. The system consisted of a skim tank, flotation unit, surge tank and in the case of the discharged stream, an earthen pond. The addition of chemicals was used to increase separation efficiency. The produced water treatment capacities of the two systems surveyed were 70,000 and 340,000 bpd respectively.

Parallel Plate Coalescers: Parallel plate coalescers are gravity separators which contain a pack of parallel, tilted plates arranged so that oil droplets passing through the pack need only rise a short distance before striking the underside of the plates. Guided by the tilted plate, the droplet then rises, coalescing with other droplets until it reaches the tip of the pack where channels are provided to carry the oil away.

In their overall operation, parallel plate coalescers are similar to API gravity oil water separators. The pack of parallel plates reduces the distance that oil droplets must rise in order to be separated; thus the unit is much more compact than an API separator. Suspended particles, which tend to sink, move down a short distance when they strike the upper surface of the plate; then they move down along the plate to the bottom of the unit where they are deposited as a sludge and can be periodically drawn off. Particles may become attached (scale) to the plate surface of the plate; then they move down along the plate surfaces, requiring periodic removal and cleaning of the plate pack.

Where stable emulsions are present, or where the oil droplets dispersed in the water are relatively small, they may not separate in passing through the unit.

The verification survey of coastal Louisiana facilities included seven plate coalescer systems which had design capacities from 4,500 to 9,000 bpd. A recent survey indicated that approximately 10% of the units in this area were plate coalescers and they treated about 9% of the total volume of produced water in offshore Louisiana waters. Both the long-term performance data and the verification survey indicated that performance of these units was considerably poorer than that of flotation units. In addition to the physical limitations, coalescers' operation and maintenance data indicated that the units require frequent cleaning to remove solids.

No plate coalescers are in use in Cook Inlet or California, either onshore or offshore.

Filter Systems (Loose or Fibrous Media Coalescers): Another type of produced water treatment system is filters. They may be classified into two general classes based on the media through which the waste stream passes.

1) Fibrous media, such as fiberglass, usually in the form of a replaceable element or cartridge.
2) Loose media filters, which normally use a bed of granular material such as sand, gravel, and/or crushed coal.

Some filters are designed so that some coalescing and oil removal take place continuously, but a considerable amount of the contaminants (oil and suspended fines) remain on the filter media. This eventually overloads the filter media, requiring its replacement or backwashing. Fibrous media filters may be cleaned by special washing techniques or the elements may simply be disposed of and a new element used. Loose media filters are normally backwashed by forcing water through the bed with the normal direction of flow reversed, or by washing in the normal direction of flow after gasifying and loosening the media bed.

Filters which require backwashing present somewhat of a problem on platforms because the valving and controls need regular maintenance and disposal of the dirty backwash water may be difficult. Replacing filter media and contaminated filter elements also create disposal problems.

Measured by the amount of oil removed, filter performance has generally been good (provided that the units are backwashed sufficiently often); however, problems of excessive maintenance and disposal have caused the industry in the Gulf Coast to move away from this type of unit, and a number of them have been replaced with gas flotation systems.

The Gulf Coast survey information indicated that when filter systems are used there is no initial pretreatment of the waste other than surge tanks. Backwashing, disposal of solids, and complex instrumentation were reported as the main problems with these units.

On the West Coast and Cook Inlet, no filter systems are in use as the primary treatment method. Filters are, however, used for final treatment in injection systems in California and several steps of filtration are used prior to seawater injection in Cook Inlet. On the West Coast, these units are preceded by a surge tank, flotation unit, and other treatment units which remove most of the oil and suspended particles. These units, when used in series with other systems, perform well.

In Wyoming a site was visited where approximately 6,600 bpd were being treated by a mixed sand media pressure filter. Earthen ponds both preceded and followed the filter unit with backwash feed being pumped from the final pond and discharged to the primary pond.

Gravity Separation: The simplest form of treatment is gravity separation. The produced water is retained for a sufficient time for the oil and water to separate. Tanks, pits, and, occasionally, barges are used as gravity separation vessels. Large volumes of storage to permit sufficient retention times are characteristic of these systems. Performance is dependent upon the characteristics of the wastewater, water volumes, and availability of space. While total gravity separation requires large containers and long retention times, any treatment system can benefit from quiescent retention prior to further treatment. This retention allows some gravity separation and dampens surges in volume and oil content.

About 75% of the systems on the Gulf Coast are gravity separation systems. The majority are located onshore and have limited application on offshore platforms because of space limitations. Properly designed, maintained, and operated systems can provide adequate treatment. A 30,000 bpd gravity system with the addition of chemicals produced an effluent of less than 15 mg/l during the verification survey.

Two of the onshore treatment systems in Cook Inlet use gravity separation with various configurations of settling tanks and pits. No gravity systems were reported to be in use on the West Coast. The four installations visited in the Texas verification study all use gravity separation tanks offshore and a combination of tanks and/or pits onshore.

The most prevalent treatment method for produced water encountered in the onshore field surveys of California, Wyoming, Texas, Louisiana and Pennsylvania onshore production sites were tanks and ponds when utilized as the single treatment process. As previously mentioned, tanks do not afford the retention times of ponds, but whether or not their primary function is separation they are effective in skimming readily removed free oil.

In California four sites were visited which utilized tankage as the single method of treatment prior to disposal by reinjection. The capacity of these systems to treat produced water ranged from 6,000 to 35,000 bpd.

In Wyoming a total of 37 production facilities were visited which utilized either tanks or ponds as the method of treatment. Of the 23 sites using tanks for treatment ranging in produced water capacity from 920 to 34,000 bpd, 11 were reinjecting for disposal and the remainder were reinjecting for secondary recovery purposes. Of the 14 sites using ponds for treatment, nine were discharging, two were reinjecting for recovery, while the remaining three both discharged and reinjected for recovery.

In Pennsylvania, where disposal by discharge is the rule rather than the exception, 11 sites were visited which utilized ponds for separation treatment ranging in capacity from 2,000 to 8,000 bpd of produced water capacity.

Distillation: In California a site was visited which utilized produced water as boiler feedwater. The boiler was fired by field natural gas and discharged condensate to the local groundwater table. The steam was utilized to heat onsite crude storage tanks and the boiler blowdown containing oil and grease residue was hauled to a Class I (California Classification) landfill site. Reported daily fuel costs for the 150 bpd facility are $70.

Chemical Treatment: The addition of chemicals to the wastewater stream is an effective means to increase the efficiencies of treatment systems. Pilot studies for a large onshore treatment complex in the Gulf of Mexico indicated that addition of a coagulating agent could increase efficiencies approximately 15% and the addition of a polyelectrolyte and a coagulating chemical could increase efficiencies 20%.

Three basic types of chemicals are used for wastewater treatment, and many different formulations of these chemicals have been developed for specific applications. The basic types of chemicals used are:

1) *Surface Active Agents*—These chemicals modify the interfacial tensions between the gas, suspended solids, and liquid. They are also referred to as surfactants, foaming agents, demulsifiers, and emulsion breakers.
2) *Coagulating Chemicals*—Coagulating agents assist the formation of floc and improve the flotation or settling characteristics of the suspended particles. The most common coagulating agents are aluminum sulfate and ferrous sulfate.
3) *Polyelectrolytes*—These chemicals are long chain, high molecular weight polymers used to assist in removal of colloidal and extremely fine suspended solids.

The results of two EPA surveys of 33 offshore facilities using chemical treatment in the Gulf Coast disclosed the following:

1) Surface active agents and polyelectrolytes are the most commonly used chemicals for wastewater treatment.
2) The chemicals are injected into the wastewater upstream from the treatment unit and do not require premixing units.
3) Chemicals are used to improve the treatment efficiencies of flotation units, plate coalescers, and gravity systems.
4) Recovered oil, foam, floc, and suspended particles skimmed from the treatment units are returned to the process system.

A similar survey of facilities in Cook Inlet, Alaska indicated that a facility uses coagulating agents and polyelectrolytes to improve treatment efficiency. Recovered oil and floc are returned to the process system.

Chemical treatment procedures on the West Coast are similar to those used in the Gulf Coast and Cook Inlet. However, there are exceptions where refined clays and bentonites are added to the waste stream to absorb the oil and both are removed after addition of a high molecular weight nonionic polymer to promote flocculation. The oil, clay, and other suspended particles removed from the waste stream are not returned to the process system but are disposed of at approved land disposal sites.

A 14,000 bpd treatment system using refined clay was reported to have generated 60 bpd of oily floc which required disposal in a state-approved site. Selection of the proper chemical or combination of chemicals for a particular facility usually requires jar tests, pilot studies, and trial runs. Adjustments in chemicals used in the process separation systems may also require modification of chemicals or application rate in the waste stream. Other chemicals may also be added to reduce corrosion and bacterial growths which may interfere with both process and waste treatment systems.

Efficiency of Treatment Systems: Table 14 gives the relative long term performance of existing wastewater treatment systems. The general superiority of gas flotation units and loose media filters over the other systems is readily apparent. However, individual units of other types of treatment systems have produced comparable effluents.

Table 15 gives the performance of existing produced water treatment systems over a 6-month to one and one-half year period of weekly and monthly sampling. The data have been divided into treatment systems according to state of location.

TABLE 14: PERFORMANCE OF VARIOUS TREATMENT SYSTEMS LOUISIANA COASTAL

Type Treatment System	Mean Effluent, Oil and Grease (mg/l)	Number of Units in Data Base
Gas flotation	27	27
Parallel plate coalescers	48	31
Filters		
Loose media	21	15
Fibrous media	38	7
Gravity separation (4)		
Pits	35	31
Tanks	42	48

Source: Reference (3)

TABLE 15: PERFORMANCE OF VARIOUS TREATMENT SYSTEMS WYOMING AND PENNSYLVANIA

State	Type of Treatment System	Mean Effluent Oil and Grease (mg/l)	Number of Units in Data Base
Wyoming	Ponds	8.2	6
	Gas flotation	10.6	2
	Sand filtration	12.5	1
Pennsylvania	Ponds	4.1	4

Source: Reference (3)

Water produced along with liquid or gaseous hydrocarbons may vary in quantity from a trace to as much as 98% of the total fluid production. Its quality may range from essentially fresh to solids-saturated brine. The no discharge control technology for the treatment of raw wastewater after processing varies with the use or ultimate disposition of the water. The water may be:

1) Discharged to pits, ponds, or reservoirs and evaporated.
2) Injected into formations other than their place of origin.

Evaporation: In some arid and semiarid producing areas, use of evaporation is acceptable, although limited in its practice. The surface pit, pond, or reservoir can only be used where evaporation rates greatly exceed precipitation and the quantity of emplaced water is small. The pit or pond is ordinarily located on flat to very gently rolling ground and not within any natural drainage channel, so as to avoid danger of flooding. Pit facilities are normally lined with impervious materials to prevent seepage and subsequent damage to fresh surface and subsurface waters. Linings may range from reinforced cement grout to flexible plastic liners. Materials used are resistant to corrosive chemically treated water and oily wastewater.

In areas where the natural soil and bedrock are high in bentonite, montmorillonite, and similar clay minerals which expand upon being wetted, no lining is normally applied and sealing depends on the natural swelling properties of the clays. All pits are normally enclosed to prohibit or impede access.

In much of the Rocky Mountain oil and gas producing area, the total dissolved solids of the produced waters are relatively low. These waters are discharged to pits and put to use for local farmers and ranchers by irrigating land and watering stock. A typical produced water system widely in use is shown in Figure 13. A cross section of the individual pit is shown in Figure 14.

FIGURE 13: ONSHORE PRODUCTION FACILITY WITH DISCHARGE TO SURFACE WATERS

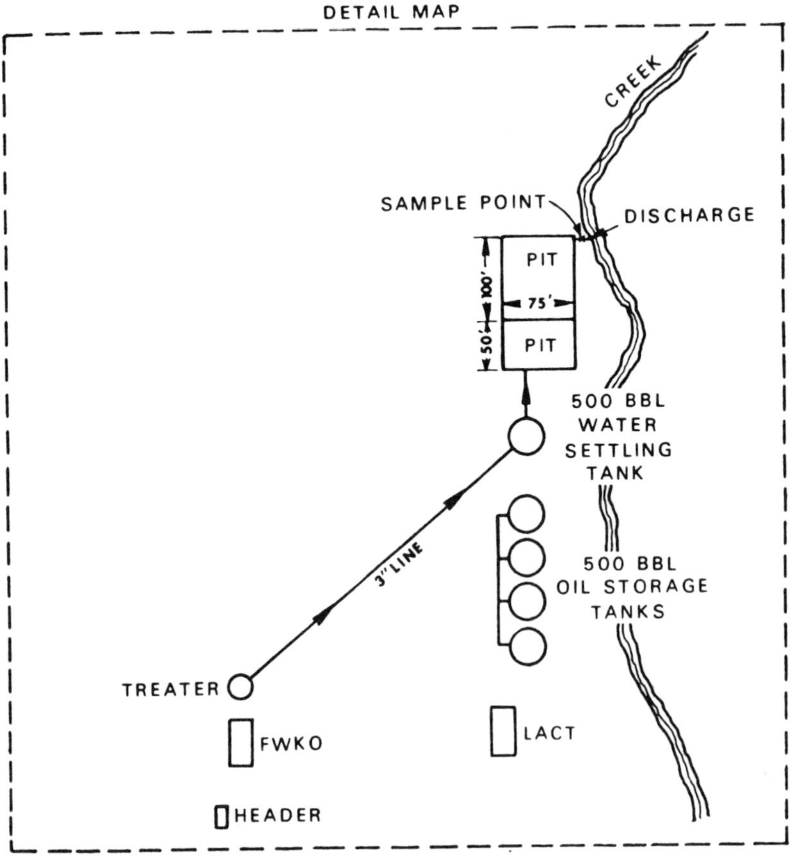

Source: Reference (3)

FIGURE 14: TYPICAL CROSS SECTION UNLINED EARTHEN OIL-WATER PIT

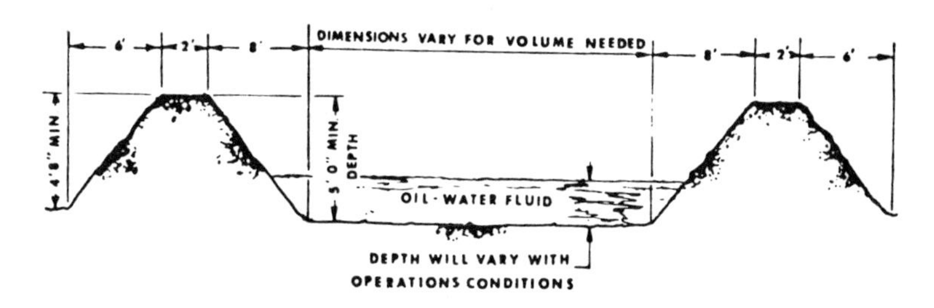

NOTE
PITS ARE EQUIPPED WITH PIPE DRAINS FOR SKIMMING OPERATIONS TO OBTAIN OIL-FREE WATER DRAINAGE

Source: Reference (3)

A producing oil field in Nevada discharges produced water to a closed saline basin. The basin contains no known surface or subsurface fresh water and is normally dry. The field contains 13 wells and produces approximately 33 barrels of brine per well per day.

Subsurface Disposal: Injection and disposal of oil field produced water underground is practiced extensively by the petroleum industry throughout the United States. The term "disposal" as used here refers to injection of produced fluids, ordinarily into a formation foreign to their origin. This injection is for disposal only and plays no intentional part in secondary recovery systems. (Injection for pressure maintenance or secondary recovery refers to the emplacement of produced fluids into the producing formation to stimulate recovery of additional hydrocarbons and is not considered end-of-pipe treatment.)

Current industry practice is to apply minimal or no treatment to the water prior to disposal. If water destined for disposal requires treatment, it is usually confined to the application of a corrosion inhibitor and bactericide; a sequestering agent may be added to waters having scaling tendencies. The amount of treatment depends on the formation properties, water characteristics, and the availability and cost of storage and stand-by wells.

Corrosion is ordinarily caused by low pH, plus dissolved gasses. Bactericides serve to inhibit the development of sulfate-reducing and slime-producing organisms. Chemicals and bactericides are frequently combined into a single commercial product and sold under various trade names.

A wide range of stable, semipolar, surface-active organic compounds have been developed to control corrosion in oil field injection and disposal systems. The inhibitors are designed to provide a high degree of protection against dissolved gasses (carbon dioxide, oxygen, and hydrogen sulfide), organic and mineral acids, and dissolved salts. The basic action of the inhibitors is to temporarily "plant" or form a film on the metal surfaces to insulate the metal from the corrosive elements.

The life of the film is a function of the volume and velocity of passing fluids. Inhibitors may be water soluble or dispersible in fresh water or brine. They may be introduced full strength or diluted. Treatment, usually in the range of 10 to 50 ppm, may be continuous or intermittent (batch or slug). Effectiveness of corrosion inhibition is determined in several ways, including corrosion coupons, hydrogen probes, chemical analyses, and electrical resistivity measurements.

Three primary types of bacteria attack oil field injection and disposal systems and cause corrosion:

1) Anaerobic sulfate-reducing bacteria (*Desulfovibrio desulfuricans*)—These bacteria promote corrosion by removing hydrogen from metal surfaces, thereby causing pitting. The hydrogen then reduces sulfate ions present in the water, yielding highly corrosive hydrogen sulfide, which accelerates corrosion in the injection or disposal system.
2) Aerobic slime-forming bacteria—These may grow in great numbers on steel surfaces and serve to protect growths of underlying sulfate-reducing bacteria. In extreme instances, great masses of cellular slime may be formed which may plug filters and sandface.
3) Aerobic bacteria that react with iron—Sphaerotilus and Gallionella convert soluble ferrous iron in injection water to insoluble hydrated ferric oxides, which in turn may plug filters and sandface. Oxygen entry into a system may also cause the formation of ferric oxide.

Treatment to combat bacterial attack ordinarily consists of applying either a continuous injection of 10 to 50 ppm concentration of a bactericide or batching once or twice a week.

Scale inhibitors are commonly used in the injection or disposal system to combat the development of carbonate and sulfates of calcium, magnesium, barium, or strontium. Scale solids precipitate as a result of changes in temperature, pressure, or pH. They may also be developed by combining of waters containing high concentrations of calcium, magnesium, barium, or strontium with waters containing high concentrations of bicarbonate, carbonate, or sulfate. Scale inhibitors are basically chemicals which chelate, complex, or otherwise inhibit or sequester the scale-forming cations.

The most widely used scale sequestrants are inorganic polymetaphosphates. Relatively small quantities of these chemicals will prevent the precipitation and deposition of calcium carbonate scale. Dimetallic phosphates or the so-called "controlled solubility" varieties are widely used by the oil industry in scale control and are preferred over the polyphosphates.

The downhole completion of a typical injection well is shown in Figure 15. A producing well is shown for comparison. Injection wells may be completed in a complicated fashion with multiple strings of tubing, each injected into a separate zone. If the disposal well is equipped with a single tubing string, and injection takes place through tubing separated from casing by packer, the annular space between tubing and casing is filled with noncorrosive fluids such as low-solids water containing a combination corrosion inhibitor bactericide, or hydrocarbons such as kerosene and diesel oil. All surface casing is cemented to the ground surface to prevent contamination of fresh water and shallow groundwater. Pressure gauges are installed on the casing head, tubing head, and tubing to detect anomalies in pressure. Pressure may also be monitored by continuous clock recorders which are commonly equipped with alarms and automatic shutdown systems if a pipe ruptures.

FIGURE 15: TYPICAL COMPLETION OF AN INJECTION WELL AND A PRODUCING WELL

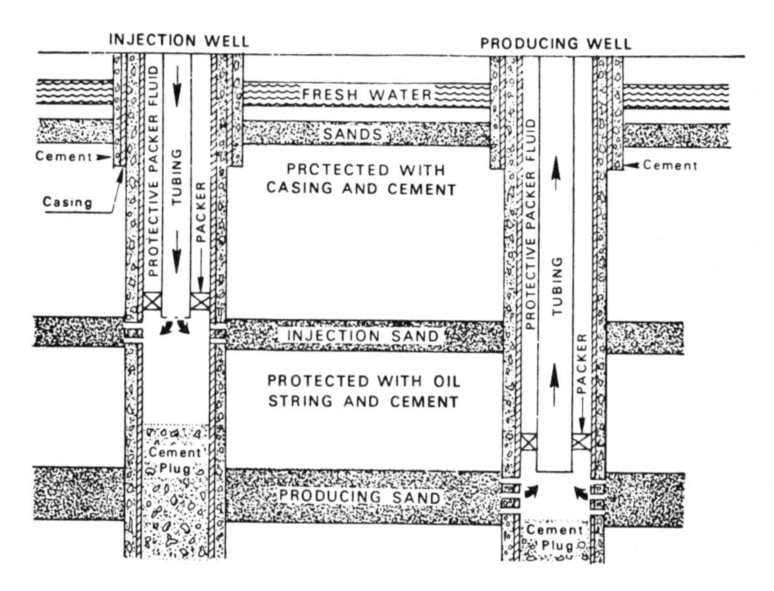

Source: Reference (3)

The injection well designed for pressure maintenance and secondary recovery purposes is completed in a manner identical to that of the disposal well, except that injection is into the producing horizon. Treatment prior to injection may vary from that applied to the disposal well inasmuch as water injected into the reservoir sandface must be as free of suspended solids, bacterial slimes, sludges, and precipitates as is economically possible. Ordinarily, selection of injection well sites poses few if any environmental problems. In many instances where injection is used for secondary recovery, the well site is fixed by the geometry of the waterflood configuration and cannot be altered.

Water for injection into oil and gas reservoirs requires treatment facilities and processes which yield clear, sterile, and chemically stable water. A typical open injection water treatment system includes a skim pit or tank (steel or concrete equipped with over-and-under baffles to remove any vestiges of nonsoluble material remaining after pretreatment); an aeration facility, if necessary to remove undesirable gasses such as hydrogen sulfide; a filtering system; a seepage-proof backwash pit; accumulator tank (sometimes referred to as a clear well or clear water tank) to retain the finished water prior to injection; and a chemical house for storing and dispensing treatment chemicals.

In the system described above no attempt is made to exclude air. Closed systems, on the other hand, are designed to exclude air (oxygen). This is desirable because the water is less corrosive or requires less treatment to make it noncorrosive. The truly "closed" system is difficult to attain because of the many potential points of entry of air into the production system. Air, for example, can be introduced into the system on the downstroke of a pumping well through worn stuffing box packing or seals.

In a few instances, closed injection (or disposal) system is used where product waters ordinarily have minimal corrosive characteristics, that is, where salt water is gathered from relatively few wells, fairly close together; where wells produce from a common reservoir; or where a one-owner operation is involved.

There are instances in which a closed input or produced water disposal system can be developed. In these systems all vapor space must be occupied by oxygen-free gas under pressure greater than atmospheric. If oxygen (air) enters the system, it is scavenged.

The "open" injection system has a much greater degree of operational flexibility than does the closed system. Among its more desirable factors are:

1) Wider range, type, and control of treatment methods.
2) Ability to handle greater quantities of water from different sources (diverse leases and fields) and differing formations.
3) Ability to properly treat waters of differing composition. This factor enables incompatible waters to be successfully combined and treated on the surface prior to injection.

The choice of a brine disposal zone is extremely important to the success of the injection program. Prior to planning a disposal program, detailed geologic and engineering evaluations are prepared by the production divisions of oil producing companies. Appraisal of the geologic reservoir must include the answers to questions such as:

1) How much reservoir volume is available?
2) Is the receiving formation porous and permeable?
3) What are the formation's physical and chemical properties?
4) What geological, geochemical and hydrologic controls govern the suitability of the formation for injection or disposal?
5) What are the short-term and long-term environmental consequences of disposal?

The geologic age of significant disposal and injection reservoirs throughout the nation ranges from relatively young rocks of Cambro-Ordovician period. Depths of disposal zones ordinarily range from only a few hundred feet to several thousand. However, prudent operators usually consider it inadvisable to inject into formations above 1,000 feet, particularly where the receiving formation has low permeability and injection pressures must be high. If the desired daily average quantity of water cannot be disposed of, except at surface pressures which exceed 0.5 pound per square inch surface gauge pressure per foot of depth to the disposal zone, particularly in shallow wells, an alternate zone is usually sought.

It is necessary to be familiar with both the lithology and water chemistry of the receiving formation. If interstitial clays are present, their chemical composition and compatibility with the injected fluid must be determined. The fluids in the receiving zone must be compatible with those injected. Chemical analyses are performed on both to determine whether their combination will result in the formation of solids that may tend to plug the formation.

The petroleum industry recognizes that the most carefully selected injection equipment means nothing if the disposed water is not confined to the formation into which it is placed. Consequently, the injection area must be thoroughly investigated to determine any previously drilled holes. These include holes drilled for oil and gas tests, deep stratigraphic tests, and deep geophysical tests. If any exist, further information as to method of plugging and other technological data germane to the disposal project is assembled and evaluated.

On the California Coast there is a definite trend for all onshore process systems which handle offshore production fluids to reinject produced water for disposal. Field investigations made in California were confined to OCS waters, with visits being made to five installations. Each of these facilities was performing some subsurface disposal; none were injecting for secondary recovery or pressure maintenance. Four of these installations were sending all or part of the produced fluids to shore for treatment.

All five installations were disposing of treated water in wells on the platform. Two were sending all fluids to shore, separating the oil and water, and then pumping the treated water back to the platforms for disposal. One installation was separating the oil and water on the platform and further treating the water so that it could be injected into disposal wells on the platform. Two of the platforms had been treating all fluids on the platform and injecting treated water. Since the total fluids produced are greater than the capacity of the disposal system, the excess treated water is being discharged overboard. Plans were being formulated to increase the capacity of the disposal system to return all produced water underground.

Produced water disposal is commonly handled on a cooperative or commercial basis, with the producing facility paying on a per-barrel basis. The disposal facility may be owned and operated by an individual, a cooperative association, or a joint interest group who may operate a central treatment or disposal system. The wastewater may be trucked or piped to the facility for treatment and disposal. Two examples of cooperative systems are operating in the East Texas Field and the Signal Hill and Airport Fields at Long Beach, California.

During major breakdown and overhaul of waste treatment equipment, it is common practice to continue production and bypass the treatment units requiring repair. This does not create a serious problem at large onshore complexes where dual treatment units are available, but at smaller facilities and on offshore platforms there may not be an alternate unit to use. Alternate handling practices vary considerably from facility to facility. The following methods are currently practiced offshore:

1) Discharge overboard without treatment.
2) Discharge after removal of free oil in surge tank.
3) Discharge to a sunken pile with surface skimmer to remove free oil.
4) Discharge of produced water to oil pipeline for onshore treatment.
5) Retention on the facility using available storage.
6) Production shutdown.

The method used depends upon the design and system configuration for the particular facility.

The following paragraphs describe some proprietary pollution control processes applicable to petroleum production operations.

Many oil field waters contain up to about 100 to 500 ppm of oil. This should be reduced to approximately 10 ppm or less if the water is to be dumped into surface streams. If this water is to be injected underground to aid in driving out the oil, as is common in secondary recovery operations, the oil content should be less than 10 ppm also.

An apparatus developed by *L.W. Jones; U.S. Patent 3,844,743; October 29, 1974; assigned to Amoco Production Company* is one for removing dispersed oil from water by contacting the oily water with sulfur to cause the oil to coalesce or agglomerate. This apparatus consists of a horizontal vessel containing a sulfur medium through which the oily water flows. The sulfur medium presents an area of solid phase sulfur to coalesce the dispersed oil.

The central part of the apparatus, as shown in Figure 16 is a horizontal cylindrical vessel **10** which has an oily water inlet **12** at the inlet end and the outlet end of the vessel has an oil outlet **14** at the top portion and a water outlet **16** at the lower end. The oily water flows from inlet **12** into a degassing chamber **18** which can be merely a vertical cylindrical vessel having a grid **20** at the top for escape of gas and an open end **22** at the lower end for the flow of water. A gas vent **24** with control valve **26** is provided in the upper side of vessel **10**, preferably directly above grid **20**.

FIGURE 16: APPARATUS FOR REMOVING DISPERSED OIL FROM PRODUCTION WASTEWATERS

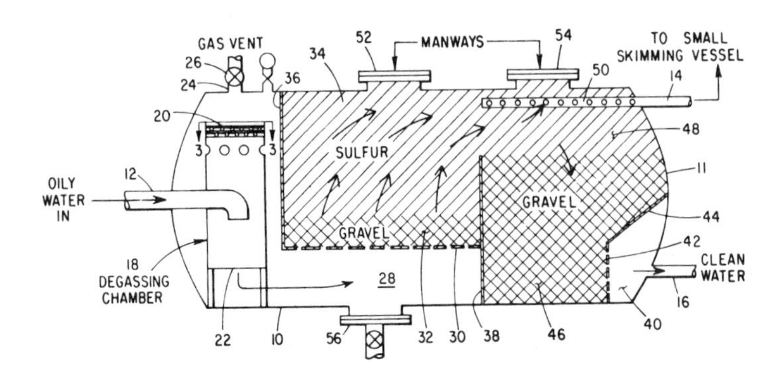

Source: U.S. Patent 3,844,743

The water flows from the degassing chamber to a water distribution chamber **18** in the lower part of vessel **10**. Immediately above chamber **28** is a perforated plate **30** which supports a gravel bed **32**. Immediately above the gravel bed is a sulfur bed **34**. The surface of sulfur is highly efficient as an oil collector. The surface of yellow sulfur is many times, at least three, more efficient in this respect than other forms of sulfur such as white sulfur, for example. Yellow sulfur is also many times, at least three, more efficient than other substances such as carbon and straw. The oil-sulfur contact angle of yellow sulfur is apparently very low for most unrefined oils and the adhesive force strong. Visual observations indicate that an oil film or sulfur can increase to a surprising thickness in the presence of flowing oily water such as water containing as little oil as 1 ppm. In contact with sulfur, the oil film appears to be abnormally cohesive.

It is believed that traces of sulfur are dissolved into the oil and induces increased oil-to-oil cohesiveness, possibly by reduced repulsive polar forces at unsaturated bonds in the hydrocarbon molecules. Bed **34** has a suface area of solid phase yellow sulfur to coalesce the dispersed oil. The purpose of the plate **30** and gravel bed **32** is merely to support the sulfur bed **32** and to provide for distribution of the oily water. A first plate **36** encloses the end of the sulfur bed **34** from the inlet side or end of vessel **10**. This causes all the water to flow through distributing chamber **28**.

A vertical baffle **38** is provided toward the outlet end **11** of chamber **28**. This baffle **38** extends, for example about half-way up through the sulfur bed. A clean water chamber **40** formed by the vessel **10** and perforated vertical wall **42** and sloping wall **44** is provided adjacent clean water outlet **16**. A gravel bed **46** is provided between vertical baffle **38** and clean water chamber **40**. Sulfur bed **34** has an auxiliary portion **48** which extends over gravel bed **46**. A perforated conduit **50** for collecting coalesced oil extends from oil outlet **14** into this portion **48** of the sulfur media. Manholes **52, 54** and **56** are provided for access to the interior of vessel **10**.

A few comments to be made concerning the operation of the device described above: Oily water is pumped into inlet **12** under pressure. If there is gas contained in the water it is separated out by degassing chamber **18** and removed through gas vent **24**. The degassed water flows into distribution chamber **28** and upwardly through gravel bed **32**. The oily water then flows up through sulfur bed **34** where the sulfur surface contacts the dispersed oil droplets and causes them to coalesce. By the time the flow of fluids reaches the auxiliary portion **48** of the sulfur media the fine drops of oil have all been coalesced. It is here that the separation of the oil and water occurs by gravity. The water flows downwardly through gravel bed **36** and out through clean water outlet **16**. The oil coalesces at the top of auxiliary bed **48** and is removed through oil outlet **14**.

It is contemplated that in the maximum efficient operation of this device there will be some water withdrawn with the oil through outlet **14**. However, as the oil at this stage has been coalesced, the oil can be readily removed from the water in a small skimming vessel exterior the vessel **10**.

The most efficient rate of the flow of oily water through the sulfur media can be easily determined. It may, of course, vary depending upon the quantity of oil in the water and how finely it is dispersed. For any flow rate, one should determine the oil content, if any, of the water flowing through clean water outlet **46**. This can be done by known laboratory techniques. If the oil has all been removed or removed to an acceptable minimum level, taking in account the intended use of the clean water, one is assured that the operation of the system is satisfactory. It is contemplated that typically the flow rate will be in the order of about 10 gallons per minute per square foot of sulfur bed **34** taken through a horizontal plane at the gravel interface.

In normal well drilling operation at an offshore site, a pressurized stream of liquefied drilling mud is introduced down the drill string as the latter rotates. The drilling mud functions both as a lubricant and as a vehicle to facilitate the cutting and removal of materials comprising the substratum. This rather heavy effluent stream carried from the well bore usually includes drilling mud, drilling cuttings, seawater and possibly oily constituents picked up from the substratum. As the drilling mud passes upwardly through the annulus defined by the rotating drill string and the bore hole, it acts as a vehicle for sand, clay, stone and other loosened solids which constitute the substratum.

These latter mentioned materials after being separated from the mud, as a matter of practicality, are normally returned to the water where they sink to the ocean floor. However, the cuttings are often coated with oily materials such as crude oil from the well bore, or other non-water-soluble constituents which make up the drilling mud mixture.

The discharge into the surrounding water of such non-water-soluble materials, can lead to a water polluting condition. Even the discharge of minor cutting amounts will tend to cause a visible discoloration at the water's surface.

Usually the drilling mud comprises essentially a water-based, flowable composition of adequate weight and chemical quality to facilitate operation under a particular set of circumstances. However, the mud is frequently compounded with a lubricant material such as diesel, crude oil, or the non-water-soluble petroleum base constituent, to facilitate the mud's lubricating characteristics.

A process developed by *L.P. Teague; U.S. Patent 3,860,019; Jan. 14, 1975; assigned to Texaco, Inc.* involves treating well drilling cuttings that normally accrue from the boring of a subterranean oil or gas well which includes circulating a drilling mud. The cuttings are introduced to a preliminary separator for removing a major part of the drilling mud. Thereafter, while yet containing a minor portion of mud, the cuttings are passed to a washer for contact with a cleaning detergent. A detergent recycling system simultaneously separates detergent from the residue of drilling mud, for reuse in the cutting cleaning process.

The apparatus which may be used in the conduct of this process is shown in Figure 17. Normally, the well bore cuttings treated according to the process are carried from well bore **4** during the drilling operation so long as the drilling mud flows. Thus, as drill string **6** is rotatably driven to urge the drill bit **7** downward, liquefied mud is forced under pressure through the drill string to exit at the lower drill bit. The mud thereby lubricates the downhole operation,

FIGURE 17: APPARATUS FOR WELL CUTTING TREATMENT AT AN OFFSHORE DRILLING SITE

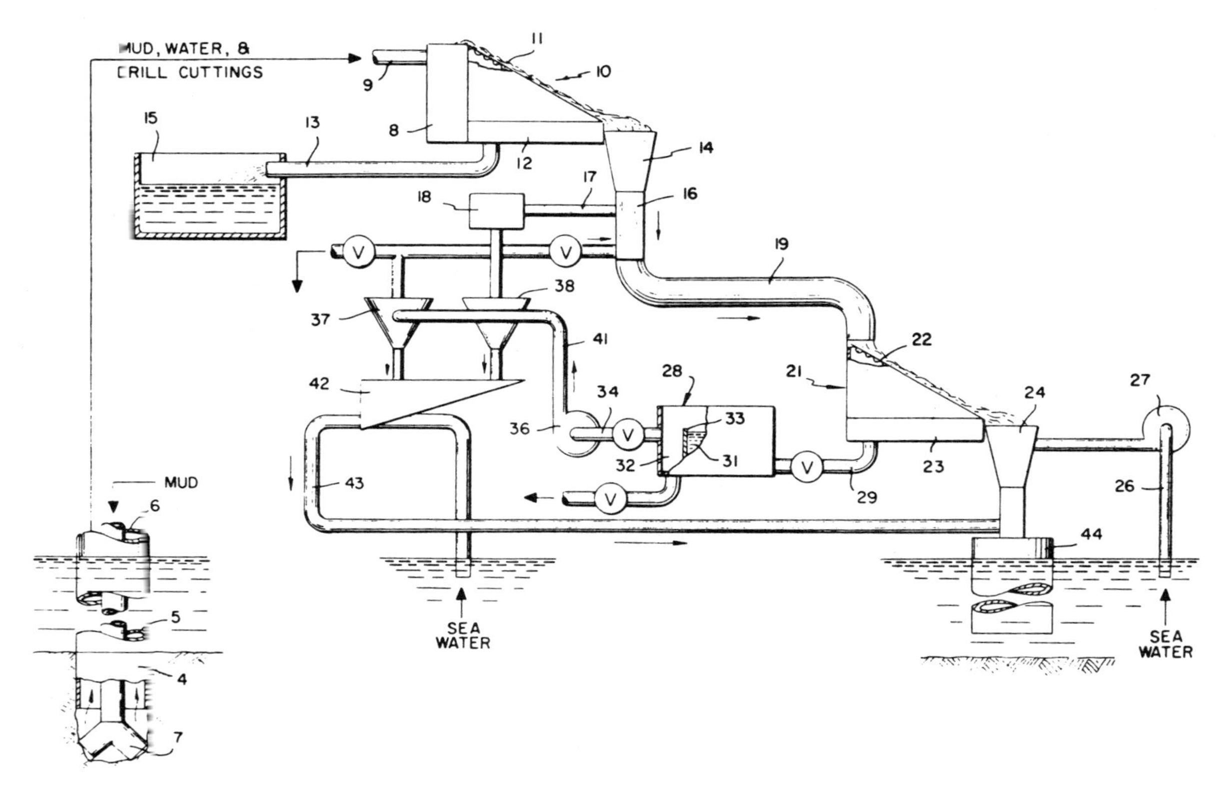

Source: U.S. Patent 3,860,019

and in passing upwardly through the annulus **5** between the drill string and the well bore wall, carries with it various forms of drilling cuttings as heretofore mentioned.

Further in regard to the drilling mud, as is generally known, the composition of the mud is usually compounded to the particular drilling situation and condition. More specifically, the weight and the chemical makeup of the mud are initially determined and subsequently altered as needed and as the drilling progresses.

The mud flow is urged under pressure from the well bore, upwardly to the drilling deck of the offshore platform, and discharged as an effluent stream by way of line **9**, into a tank **8** that is ancillary to shaker **10**. From the tank the mud mixture overflows onto the perforated face **11** of shaker **10**.

The shaker as shown, comprises a vibratory or stationary type separator having a tilted screen working surface **11** upon which the mud mixture overflows from the tank. The mesh of the screen utilized on the shaker face is variable, being contingent on the characteristics of the substratum being drilled and the type of drilling cuttings being carried by the mud flow.

In the shaker, the liquefied mud vehicle will by and large traverse the screen openings and is received in an underpositioned pan **12**. The latter is communicated by a conduit **13**, with a mud storage tank **15**. The remainder of the mixture deposited on the shaker face, and which does not pass through the screen, will comprise essentially an aggregate of solids being of sufficient size to remain at the screen surface. Solid matter, through the screen's vibratory action or through gravity flow, advances along the screen face to be discharged at the lower side thereof.

A collector **14** communicated with the shaker face receives the stream of drilling cuttings which in essence comprises a conglomerate of solid matter as well as some liquid. This flowable mass further embodies the previously mentioned non-water-soluble, oily base constituents which normally cling to the cuttings. A wash chamber **16** is communicated with the collector discharge outlet to receive a stream of unprocessed drilling cuttings. Wash or spray chamber **16** includes a compartment adapted to receive the downwardly passing drilling cuttings, with means in the compartment to retain the cuttings sufficiently long to be brought into contact with the liquid detergent.

The wash chamber, which in this arrangement is a spray chamber, further includes a spray nozzle system **17** disposed thereabout and appropriately arranged to deliver detergent streams against the cuttings. The spray nozzle system is communicated with a pressurized liquid detergent source represented by reservoir **18**.

Toward cleaning or scouring the cuttings of oily matter, the cuttings as an alternative can be immersed in a bath rather than being sprayed. The apparatus used in this latter step will be adapted in accordance with the consistency and the volume flow of cuttings as well as with other features of the drilling process.

Within the spray chamber, detergent is brought into contact with the cuttings under sufficient pressure and/or turbulence to remove substantially all of the extraneous matter clinging thereto. An elongated conduit **19** directs a stream of liquid detergent and drilling cuttings from the spray chamber whereby to physically expose the cuttings to the cleaning and separating action.

For this use, and toward achieving the necessary scouring and cleaning function, the detergent liquid can include any of a number of commercial solutions as for example, a biodegradable phosphate-free detergent.

Conduit **19** is connected at the discharge end thereof to a second separator **21**. The latter, as in the instance of shaker **10**, is a vibratory unit having a screen-type face **22** across which the detergent and cuttings flow is directed. The mesh size or openings of the screen face **22** are usually smaller than the mesh of screen face **11**, and of a sufficient size to pass the liquid detergent therethrough and into shaker reservoir **23**. The remaining cuttings stream, substantially free of detergent and other liquid, falls from screen top **22** and into a discharge chamber **24**.

This chamber includes a receptacle to receive and retain the flow of cleaned cuttings for further cleaning. The receptacle **24** in the instant arrangement is communicated with seawater drawn from the immediate area by conduit **26** and pump **27**, or from an alternate source of water. After further cleaning by contact with seawater, the cleaned cuttings are discharged into a downcomer **44**. The member comprises in its simplest form an elongated tubular conduit that extends downwardly beneath the water's surface terminating short of the sea floor. Cuttings deposited at the upper end thereby are directed toward the floor where they tend to settle without the concern of prompting a water polluting situation on the surface.

Solution, including detergent separated from the drilling cuttings within separator **21**, is received in the reservoir. The latter includes an outlet communicated with the inlet of skimmer tank **28** by a valved connecting line **29**. The skimmer tank embodies a first compartment **31** into which the detergent is fed and into which additional detergent can be added if such addition is required for reconstituting the material.

A second compartment **32** is communicated with the first compartment across a transverse panel **33**. Compartment **32** is provided with an outlet to receive detergent in valved line **34**, which in turn is communicated with the suction of detergent pump **36**. The discharge of the pump is communicated with one or more hydrocyclone units **37** and **38** or similar fluid separating units, by way of line **41**. The function of the units **37** and **38** is to provide a final separation of detergent from any remaining materials in the flow stream.

The hydrocyclone units function to centrifuge detergent from any remaining mud, water, and/or other fluidized or particulated components. The separated and cleaned fine solids pass upwardly into manifold **42** and are carried by line **43** into conduit **44**. The detergent material, essentially free of solids, discharges into wash chamber **16** to again contact the incoming mud and cuttings flow.

Open-ended downcomer conduit **44**, as noted herein is disposed in the body of water, normally depending from the offshore drilling platform. The conduit **44** preferably is positioned with its lower open end spaced from the floor at the offshore location, or provided in the alternate with openings formed about the lower end. The conduit or caisson upper end is open to the atmosphere and disposed in alignment with the discharge opening of cuttings collector **24**. Thus, in the course of this process, substantially clean cuttings are fed into the caisson upper end. The clean cuttings thus enter the water and flow downwardly by gravity through the caisson, to be deposited at the ocean floor.

Use of the disclosed method for treating drilling cuttings results in a cleaner operation as well as a more economical one. The method serves to maintain a nonpolluting condition at the offshore production or drilling site and also permits maximum recovery of both drilling mud and washing detergent for the subsequent reuse of both items.

An apparatus developed by *M.O. Stearns and J.A. Gill; U.S. Patent 3,901,254; Aug. 26, 1975; assigned to NL Industries, Inc.* is an apparatus for cleaning oily mud from cuttings at offshore drilling locations and disposing of the cuttings without the formation of an oil slick. The apparatus comprises a cuttings washer in which the cuttings are sprayed with a wash liquid and are then conveyed to a partially submerged downpipe and discharged therein below the water level, whereby the cuttings fall to bottom, releasing their oily coating which rises within the downpipe to a submersible pump therein which pumps the released oil to a receiver for optional recycling into the mud system.

Figure 18 shows such an apparatus on the platform of an offshore drilling location, which is supported by a number of vertical supports of which one, **11**, is indicated in the figure. **12** indicates the downpipe, which is conveniently welded or otherwise attached to the support leg **11**. The downpipe is conveniently of 20-inch diameter steel pipe and should extend about 30 feet below the normal water level, provided of course that the water depth at the drilling locality permits this. Otherwise, it should extend to within not closer than about 10 feet of the ocean or lake bottom. The downpipe **12** should also extend sufficiently above the normal water level so as to be above the water regardless of tidal and normal wave action, for which 8 or 10 feet normally suffices. Of course, in estuarine and like locations where large tidal action occurs suitable allowance should be made therefor.

The encased drilling string **13** is shown diagrammatically, with the flow line **14** indicating the conduit in which the drilling fluid with its burden of cuttings is conducted to vibrating screens **15**, which shake a substantial portion of the oily mud from the cuttings, from which the cuttings are discharged into a cuttings washer **17**, which conveniently takes the form of a relatively small vibrating screen over the top of which is arranged a series of spray nozzles, supplied with wash liquid from a header **22**. The cuttings are thus washed with the wash liquid, a large proportion of which falls through the screen carrying with it a portion of the oil mud from the cuttings.

However, even with the spray action and vibratory screen action to which the cuttings are subjected in the cuttings washer **17**, not all of the oily drilling fluid is necessarily removed from the cuttings. In any case, some of the wash liquid, to be described in detail hereinbelow, amalgamates with the residual oil on the cuttings and renders it more readily susceptible to eventual dislodgement when immersed in the water, generally seawater, in the downpipe **12**. The cuttings are simultaneously washed and propelled to the discharge end of the cuttings washer, when they fall into flume **24** and are conveyed by the action of gravity into the downpipe **12**. Water from the body of water surrounding the platform **10** or from the separator **29** may be pumped to the flume **24** to assist in conveying the washed cuttings to the downpipe **12**.

The flume **24** terminates and discharges the cuttings below the water level and below the pickup level of the submersible pump **25**. The cuttings fall through the downpipe **12**, and eventually accumulate on the sea bed or lake bed **26** as a tailings pile **27**. In their passage through the downpipe, however, and in view of the fact that the cuttings are essentially water-wettable, and more particularly aided by the amalgamated wash liquid carried by the cuttings, the oil together with wash liquid is displaced from the cuttings by the water and rises upwardly within the downpipe to form an oil layer adjacent to the pump **25**. The latter pumps the top layer of liquid within the downpipe, which as a practical matter includes both oil and water, and pumps it through conduit **28** to the oil water separator **29**. The oil freed of any water is then pumped from the separator **29** through conduit **30** to the oil receiving tank **31**.

FIGURE 18: POLLUTION-FREE WELL CUTTINGS DISPOSAL APPARATUS

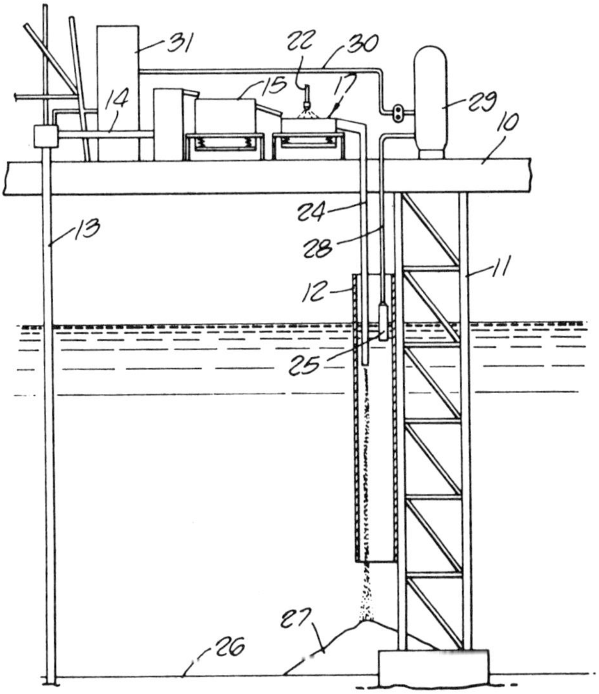

Source: U.S. Patent 3,901,254

A process developed by *M. Guillerme, J. Gratacos, A. Sirvins and B. Tramier; U.S. Patent 4,035,289; July 12, 1977; assigned to Societe Nationale Elf Aquitaine (Production), France* involves purification of the effluents from mineral oil drillings, carried out on land, for permitting the discharge of the residual waters into the natural surroundings. First, microorganisms are cultivated on a volume of the effluent, in order to make them capable of digesting the impurities of the effluent. The culture, thus obtained, is then developed in the mass of the effluent to be treated until the BOD is reduced to the required limit. This treatment is preferably followed by a flocculation with chemical agents in the known manner. This antipollution treatment permits the regulations in force to be applied in all locations where the drillings are undertaken, whatever may be the auxiliary fluids used in the drilling operation.

Petroleum Storage

Crude oil, intermediate, and finished products are stored in tanks of varying size to provide adequate supplies of crude oils for primary fractionation runs of economical duration, to equalize process flows and provide feedstocks for intermediate processing units, and to store final products prior to shipment in adjustment to market demands. Generally, operating schedules permit sufficient time for settling of water and suspended solids (8).

VAPOR EMISSIONS

A major hydrocarbon emission source in the petroleum industry is tankage. Storage emissions depend on diurnal temperature and pressure changes, filling operations, volatilization, solar radiation, and mechanical condition of the tanks (10).

Proper design of storage tanks will control hydrocarbon emissions greatly. There are five basic types of storage tanks used in the petroleum industry. These are: fixed roof, floating roof, internal floating cover, variable space, and pressure. The applicability of these tanks largely depends on the volatility of the stored liquid. Table 16 shows the type of tank generally used for storing certain volatile petroleum products.

In order to calculate the hydrocarbon emissions from petroleum storage, the following assumptions are used:

1) Storage capacity is one month for feed and products.
2) Only crude and gasoline storage will result in significant hydrocarbon emissions. Light fuel oil storage will result in a small hydrocarbon emission.
3) Heavy fuel oil storage and pressurized storage of high volatility products will result in negligible emission.
4) Crude, gasoline, and fuel oils will be stored in floating-roof tanks.

Hydrocarbon emission factors for floating roof tanks are the following:

crude oil	0.029 lb/day-10^3 gal
gasoline	0.033 lb/day-10^3 gal
light fuel oil	0.0052 lb/day-10^3 gal

From a hydrocarbon emissions standpoint, any storage facility which is flexible enough to retain all vapors emitted from the stored hydrocarbon at an economically feasible level is the system most desirable. The fixed roof tank is the only tank mentioned above which is not designed for containing vapors emitted. Usually, in a fixed roof tank, the vapors are vented to the atmosphere through a pressure-vacuum vent. The variable space-type storage tank is designed to control normal diurnal breathing losses and small filling operations. Large changes in the vapor content cannot be handled by the variable space tank alone and thus vapors will have to be vented to the atmosphere. Pressure vessels are used to hold highly volatile petroleum products under pressure. The shape of the pressure vessel depends on the pressure required. Spheres can be operated at pressures up to 217 psi; spheroids, up to 50 psi; noded spheroids, up to 20 psi; and plain or noded hemispheroids, up to 75 psi and 2½ psi, respectively (10).

TABLE 16: NATURE OF PRODUCT STORAGE AT REFINERIES

Product	True Vapor Pressure (psia at 60°F)	Types of Storage Tanks	Approximate Quantity Stored (10^6 bbl)
Fuel gas	—	Cryogenic, pressurized	—
Propane	105	Pressurized	—
Butane	26	Pressurized	—
Motor gasoline	4-6	Variable space, fixed roof, floating roof	204
Aviation gasoline	2.5-3	Variable space, fixed roof, floating roof	14
Jet naphtha	1.1	Variable space, fixed roof, floating roof	18
Jet kerosene	<0.1	Fixed roof	31
Kerosene	<0.1	Fixed roof	46
No. 2 distillate	<0.1	Fixed roof	346
No. 6 residual	<0.1	Fixed roof	346
Crude oil	2	Variable space, fixed roof, floating roof	137

Source: Reference (10).

Floating Roof Tanks

The floating roof tank is the most commonly used tank for controlling hydrocarbon emissions. Modern designs include pontoon deck floating roofs, double-deck floating roofs, and trussed-pan floating roofs. The major concerns in design of the roof are structural support and reduction of heat conduction due to solar radiation. The floating roof is constructed about eight inches shorter in diameter than the inside tank diameter. The space between is usually sealed by vertical shoes (metal plates) connected by braces to the floating roof. A fabric seal is also included to reduce hydrocarbon emissions. The space between the roof and the wall is the greatest source of emissions from a floating roof tank. Frequently an additional secondary seal is added to act as a wiper reducing the wicking action associated with floating roof tanks. Another sealing device is a flexible tube which floats on the liquid surface and keeps contact between the roof and the tank wall. Floating roof tanks are about 91% efficient in controlling hydrocarbon emissions from gasoline storage.

Fixed Roof Tanks

In order to limit emissions from fixed roof tanks, floating covers have been designed. One type is a floating plastic blanket. The blanket is constructed with plastic floats underneath and custom manufactured so that only a one-inch gap remains around the periphery. A skirt is placed above this gap to further eliminate fugitive vapors. Another new technique in sealing fixed roof tanks is a floating microsphere blanket. The microspheres are made of plastic resins less dense than the liquid petroleum product. The microspheres entirely cover the liquid surface and their fluidity gives the added advantage of being able to flow around internal tank parts.

Vapor Recovery Systems

With present day prices of petroleum products increasing, more sophisticated systems for recovering hydrocarbon vapors are becoming economically feasible. One system employed is an integrated vapor recovery system. The vapor recovery system is a closed system which is set up to recover hydrocarbon vapors emitted from storage facilities and also the loading facilities. A typical vapor recovery system including a vapor saver is shown in Figure 19. The variable space tank included is designed to control breathing losses and small vapor changes within the system. The vapor recovery unit which handles large vapor volume changes such as during loading operations or periods of drastic ambient temperature or pressure changes includes a compressor-refrigeration system.

Vapor recovery units can liquefy hydrocarbon vapors by several principles which include compression-refrigeration, absorption, and adsorption. They also can employ a combination of these principles. The efficiency of vapor recovery units typically ranges from 90% to 95%, depending upon the composition and concentration of the hydrocarbon vapors processed. Vapor recovery units are manifolded into the vapor collection systems of tankage and loading operations for the reliquefaction of hydrocarbon vapors into product. Vapor recovery systems are quite expensive as compared to floating roof tanks, and only give a little greater efficiency in recovering vapors.

FIGURE 19: COMPLETE VAPOR RECOVERY SYSTEM

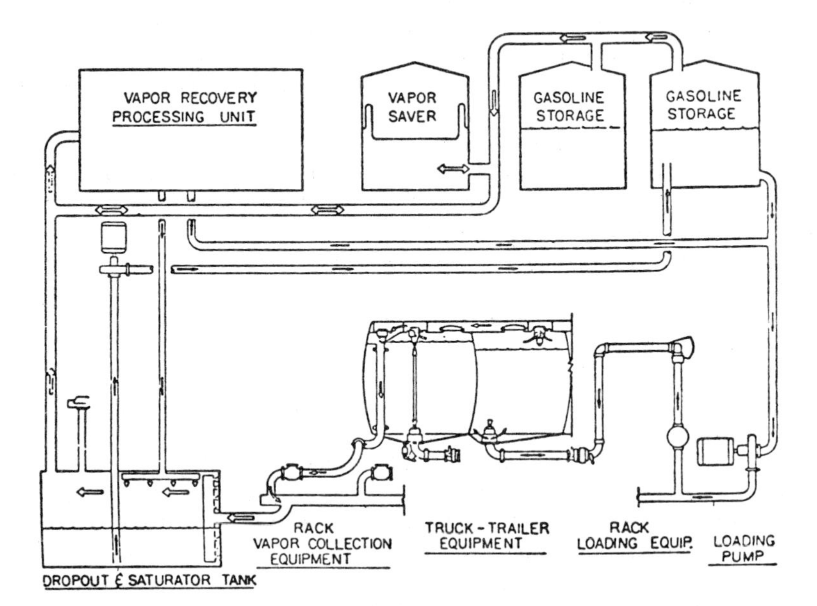

Source: Reference (10).

Another possible control technique is maintaining wet scrubbers or condensers on the vents of fixed roof tanks. The wet scrubbers can be bubble-cap tray towers, packed towers, spray towers, or Venturi scrubbers. These types of scrubbers are shown in Figure 20. The common absorbents for organic vapors are water, mineral oil, nonvolatile hydrocarbon oils, and aqueous solutions (e.g., solutions of oxidizing agents, sodium carbonate, or sodium hydroxide). If water is used, the hydrocarbon-rich water is sent to a closed wastewater stream to be treated at the wastewater treatment facility. The other absorbents can be regenerated, with the collected hydrocarbons.

Activated carbon is an adsorbent that can be used for hydrocarbon emissions. The adsorbed hydrocarbons are removed from the carbon by steam stripping and then recovered by decantation or distillation. Costs of activated carbon adsorbers are high, but the recovery of valuable hydrocarbons enhances the feasibility of the operation.

Storage Tank Maintenance

Heat from solar radiation causes problems by increasing hydrocarbon boil-off. Painting tanks and proper tank design reduce the radiation effect. The paints chosen are those that best reflect solar radiation. Table 17 lists the effectiveness of various paints on reflecting heat. Proper tank design includes a double-deck pontoon-type floating roof or trussed floating roof to avoid direct warm metal-liquid contact. Tank diameter also affects the amount of hydrocarbon emitted. A smaller tank diameter will have less emissions.

FIGURE 20: HYDROCARBON VAPOR SCRUBBERS

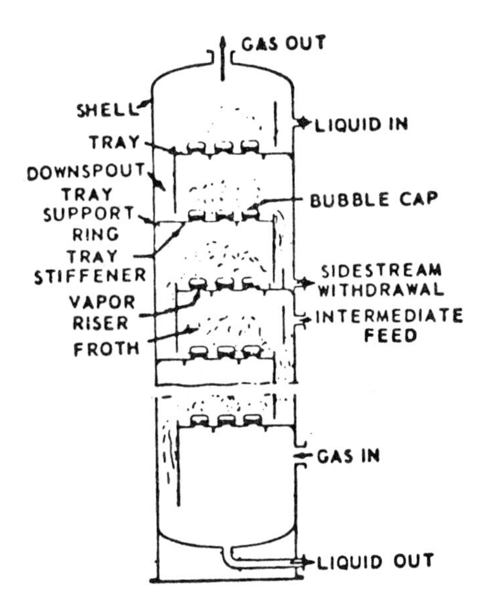

Schematic diagram of a bubble-cap tray tower.

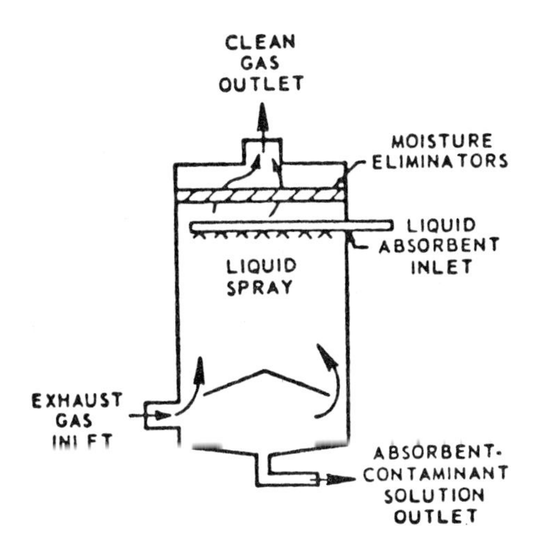

Spray tower.

(continued)

FIGURE 20: (continued)

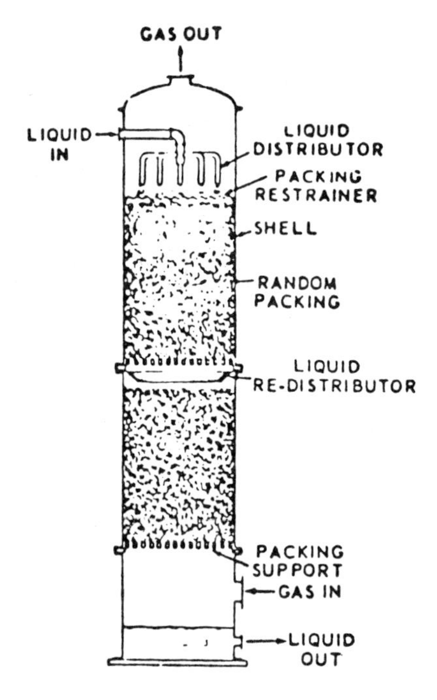

Packed tower.

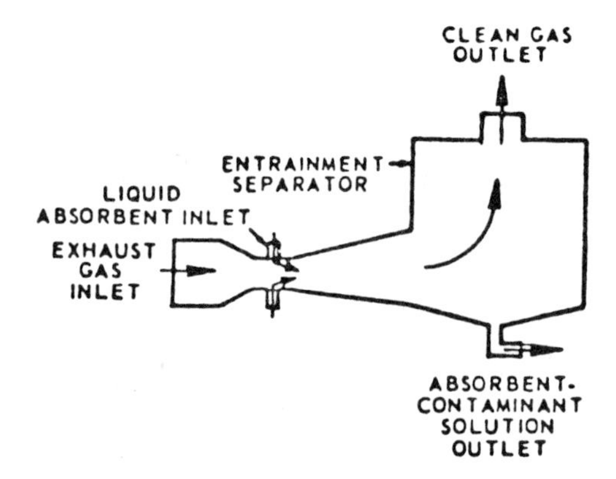

Venturi scrubber.

Source: Reference (10)

TABLE 17: RELATIVE EFFECTIVENESS OF PAINTS IN KEEPING TANKS FROM WARMING IN THE SUN

Color	Relative Effectiveness as Reflector or Rejector of Heat, %
Black	0
No paint	10.0
Red (bright)	17.2
Red (dark)	21.3
Green (dark)	21.3
Red	27.6
Aluminum (weathered)	35.5
Green (dark chrome)	40.4
Green	40.8
Blue	45.5
Gray	47.0
Blue (dark Prussian)	49.5
Yellow	56.5
Gray (light)	57.0
Aluminum	59.2
Tan	64.5
Aluminum (new)	67.0
Red iron oxide	69.5
Cream or pale blue	72.8
Green (light)	78.5
Gray (glossy)	81.0
Blue (light)	85.0
Pink (light)	86.5
Cream (light)	88.5
White	90.0
Tin plate	97.5
Mirror or sun shaded	100.0

Source: Reference (10)

Proper maintenance practices help to eliminate hydrocarbon emissions. Particular trouble spots are leaky and poorly regulated vents on fixed roof tanks and leaky seals on floating roof tanks. Maintaining properly painted tanks helps in eliminating emissions. Proper scheduling such as pumping liquids into storage tanks during cool hours and withdrawing liquids at hotter times and maintaining short periods between pumping operations should be followed.

WASTEWATER CHARACTERISTICS

Wastewaters associated with storage of crude oil and products are mainly in the form of free and emulsified oil and suspended solids. During storage, water and suspended solids in the crude oil separate. The water layer accumulates below the oil, forming a bottom sludge. When the water layer is drawn off, emulsified oil present at the oil-water interface is often lost to the sewers. This waste is high in COD and contains a lesser amount of BOD_5. Bottom sludge is removed at infrequent intervals. Additional quantities of waste result from leaks, spills, salt "filters" (for product drying), and tank cleaning.

Intermediate storage is frequently the source of polysulfide bearing wastewaters and iron sulfide suspended solids. Finished product storage can produce high BOD_5, alkaline wastewaters, as well as tetraethyllead. Tank cleaning can contribute large amounts of oil, COD and suspended solids, and a minor amount of BOD_5. Leaks, spills and open or poorly ventilated tanks can also be a source of air pollution, through evaporation of hydrocarbons into the atmosphere.

DESIGN TRENDS

Many refineries now have storage tanks equipped to minimize the release of hydrocarbons to the atmosphere. This trend is expected to continue and probably accelerate. Equipment to minimize the release of hydrocarbon vapors includes tanks with floating-roof covers, pressurized tanks, and/or connections to vapor recovery systems. Floating-roof covers add to the wastewater flow from storage tanks. Modern refineries impose strict Bottom Sediment and Water (BS&W) specifications on crude oil supplies, and frequently have mixed-crude storage tanks; consequently, little or no wastewater should originate from modern crude storage. Another significant trend is toward increased use of dehydration or drying processes preceding product finishing. These processes significantly reduce the water content of finished product, thereby minimizing the quantity of wastewater from finished product stroage (8).

Petroleum Transportation

The proper design, operation and maintenance of petroleum transportation facilities have become increasingly vital with the advent of the Alaskan pipeline, supertankers and superports. Down the line from those rather mercurial problems, attention is focussing even at the point of delivery—at the service station hose where vapor emissions are becoming of increasing concern.

AIR POLLUTION CONTROL

Air pollution by petroleum products is one of the most urgent problems of our present society. Freedom of travel in daily activities is directly affected by the availability of gasoline for motor vehicles. For example, approximately 2.2 billion gallons per year of motor vehicle gasoline are consumed within the San Francisco Bay Area air pollution control district alone. Marketing of such gasoline involves the transfer of gasoline from one container to another, as for example, from a refinery storage tank to a bulk handling motor vehicle tank truck, thence from the tank truck to an underground storage tank at a service station, and thence from the underground storage tank to an automobile gasoline tank. The transfer of liquid gasoline from one container to another container produces gasoline rich vapors which are displaced into the atmosphere as the container is filled. In the Bay Area district it is estimated that 75 tons per day of gasoline enter the district area atmosphere.

Air pollution in the form of haze and smog formation includes breakdown of hydrocarbons of a type found in gasoline vapor. Gasoline marketing operations involving transfer of gasoline from one container to another container require careful consideration if improvements in air quality are to be achieved.

Vapor Control

Hydrocarbon emissions from transport loading operations are generally controlled by the use of a vapor collection device manifolded into a vapor recovery unit. The transport vehicle may be a tank truck, rail car, barge, or marine vessel.

The type of vapor collection system installed depends on how the transport vehicle is loaded. If the unit is top loaded, vapors are recovered through a top loading arm. Product is loaded through a central channel in the nozzle. Displaced vapors from the compartment being loaded flow into an annular vapor space surrounding the central channel and in turn flow into a hose leading to a vapor recovery system.

If the transport is bottom loaded, the equipment needed to recover the vapor is considerably less complicated. Vapor and liquid lines are independent of each other with resultant simplification of design. Product is dispensed into the bottom of the transport and displaced vapors are collected from the tank vents and returned to a vapor recovery unit.

Bottom loading vapor recovery has many advantages over top loading vapor recovery. Bottom loading generates much less vapor, generates almost no mist and is safer from a static electricity point of view.

The vapor collection efficiency of loading controls is in excess of 95%. However, the overall emission reduction is also dependent on the efficiency of the vapor recovery unit. A 90% efficient vapor recovery unit would make a loading control system 85% efficient (10).

A system which was developed by *L.T. Cavallero and W.J. Elnicki; U.S. Patent 3,817,687; June 18, 1974; assigned to Aer Corp.* is a pollution-free system for disposing of gasoline vapors at a tank truck loading station. A booster pump energized in response to pressure within a vapor saver tank which receives the truck tank vapors feeds the vapors from the vapor saver tank through a control valve to the supply line of the hydrocarbon oxidizer. Means responsive to the temperature in the oxidizer operate the control valve to regulate the vapor flow to the oxidizer to maintain the temperature therein below a predetermined temperature. Various safety devices are incorporated to ensure the safety of both the installation itself and of personnel working at the installation.

Such a system is shown in Figure 21. The system **10** is adapted to be installed at a station at which gasoline tank trucks such, for example, as a truck having a tank **12**, are to be refilled with gasoline for distribution. While there is illustrated only one tank truck in the drawings, it will readily be appreciated that there are facilities for handling a multiplicity of such trucks at the normal refilling station. At such a station, gasoline is fed into the tank **12** through a line **16** leading into the refill fixture **14** on the tank. At the same time, vapor from within the tank **12** is forced outwardly through a line **18** and past a check valve **19**. A gate valve **20** is adapted to be opened to pass the vapor to a vapor saver tank **22** of a type known in the art. Further as is known in the art, the tank **22** includes a bladder **24** which expands much in the manner of a balloon as the vapor pressure builds up within the tank **22**.

FIGURE 21: APPARATUS FOR VAPOR INCINERATION AT A TANK TRUCK LOADING STATION

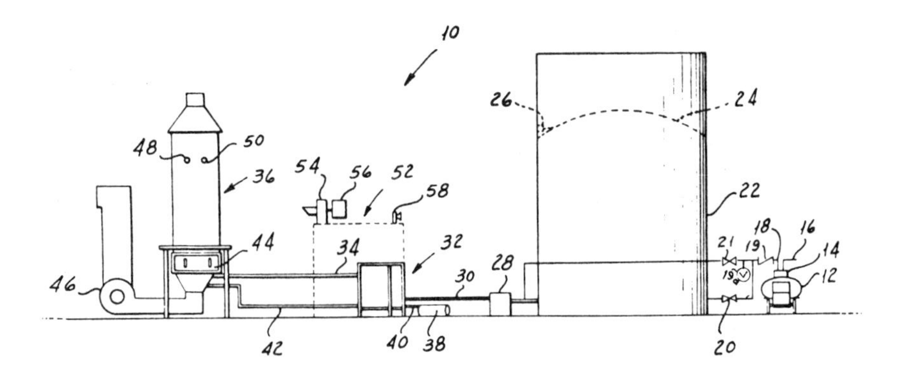

Source: U.S. Patent 3,817,687

The tank **22** is provided with a switch **26** adapted to be operated by the bladder **24** when the pressure within the tank reaches a predetermined value to activate a booster pump **28** to feed vapor from the tank **22** through a line **30** leading to a piping rack, indicated generally by the reference character **32**. In some instances it may be desirable to eliminate the tank **22** and to feed the vapor from the tank **12** directly to the booster pump **28**. In such a case the vapor comes from the truck through line **18** and past a check valve **19** to a specially designed valve **21**. The valve **21** is activated by a specially designed pressure switch **19a**. As the valve **21** is activated the booster pump **28** is also activated to feed vapor through a line **30** leading to a piping rack, indicated generally by the reference character **32**.

The specially designed switch **19a** is known to the art, is manufactured by The Foxboro Company, and is known as Model 43A Controller. The arrangement is such that it continuously detects the difference between an actual measurement and its desired value and converts this difference to an air signal between 3 pounds and 15 pounds per square inch. It is this air signal which operates the valve. The actual measurement, the desired measurement, and the output signal are indicated on a controller. The arrangement is such that an air supply passes through a regulator, then through a pneumatic controller, and then to the control valve. The valve **21** is also known to the art and is manufactured by Fisher Controls Company. By way of example and not by way of limitation, the valve is a rotary butterfly which can maintain bubble-tight shutoff. As the valve rotates to open position, it actuates a rotary switch (not shown) to make electrical contact to activate the booster **28**. Vapor from the rack **32** is carried by a pipe **34** to the supply line of a hydrocarbon oxidizer indicated generally by the reference character **36**, whereat the vapor is ignited and burns. Any suitable type of pilot fuel can be employed, for example, as butane contained in a tank **38**. A line **40** conducts pilot fuel from tank **38** to the rack **32**. Another pipe **42** leads from the rack **32** to the pilot line of the oxidizer **36**.

Oxidizer **36** includes a door **44** which affords access to the interior of the oxidizer. A combustion air fan **46** is adapted to be activated in a manner to be described hereinbelow to supply combustion air to the oxidizer **36**. A pair of thermocouples **48** and **50** sense the temperature within the stack of the oxidizer. The system includes a control house indicated generally by the reference character **52**, which houses the control panel (not shown) and associated equipment for the system.

A purging fan **54** driven by a motor **56** is mounted on housing **52** to ensure that the interior of the housing is at all times free of combustible vapors. A horn **58** is energized each time the system starts up to warn personnel in the area.

More specifically, the booster **28** may be, for example, a three-quarter horsepower turbine rated at 100 cubic feet per minute at eight inches driven by a three quarter horsepower explosion proof motor. The pressurizing or purge fan **54** may be any suitable type of blower driven by a one-third horsepower explosion proof motor **56**. The supply fan may, for example, have a capacity of 6,500 cubic feet per minute driven at 1,368 rpm by a five horsepower motor which also should be explosion proof.

A scheme developed by *G.P. Canevari and W.M. Cooper, Jr.; U.S. Patent 3,850,206; November 26, 1974; assigned to Exxon Research and Engineering Co.* is one in which a layer of an aqueous foam is applied to the surface of crude oil during loading into vessels or tankage thereby preventing loss of hydrocarbon vapor and the resultant air pollution as the tank is filled.

Figure 22 shows the application of this scheme to tanker loading at three different stages of the loading process.

FIGURE 22: DIAGRAMS SHOWING THE USE OF FOAMED VAPOR BARRIERS IN LOADING OF CRUDE OIL INTO TANKERS

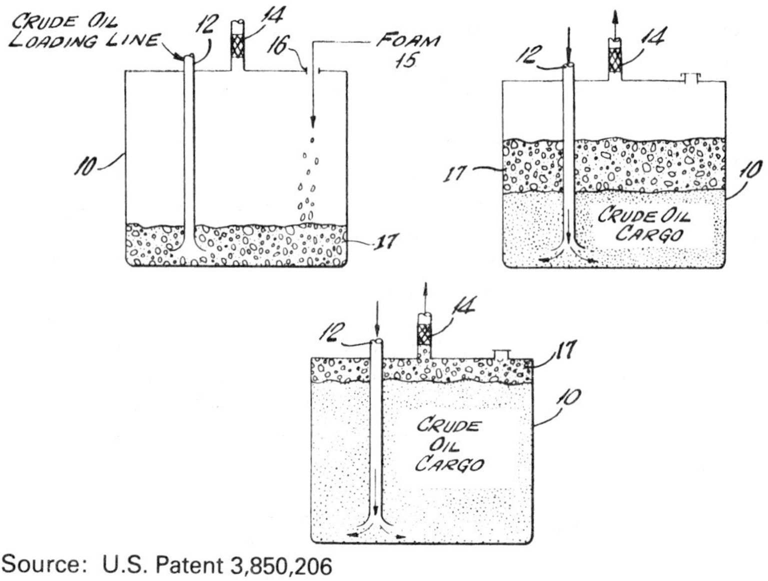

Source: U.S. Patent 3,850,206

The upper sketch illustrates an empty tanker compartment **10** prior to the admission of oil thereto. Foam **15** will be ordinarily applied through existing openings **16** used for the introduction of cleaning apparatus. A layer **17** of approximately 3 to 12 inches, preferably about 6 inches of foam, is applied to the compartment. Although 3 to 12 inches of foam is sufficient to prevent significant vapor loss, more foam may be needed in order to assure complete coverage of the surface when the tanker compartments are filled nonuniformly owing to the internal structural members. Oil is introduced beneath the foam layer through the loading line **12** as shown in the upper right-hand sketch. As crude oil enters, the foam layer **17** rises and conforms to the irregularities present in all tanker compartments, sealing off effectively the surface of the crude which is being introduced. Hydrocarbon vapors are chemically dissimilar to the foam and do not readily pass through it. The vapor space is continually vented through outlet vent **14** as the liquid level rises. Since essentially no hydrocarbon reaches the vapor phase, none is vented during loading operation, requiring no recovery facilities and avoiding localized air pollution hydrocarbon vapors.

The bottom sketch in the figure illustrates the tanker compartment when the loading process is being completed. If the loading process is continued any foam which passes out the vent **14** will be broken by contact with a foam breaking device located in or near the vent but, if not, the foam will naturally break over a relatively short period of time so that when the cargo is unloaded it will contain only very minor traces of water and the foam forming compound, which will enter the refining process and effectively disappear. A typical foam breaking device is a mesh screen comprising a hydrophobic material. As an alternative, injection of a defoaming agent, e.g., silicones may be employed.

The foaming materials which can be used include any of a number of commercially available foam forming materials, for example, a 6% aqueous solution of Aerowater (a trademark of National Foam Co.) has been found to be effective. Others which may also be used, include fatty acid salts, e.g., sodium laurate, sodium myristate, and dodecyl sodium sulfate.

It has been discovered that, while many foam solutions may be satisfactorily applied to crudes handled at relatively low ambient temperatures, these foams are not sufficiently stable in contact with naturally warm or heated crudes. In order to retain their integrity for the entire loading process, which may take as much as 20 hours, it has been discovered that small amounts of water thickeners may be added to the foam forming material before the foam is generated in order to strengthen the foam so that it will retain its effectiveness during the entire loading period. An example of such an application is in the use of 0.25% of a commercial water thickener (Kelzan produced by Kelco Co. of Clark, N.J.) to the 6% aqueous solution of Aerowater foam which has been mentioned above.

Other water-soluble polymers could also be used to increase the viscosity of water, e.g. polyacrylamide, polyoxyethylene and sulfonated polystyrene. A substantial improvement in durability was obtained by adding the water thickener, which apparently operates to effectively prevent drainage of water from the bubble films which comprise the foam structure. However, the amount of water thickener added must be carefully selected in order to improve the stability of foam. It has been found that an excessive amount of water thickener will result in an embrittlement of the foam and a loss of effectiveness since the foam in bulk may be broken. In contrast, most foams fail by drainage of water from the bubbles under the influence of gravity which causes the film to thin out until the bubbles rupture. Thus, an optimum amount of water thickener should be applied.

A system developed by *A. Kattan and J.E. Gwyn; U.S. Patent 3,979,175; September 7, 1976; assigned to Shell Oil Co.* is one for preventing gasoline vapors contained in air vented from gasoline storage tanks from entering the atmosphere. It includes a conduit to pass gasoline vapor-laden air to a bed of adsorbent for gasoline and, when the adsorbent approaches saturation, subjecting it to back-flushing, with or without heat, to an extent and for a time adequate to remove enough gasoline from the adsorbent to restore it to a regenerated condition, and incinerating the gasoline removed from the adsorbent bed.

Figure 23 shows such a system. The line **1** in the drawing represents grade below which is buried a filling station storage tank **2**. There is also illustrated a tank delivery truck **3** provided with a liquid delivery hose **5** for delivering liquid gasoline through a suitable conduit **6** to join the volume of gasoline **7** maintained in the tank **2**. There is also illustrated a conduit **8** for delivering gasoline through a pump **10** and a filling station island **11** for delivery through a hose **12** to an automobile or other vehicle **13**. The equipment thus far described is the conventional equipment that already exists in virtually every filling station or is periodically associated with every filling station.

FIGURE 23: SHELL OIL CO. VAPOR RECOVERY AND DISPOSAL SYSTEM FOR GASOLINE STORAGE TANKS

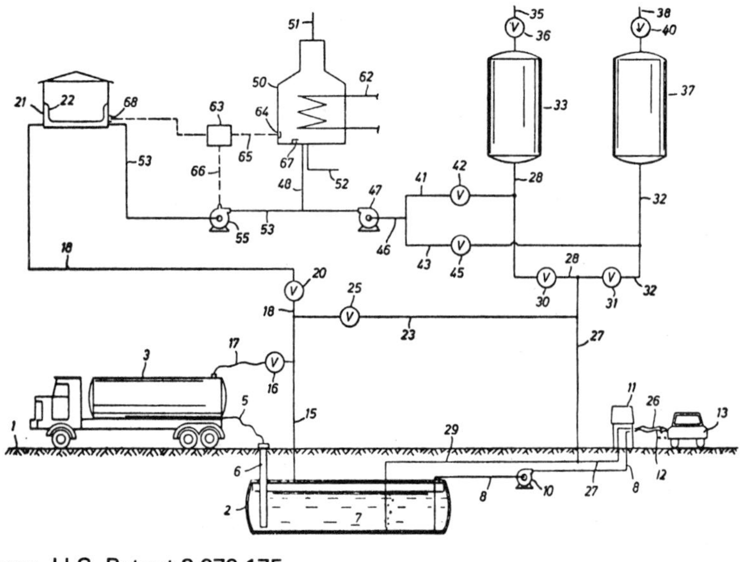

Source: U.S. Patent 3,979,175

In accordance with the present process, the vapor space in the storage tank **2** may be connected via pipe **15**, valve **16** and conduit **17** to the vapor space in the tank on delivery truck **3**. Accordingly, when liquid is delivered from delivery truck **3** to storage tank **2**, if valve **16** is open the vapor-laden air in the vapor space of vessel **2** passes through line **15** and is drawn into the tank of truck **3** via line **17** as the level of liquid in the tank of the truck **3** diminishes. Any excess vapor-laden air from the vapor space or tank **2** passes on through line **15** to one or another of the vapor recovery systems to be described hereinafter. The lines **5, 15** and **17** constitute a closed or a tight system, and the vapor-laden air passing through line **15** will be substantially saturated with gasoline vapors.

When the volume exchange system is not employed, valve **16** may be closed and the vapor-laden air from tank **2** may be passed through two alternative routes. The first route is through line **18** and valve **20** when valve **20** is open which causes the vapor-rich air to pass into storage vessel **21** illustrated with a flexible diaphragm **22** that is capable of expanding the volume of vapor-laden air within the vessel **21** without loss to the atmosphere. The vapor-rich air in vessel **21** will be disposed of as explained in more detail hereinafter.

Still another alternative route for the material flowing through line **15** is through line **23** and valve **25**. When valve **25** is open, the vapor-laden air passing through line **23** has the vapor removed therefrom as will be described in more detail hereinafter.

When filling a vehicle fuel tank, a means will be provided for collecting the vapor-laden air expelled from the vehicle storage tank, which means is not shown herein, and the collected vapor-laden air is passed through flexible hose **26** and into line **27**. The material in line **27** can be passed through one of several alternative routes. If a closed system or a tight system exists between tank **2** and vehicle **13**, the vapor-laden air is passed through line **27** and line **29** and is introduced beneath the surface of the gasoline in tank **2**. This volume exchange provides for vapor-rich air to enter tank **2** rather than ordinary air, and it thereby avoids gasoline losses by introducing most of the air vented from the vehicle tank into the underground storage tank. The vapor-laden air introduced into tank **2** through line **29** is saturated as it bubbles through the gasoline to avoid formation of explosive mixtures in tank **2** and disposed of via line **15**. Alternative methods for saturating this vapor-laden air may be employed.

If an open system exists between vehicle **13** and line **27**, it will not be economical to use line **29** and volume exchange because too much gasoline would be consumed saturating the excessive amounts of air drawn from the fuel tank of vehicle **13**; thus, the excess vapor laden air is passed through line **27** and then either through line **20** to the adsorbent bed in vessel **33** or through line **32** and the adsorbent bed in vessel **37**. Valves **30** and **31** will be opened and closed appropriately so that one or the other of vessels **33** and **37** will be in use. In passing through the bed of adsorbent, the gasoline vapors are removed from the air, and vapor-free air passes through the vent to the atmosphere. When the vessel **33** is in the adsorption mode, valve **31** in line **32** will be closed to prevent the flow of vapor-laden air therethrough. However, when the adsorbent bed in vessel **33** becomes sufficiently saturated with gasoline to be regenerated, the valve **30** will be closed and the valve **31** will be opened so that vapor-laden air may pass through line **32** and into vessel **37** which also contains a bed of adsorbent for gasoline vapors. The resultant gasoline-free air is vented through vent **38** and valve **40** to the atmosphere. When desired, the vapor-saturated air passing through line **15** and line **23** may also be passed to one or other of the adsorbent beds contained in vessel **33** or **37**.

For the sake of illustration, it will be presumed that the adsorbent bed in vessel **33** is in the adsorption mode and the adsorbent bed in vessel **37** has been regenerated. When sufficient gasoline has been adsorbed in the bed in vessel **33** so that regeneration of that bed is appropriate, valve **30** will be closed and valve **31** will be opened. At that point valve **42** in line **41** will be opened and the blower **47** will be placed in operation so that a flow of air passes through line **35** to backwash the adsorbent bed in vessel **33** with air, through blower **47** and ultimately into line **48** which discharges into the furnace **50**. In the furnace **50** sufficient air added through line **52**, and perhaps catalyst is employed, so that the gasoline in the air entering the line **48** is burned to essentially innocuous material such as carbon dioxide and water vapor which are vented through the line **51**.

When the adsorbent in vessel **37** approaches saturation, regeneration in that vessel is effected by closing valves **31** and **42** while valves **30** and **45** are opened so that air backwashes the bed in vessel **37**, desorbing the gasoline from the bed and regenerating it. The backwash air passes through line **32**, line **43** and line **46**, through the blower **47** and the before-mentioned line **48** so that the gasoline vapors are disposed of in furnace **50**.

It is evident that either all or part of the vapor-rich air expelled from storage tank **2** when it is being loaded may be passed into the adsorbent bed in vessel **33** or vessel **37** when a separate vessel for saturated vapors is not employed. In one embodiment of this process a volume of vapor-rich or saturated air is stored in a storage tank **21** in order to effect more complete regeneration of the adsorbent in vessels **33** and **37**. The use of the vapor-rich air for this purpose is as follows.

When regenerating the adsorbent bed, for example the bed in vessel **37**, air is drawn into the bed through line **38** to backwash the bed and desorb gasoline vapors which are passed ultimately to the furnace **50**. When regeneration is begun, a high concentration of gasoline vapors is in the backwash air and combustion of this stream offers no difficulty. However, as the regeneration proceeds the concentration of gasoline in the backwash air becomes smaller and smaller until ultimately the mixture entering furnace **50** through line **48** does not contain enough gasoline to support combustion. At this point the regeneration of the adsorbent in vessel **37** must stop whether a sufficient amount of gasoline has been desorbed or not. In accordance with this embodiment of the process, when the gasoline contained in the backwash air is not sufficient to maintain combustion, the blower **55** is turned on so that gasoline-saturated air passes from container **21** through line **53** to join the air-gasoline mixture in line **48**. Since the air passing through line **53** is saturated or very enriched with gasoline vapors, it increases the concentration of gasoline in the combined stream in line **48** to the extent that combustion may be effected. Thus, the backwashing of the adsorbent in bed **37** may be continued long after the amount of gasoline desorbed from that bed is sufficient to support combustion. This may increase the capacity of the adsorbent bed in vessel **37** between successive regenerations.

It is also possible to operate the furnace when storage tank **2** is being filled to dispose of some of the vapor issuing from it while the rest is taken care of in one or the other of the adsorbent beds in vessels **33** and **37**. In this embodiment, during the loading of the storage tank **2** the vapor expelled from it passed through line **15**, line **23**, line **27** and one or the other of the lines leading to vessel **33** or **37**. For purposes of illustration, vessel **33** will be employed to adsorb the excess vapors. In this embodiment the valve **42** is opened and the blower **47** is operated. In this embodiment the vapor-saturated air is drawn through the blower **47** and discharged into the furnace **50** to be disposed of by combustion. Any excess vapor-laden air that is beyond the capacity of the blower **47** to handle simply passes through line **28** into the vessel **33** wherein the gasoline is removed from the air and pollutant-free air is discharged through the vent **35**. The blower **47** may be turned off when storage tank **2** is filled, or it may remain in operation until the adsorbent in vessel **33** is regenerated as indicated by insufficient gasoline in the vapor-air mixture passing through line **48** to support combustion.

Finally, the gasoline disposed of in the furnace **50** need not be washed. A coil, schematically represented as **62**, may be employed to heat water, to generate steam, or even to provide hot air which may be employed in internal coils in the vessels **33** and **37** to aid in the regeneration of the adsorbent contained therein.

One suitable control for regeneration includes a control center **63** having suitable conventional means to receive impulses characteristic of specific conditions and to produce a suitable control impulse in response thereto. As shown herein, the control center **63** is connected via line **65** to a suitable sensing means **64** that is capable of sensing a flame or of measuring a high temperature. When a flame from burning the gas entering the furnace **50** via line **48** exists, the control center **63** will maintain blower **55** inoperative via line **66**.

When the flame is extinguished, however, blower **55** is turned on and vapor-saturated air is introduced through line **53** into line **48**. Igniter **67** will then cause the mixture from line **48** to burn until the volume of vapor-rich gas in vessel **21** is exhausted, as indicated by sensor **68**, at which time control **63** will turn off blower **55** as well as blower **47** by conventional means not shown. It is evident that many conventional control circuits may be employed to introduce vapor-rich air when backwash gas will no longer support combustion and to discontinue operation of all blowers when all combinations of backwash gas and vapor-rich air that will support combustion have been exhausted.

The solid adsorbents useful in this process may be any of those known to the art which are capable of adsorbing gasoline vapors from a mixture of gasoline vapors and air. Examples of suitable adsorbents are activated carbon such as charcoal, silica gel and certain forms of porous minerals such as alumina, magnesia, etc., which are known to selectively adsorb gasoline vapors from air.

A system developed by *G.E. Wengen; U.S. Patent 3,995,440; December 7,*

1976; assigned to Sun Oil Co. of Pennsylvania is a system for preventing substantial pollution of the atmosphere by benzene vapors displaced from a tank truck during loading of the truck with benzene. The benzene is loaded into the truck at its ambient temperature, and benzene vapors displaced from the truck are cooled to condense them to a liquid state. The system for cooling the benzene vapors is one in which natural gas at high pressure is expanded to a lower pressure, and the cooling effect which takes place during the expansion is utilized to cool and condense the benzene vapors.

Recently enacted Environmental Protection Agency regulations governing loading of volatile substances into tank trucks specify that a vapor control system is required for substances which have a vapor pressure exceeding 1.5 psia at loading temperature and wherein the loaded volume exceeds 20,000 gallons per day. The present system was designed to comply with EPA standards during the loading of benzene into tank trucks.

Figure 24 is a flow diagram of the present system wherein the flow of natural gas is indicated by a solid line, the flow of coolant is illustrated by a dashed-dotted line, and the flow of benzene is illustrated by a dotted line. A tank truck **10** is shown being filled from a benzene supply line **12**. The tank truck has a sealed hatch on its top, not shown in detail, which prevents benzene vapors from escaping directly into the atmosphere. The sealed hatch has a vapor removing pipe **14** attached thereto such that as benzene vapors are displaced from the tank truck during filling, they are directed into pipe **14** where they are transported to a condenser **16**. They are cooled in the condenser, in a manner as will be explained later, and are condensed into a storage drum **18**. In an alternative embodiment, the condensed liquid might be pumped directly to benzene tanks in the refinery.

The cooling system operates as follows. Natural gas is received at the refinery over a supply pipe **20** at a very high pressure, ordinarily 500-600 psi. The pressure of the natural gas must be reduced to a lower pressure, generally 100-150 psi, by a pressure break down station **21** before it enters a general distribution system in the refinery. A tremendous amount of heat is absorbed during this expansion because of the heat of expansion of the natural gas, and ordinarily the potential work of this cooling effect is disregarded. The system takes advantage of this cooling effect for both supplying power to the system and cooling the benzene vapors. The natural gas is expanded through a turbine **22** which drives a coolant pump **24**. The expanded, cooled natural gas is directed via a pipe **26** to a heat exchanger **28**. The natural gas is then returned via a pipe **30** to the low pressure side of the break down station **19** where it enters the general distribution system **31** for the refinery.

Most of the coolant pressure after the pump **24** is dissipated across a partially opened plug disc globe valve **32**. The coolant, which in the preferred embodiment is kerosene, then flows into cooler **28** where its temperature is substantially reduced, and continues via pipe **34** to condenser **16** where it is utilized to cool the benzene vapors. The kerosene coolant is then direced by a pipe **36** to an insulated holding tank **38** which is vented at **40** to provide kerosene at atmospheric pressure to the pump **24**.

The system is designed to be automatically self-controlled, with operator assistance required only at start up. The coolant circulation is to be maintained at all times, and thus there is no need to start and stop the system for each truck loading. The system includes three self-regulating controls. A turbine governor control **50** includes an rpm meter which controls the opening of a valve to maintain the rpm of the turbine **22** steady at some predetermined value. A pressure control system includes a pressure transducer **44** and a control valve **46** which function together to control the amount of natural gas bypassing the cooler **28**, and accordingly control the pressure in the cooling loop beyond the turbine. A temperature control system includes a temperature transducer **48** and a control valve **50** which function together to maintain the temperature of the coolant at a given value. Because of the nature of these controls, the system can run continuously, even when no vapor is flowing through condenser **16**, as the combination of the temperature/pressure control systems admits only enough gas to the cooler to maintain the coolant temperature at a selected value.

FIGURE 24: SUN OIL CO. VAPOR CONTROL SYSTEM FOR BENZENE LOADING OPERATIONS

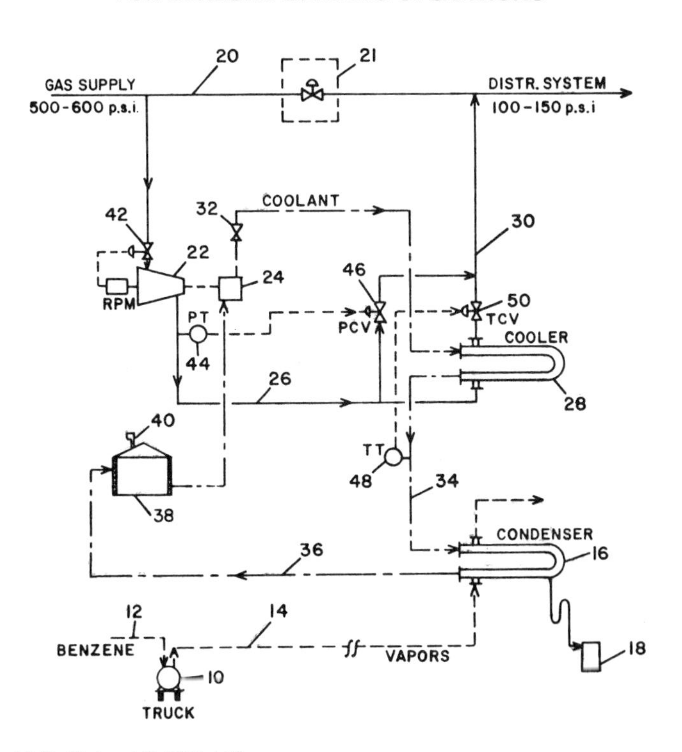

Source: U.S. Patent 3,995,440

By way of specific example, one embodiment was designed with the following parameters. All of the natural gas piping was two inches, insulated before the cooler **28** to keep heat absorption to a minimum, and uninsulated after the cooler to allow the natural gas to warm before it enters the general distribution system of the refinery. The gas is received from a supplier at a pressure of approximately 500-600 psi, and the pressure is reduced to 100-500 psi before it enters the general distribution system. The turbine **22** is a small (15 hp) noncondensing gas expansion turbine with a constant speed governor, as is available from Coppus Engineering. The coolant pump **24** is designed for a flow of 61 gpm at 140 psig discharge pressure. This high flow and discharge pressure is required to consume enough driver power to ensure a sufficient flow of natural gas for cooling.

The expansion of the natural gas through the turbine gives a gas temperature of approximately 0°F, and the cooled gas flow is regulated to deliver the kerosene coolant at 55°F temperature into the condenser. The kerosene cooler and the vapor condenser are standard Brown Fin Tube units. The cooler is an all-steel, type 1JH24, 15 foot long cooler, while the condenser is an all-steel, type 1JH36, 20 foot long condenser. Gas streams flow on the fin sides of both exchangers, while kerosene circulates on the tube sides of both exchangers. These exchangers have the inherent flexibility of adding capacity by simply stacking on additional sections.

The temperature control system maintains the circulating kerosene coolant at 54°-55°F. The kerosene holding tank has a capacity of 100 gallons, and the storage drum is a 55 gallon drum. The kerosene circulation lines are all 2 inch insulated piping. The air-benzene vapors displaced from the tank truck have a pressure of approximately 3 inches (H_2O), and the vapors vented into the atmosphere are cooled to approximately 65°F and have a pressure of approximately 1 inch (H_2O). In the designed embodiment, each control loop would use a feed into the transducer appropriate to that application, i.e., speed measurement could be electrical, pressure measurement would be by a direct pressure feed to the transducer, etc.

Conventional transducer control systems would then be used to convert the input signal into an output air pressure signal (usually 3 to 15 psig) sent to the appropriate control valve. For example, the rpm sensor (probably electromagnetic) would feed an electrical signal proportional to rpm into the transducer, where the difference between the process and set point signals would generate an air pressure to the control valve to regulate incoming steam into the turbine. There are new transducers available which operate completely electronically. They receive an input signal appropriate to the variable being measured, and send an electric output signal to the valve. Since the control loops are intended to be "local" control (controlling on location rather than transmitting signals to and from a control room), these electronic transducer controllers probably would not be used. They have a big advantage in applications where the unit is located a sizeable distance from its control room.

A system developed by *W.B. Hansel; U.S. Patent 3,996,975; December 14, 1976; assigned to Sun Oil Co. of Pennsylvania* is particularly applicable for use on a hydrocarbon tank truck which receives the vapors from the underground tank being filled, so that the pressure built up during filling can be released into the atmosphere through a filter in order to minimize the hydrocarbons released into the atmosphere when the connections to the underground tank are uncoupled. The system includes a pressure relief line connected to a float valve located inside a tank compartment on one end and to a filter on the other end. The filter can be made out of an adsorbent material which permits it to be cleaned and used over again.

Figure 25 shows the application of this system to a standard tank truck equipped to receive the vapors displaced from an underground storage tank and having three individual compartments. The tank truck has a vapor recovery line **10** which is connected to the upper portion of each compartment of the tank truck and to vapor recovery hose **11** through a swivel fitting **12**. The bottom of each tank compartment is connected to a dispensing control panel **13** (by means not shown). Dispensing hose **14** is connected to dispensing panel **13** on one end and is connected to fillpipe **15** of underground tank **16** through dry break connection **17**. When filling underground tank **16**, the gasoline flows from each compartment through dispensing panel **13**, dispensing hose **14**, and into the tank **16** through fillpipe **15**. The vapors in tank **16** flow back into the compartments of the tank truck through the vapor recovery hose **11** and the vapor recovery line **10**.

FIGURE 25: VAPOR CONTAINMENT SYSTEM FOR TANK TRUCK FILLING OPERATIONS

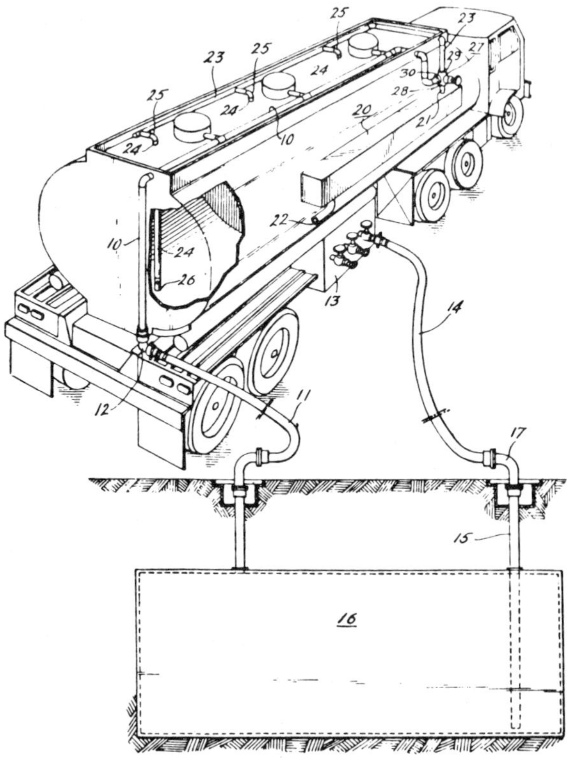

Source: U.S. Patent 3,996,975

A vapor collection device, such as filter **20**, is mounted on the side of the tank truck, and has an intake **21** and an exhaust **22**. The filter is preferably of the type which can adsorb the vapors so that they can easily be removed to permit reuse of the filter, such as charcoal. Intake **21** is connected to a liquid level sensitive valving system located in each tank compartment through pressure relief line **23**. Each valving system includes a vertical line **24** which extends from the bottom of each compartment through the top of the compartment, where it is connected to pressure relief line **23** through elbow fittings **25**. At the end of each vertical line **24**, which is located at the bottom of each compartment, a float valve **26** is installed. Near the point where pressure relief line **23** connects intake **21** of filter **20**, a three-way diverting valve **27** is placed. Ports **28, 29** and **30** are connected to intake **21**, relief line **23**, and vapor return line **10**, respectively, so that when valve **27** is in first position, flow between intake **21** and pressure relief line **23** is permitted and in a second position, flow between intake **21** and vapor return line **10** is permitted.

When the tank truck is filled with gasoline, valve **27** is placed in the first position and float valve **26** remains closed, thereby preventing the exit of any vapors. During filling of underground tank **16**, the vapors flow back into each compartment through vapor return hose **11** and vapor return line **10**. Each float valve **26** opens once the liquid level in its respective tank compartment falls below the level of the float, thereby releasing any excess pressure into the atmosphere through filter **20**.

A process developed by *J.H. Hirt; U.S. Patent 4,009,985; March 1, 1977; assigned to Hirt Combustion Engineers* is one for abatement of excess gasoline vapor emissions which occur during transfer of service station liquid gasoline from one storage tank to another storage tank, such as the transfer of gasoline at a gasoline service station from a gasoline storage tank truck to underground gasoline storage tanks and also from underground gasoline storage tanks to an automobile or vehicle storage tank through a gasoline pump station.

Vent outlet pipes of the underground storage tanks are manifolded to a common vent pipe, where the vapor pressure is sensed and upon reaching a predetermined vapor pressure, (normally slightly below atmospheric) vapors are directed along a path from the vent pipe to a burner means. Gasoline vapors are directed to the burner by means of suction produced by an ejector using compressed air and an air-fume mixer so that substantially complete combustion of the resulting vapor air mixture will occur in the burner means. The burner is automatically ignited and burns the vapor-air mixture whenever preselected vapor pressure conditions occur in the common vent pipe during transfer of liquid gasoline between such tanks.

Figure 26 shows a suitable form of apparatus for conduct of such a process. As shown there, a gasoline pump **10** is provided with a dispensing hose **11** having a nozzle **12** for insertion into a fill pipe **14** of an automobile gasoline tank **15**. Gasoline pump **10** is connected by a gasoline line **17** to an underground liquid gasoline storage tank **18** which is shown as being partially filled with liquid gasoline at a level indicated at **19**. The space above the liquid gasoline level **19** contains air and gasoline vapors and the volume of the space changes as liquid gasoline enters or is withdrawn from tank **18**. Also connecting the storage tank **18** with the pump island is a gas vapor line **22** having an opening **23** at the top of tank **18** and in communication with gasoline vapors above the liquid level in tank **18**. Vapor line **22** lies alongside dispensing hose **11** and at dispensing nozzle **12** may enter and communicate with fill pipe **14** and automobile gasoline supply tank **15**.

FIGURE 26: APPARATUS FOR ABATEMENT OF GASOLINE VAPOR EMISSIONS IN LOADING OPERATIONS

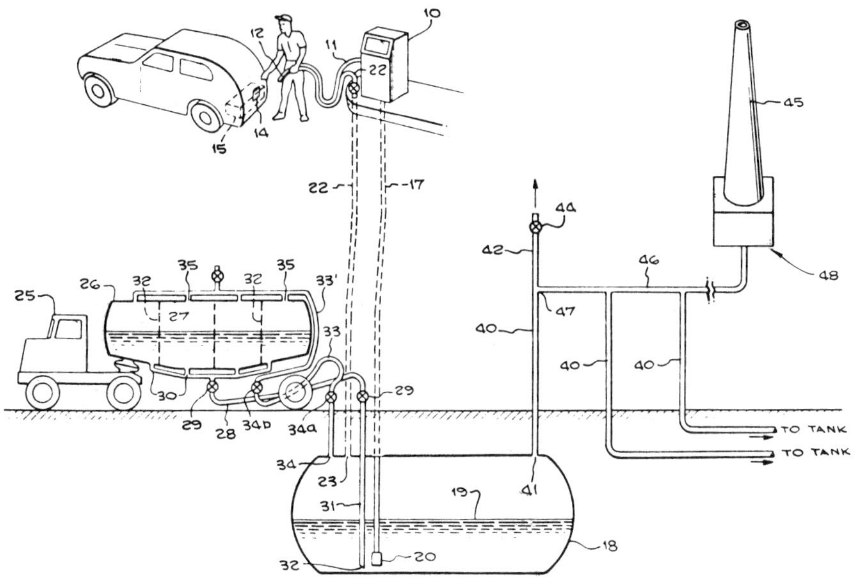

Source: U.S. Patent 4,009,985

Means for preventing excess gasoline vapors from being discharged into atmosphere at the automobile tank includes a seal means (not shown) which is effective to provide a vapor tight connection between fill pipe **14** and nozzle **12** which has a suitable check valve. Vapor line **22** provides a closed path sealed at the fill pipe **14** for communication of gas vapors between the spaces above the liquid levels in automobile supply tank **15** and underground storage tank **18**. There is thus provided a closed vapor tight circulation system for liquid gasoline from storage tank **18** to automobile tank **15** and a vent system for communication and passage of vapors between vapor space above the level of the gasoline in automobile tank **15** and above the liquid level **19** in the storage tank **18**.

The figure also illustrates a similar closed and vapor tight sealed gasoline liquid and vapor circulation system between underground storage tank **18** and a gasoline delivery truck **25** having a truck tank **26** in which the level of gasoline is generally indicated at **27**. The delivery tank **26** is connected by a fill hose **28** to underground tank **18**. Hose **28** has suitable manifold connections at **30** with compartments **32** defined by partition walls or bulwarks provided in the delivery tank **26**. The ends of fill hose **28** are connected to suitable valves **29** at the manifold and at tank fill pipe **31** which has a bottom end **32** located close to the bottom of storage tank **18** for submerged filling. Sealed valve connections are provided to avoid loss of liquid or loss of gasoline vapors at ends of hose **28**. Flow of liquid gasoline from tank **26** to storage tank **18** is normally by gravity drop when the valves are opened.

Delivery tank **26** is also connected to a bulk gasoline vapor line **33** which has an inlet opening **34** at the top of tank **18** and which includes a suitable check or block valve **34a**. Line **33** also has a connection to a truck vapor line **33'** through suitable valve **34b**. Vapor line **33'** has a plurality of connecting openings **35** each communicating with one of the compartments of delivery tank **26**. Thus between delivery tank **26** and the underground storage tank **18** there is provided a closed vapor tight liquid and vapor communication system for transfer of liquid gasoline and passage of gasoline vapors from the delivery tank **26** to the storage tank **18**. Preferably vapor lines **22** and **33** enter storage tank **18** adjacent one end of the tank.

Transfer of gasoline liquids and vapors under the two situations described above, that is storage tank to automobile supply tank and delivery tank truck to storage tank occurs under a closed vapor tight sealed system which prevents loss of gasoline vapors to atmosphere. However, the system must be safely operable under many different conditions of temperature, pressure and volumes of liquid gasoline and gasoline vapors which affect release of gasoline vapors from the closed system. For this purpose, storage tank **18** is provided with a tank vapor vent pipe **40** which has a vent opening **41** at the top of tank **18** preferably at the end of the tank opposite the entry of vent lines **22**, **33**. At the top of vent pipe **40** is an outlet opening **42** for release of vapors to atmosphere under certain extreme conditions wherein some gasoline vapors may be vented. A pressure-vacuum and blow-off relief valve **44** may be provided at opening **42**, valve **44** being operable at −4.6 inches WC and +5.7 inches pressure and blow-off pressure at +12 inches WC.

It will be understood that while only one underground storage tank has been shown, a gasoline service station may have three or more tanks each for a different type of gasoline. The fill and outlet pipes for each tank may be arranged as described for tank **18**. The vent pipes for each tank are joined at a vent header pipe **46**.

The system contemplates means for control and abatement of vapor emissions from a gasoline service station equipped as described above wherein excess gasoline vapor emissions at vent pipes of one or more underground storage tanks are controlled and abated by directing such vent gases under certain pressure conditions, to an incinerator means where substantially complete combustion of the hydrocarbons in the vent gases occurs. The control and abatement means in this example is arranged to operate in two vent gas pressure disposal stages normally encountered at a service station, that is, where disposal pressures may be relatively low, 0 inches to −0.5 inches WC as from a delivery tank truck to the underground storage tank and when disposal pressures are relatively high, −0.5 inches WC or above, as during the dispensing of relatively small quantities of liquid gasoline to an automobile tank through the gasoline service station pumps.

The two disposal pressure stages may involve the handling of liquid gasoline and vapors at different pressures, temperatures, and volumes. The vented gasoline vapors of each stage require preset amounts of air in order to achieve complete combustion of that preset amount of vented gases and the reduction to a minimum of unwanted hydrocarbon type air pollutants.

The two-stage control and the abatement means of this process includes a pressure-volume and blow-off relief valve **44** in tank vent pipe **40** and the transfer of vent gases from the storage tank **18** to a combustion or fume incinerator device generally indicated at **45**. Vent header line **46** is connected at **47** to vent pipe **40** between tank **18** and valve **44**. Depending upon the selected pressures for actuation of pressure switches, vent gases will be transferred through vent header line **46** to a gasoline vapor-air mixing system generally indicated at **48**. The vapor-air system **48** and incinerator device **45** are designed and arranged to provide proper vapor-air mixtures to incinerator device **45** so that device **45** will destroy at least 90% and as much as 99.9% of the hydrocarbons supplied to it from the vent gas vapors.

An apparatus which was developed by *L.W. Pollock and G.H. Dale; U.S. Patent 4,010,779; March 8, 1977; assigned to Phillips Petroleum Co.* is an apparatus for recovering vapors expelled from a tank during filling. The apparatus includes a conduit arrangement for selectively conducting vapor to one of a storage tank or a vapor liquefying means. Vapor conducted to the liquefying means is contacted with refrigerated liquid for liquefaction by condensation and/or absorption. Periodically, liquid and liquefied condensable portions of the vapor are discharged from the liquefying means and are replaced with fresh liquid. A pressure controller determines the flow path of the vapor to the storage tank or the vapor liquefying means wherein only a portion of the vapor is conducted to the vapor liquefying means.

Figure 27 shows the details of such an apparatus. The reference numeral **1** designates generally a vapor recovery apparatus which includes a conduit **2** which is operable to receive vapor therein for conducting same to vapor liquefaction means **3** via a conduit **4** connected to the conduit **2**. The conduit **2** is also connected to a storage tank **5**. Means **7** is operably associated with the conduit **2** and **4** to ultimately regulate the flow of vapor to the liquefaction means **3** whereby only a portion of expelled vapor flows to the liquefaction means **3**.

In the illustrated structure the storage tank **5** includes pump means **8** which is operable to pump liquid such as gasoline from the storage tank **5** through a conduit **9** to a dispensing unit **10**. The dispensing unit **10** is of conventional form, such as one commonly used at a gasoline filling station, and has a flexible conduit **11** with a filler nozzle **12** which is adapted to be received in an opening of a tank **14** such as a fuel tank of an automobile **15** or the like. The nozzle **12** is equipped with means (not shown), as is known in the art, to seal the opening of the tank **14** whereby vapor displaced by the gasoline introduced into the tank **14** flows into the conduit **2** which communicates with the interior of the tank **14**. As described above, the conduit **2** is connected to both the liquefaction means **3** and the storage tank **5**. The storage tank **5** is of stan-

FIGURE 27: PHILLIPS PETROLEUM CO. VAPOR RECOVERY SYSTEM FOR TANK FILLING OPERATIONS

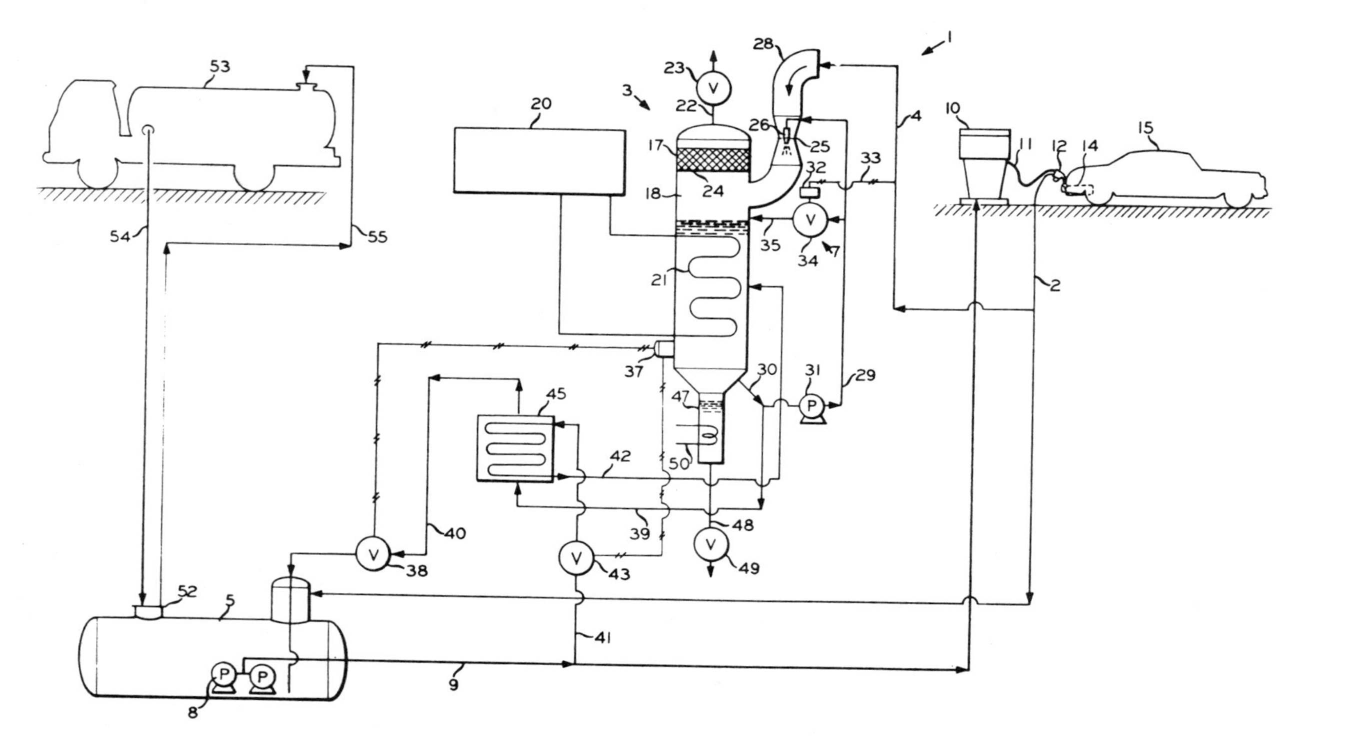

Source: U.S. Patent 4,010,779

dard construction and, as shown, is the tank used to store gasoline underground at a filling station.

Any suitable liquefaction means **3** can be used and, as shown, some include a vessel **17** having a chamber **18** therein. Preferably, refrigeration means are in heat transfer relation with the contents of the chamber **18** and, as shown, the refrigeration means includes a refrigeration unit **20**, of generally standard construction which has a cooling coil **21** positioned in the chamber **18** and operable to remove heat from the liquid therein. The vessel **17** is provided with a vent **22** preferably with a pressure relief valve **23** to permit venting of nonliquefiable gases such as air from the chamber **18** to atmosphere at a predetermined pressure as determined by the relief valve **23**. As shown, although not required a suitable mist-eliminating device **24** is secured in the chamber **18** and positioned adjacent the vent **22** and is operable to prevent mist from being discharged through the vent **22**. Any suitable mist-eliminating device can be used such as wire mesh screens or the like.

The liquefaction means **3** includes means for contacting vapor, provided by the conduit **4** from the conduit **2**, with liquid, preferably liquid from the chamber **18**, to effect liquefaction of the vapor. The liquid will generally be the same filling liquid from which the vapors evolved. Any suitable contact means can be provided and, as shown, this includes a venturi **25** having a liquid dispersion nozzle **26** opening thereinto. The venturi **25** is part of a neck or conduit **28** which connects the conduit **4** to the chamber **18**. The nozzle **26** is connected to a source of liquid and, as shown, a conduit **29** is connected to a lower portion of the chamber **18** by a discharge conduit **30**. A pump **31** is connected to the conduit **30** and the conduit **29** to circulate liquid from the chamber **18** to and through the nozzle **26**. Preferably, the nozzle **26** is directed to effect cocurrent flow of the liquid dispensed therethrough and the vapor flowing through the venturi **25** and preferably is of a type to atomize or otherwise disperse the liquid as same is discharged. The chamber **18** is adapted to contain liquid, as described above, which is substantially the same liquid as stored in the tank **5** and which is cooled by the refrigeration unit **20** and cooling coil **21** whereby the chilled liquid effects cooling of the vapor flowing through the venturi **25** and most or all of the hydrocarbons in the vapor are liquefied by condensation and/or absorption.

The means **7** is operable to control the amount of chilled liquid passing through the venturi and hence to control the amount of vapor which is drawn through the venturi **25** and the amount of vapor which returns to the storage tank **5** via conduit **2**. Any suitable control means can be used but preferably it will operate on pressure sensed in the conduit **4**. In the form shown, a pressure controller **32** ensures pressure in the conduit **4** with the signal being transmitted via conductors **33**. The controller **32** is operably connected to a valve **34** which is connected in a conduit **35**. The conduit **35** is connected to the chamber **18** and the conduit **29** and forms a bypass for liquid flow from the conduit **29** to the chamber **18** without flowing through the nozzle **26**.

The valve **34** can be the on/off type as would be the controller **32** but a proportioning type valve **34** and controller **32** is preferred to regulate flow volume through the conduit **35** and thereby the flow volume through a nozzle **26** in response to pressure changes in the conduit **4**. In operation of a proportioning type valve and controller, an increase in pressure in the conduit **4** will effect a reduced flow through **35** and a higher flow of liquid through the nozzle **26** to handle the increased vapor load. A decrease in pressure in the line **4** will correspondingly effect higher flow volume of vapor through the conduit **2** to the storage tank **5**. Flow of vapor through the conduit **4** is induced by pressure differential between the chamber **18** and the conduit **2** which can be changed by the amount of liquid dispersed through the nozzle **26** in venturi **25** which liquefies the vapor thereby effecting a greater pressure differential and higher flow rates of vapor through the conduit **4**.

In the liquefaction of hydrocarbon vapors such as gasoline vapors, normally the vapors expelled are of the lighter hydrocarbons and during liquefaction of same, the hydrocarbon liquid in the chamber **18** becomes enriched with the lighter hydrocarbons and preferably same is replaced by fresh gasoline such as from the storage tank **5**. As shown, a liquid level controller **37** senses the level of hydrocarbon in the chamber **18** and is operably connected to a valve **38** which is connected to a discharge of the chamber **18** which is shown as the discharge **30** and to the storage tank **5** whereby at a predetermined level of liquid in the chamber **18** the valve will open allowing discharge of a portion of the liquefied so as to maintain a relatively constant level of liquid.

The valve **38** can be a timed valve where, at a predetermined interval, it will open and permit discharge of liquid from the chamber **18** for return to storage tank **5** whereby the level controller **37** would be operable to override the timed operation of the valve **38** in the event the chamber **18** becomes overfull before the valve **38** is opened by timed operation. A conduit **39** is connected to the discharge **30** and is in turn connected to the valve **38** by a conduit **40** which connects the valve **38** also to the tank **5**. When chamber **18** is essentially empty, valve **38** closes in response to operation of the level controller **37** thereby maintaining a predetermined quantity of liquid in the chamber **18**. Fresh gasoline is supplied from the tank **5** through the conduit **9** and a conduit **41** which is connected to the chamber **18** by a conduit **42**. Preferably the valve **43** is operably connected to the level controller **37** and at a predetermined level during discharge of liquid from the chamber **18** the valve **43** is opened, at about the same time the valve **38** is closed to permit filling of the chamber **18** with gasoline from the tank **5**. When the level of fresh gasoline reaches a predetermined level during filling, the level controller will close the valve **43**. The level after filling is preferred to be such as to maintain the cooling coils submerged. Alternately the valve **43** can be a timed valve, which would be timed to open after valve **38** closes and permits flow of a suitable amount of gasoline from the tank **5** to the chamber **18** after it has been discharged of the enriched gasoline.

To improve the operating efficiency of the liquefaction means **3** a heat exchanger **45** is provided to effect heat transfer between relatively warm liquid supplied from the storage tank **5** and the relatively cold liquid being returned to the storage tank **5** from the chamber **18**. The heat exchanger is connected to the conduits **41** and **42** for flow of fresh liquid from the storage tank **5** therethrough and to the chamber **18** which liquid is in heat transfer relation with enriched liquid being discharged from the chamber **18** through the conduit **39** and into the conduit **40** for return to the storage tank **5** whereby the fresh liquid is cooled before being charged into the chamber **18**. Preferably, the heat exchanger **45** has a liquid holdup capacity of about the volume of liquid discharged from the chamber **18** on each cycle with the valve **38** closing, during operation, after a volume of liquid essentially equal to the liquid holdup capacity has been discharged.

Preferably the liquefaction means **3** is provided with a water trap and, as shown, a water trap **47** is positioned adjacent the lower portion of the chamber **18** and has a discharge **48** connected thereto for discharge of water. A valve **49** is connected to the discharge **48** for selective discharge of the water. It is to be noted that a liquid level controller (not shown) as is known in the art can be used to open and close the valve **48** to maintain a predetermined level of water. Such controller usually functions on a difference in specific gravity of the liquids. Preferably a heater **50** is provided to maintain the water in a liquid state in the event that the liquid in the chamber **18** is maintained at a temperature at or below the freezing point of water to thereby assure discharge of the water through the discharge **48**.

The storage tank **5** has a filler **52** which is connected to a transport truck tank **53** or the like by a conduit **54** for filling the tank **5** with gasoline. A second conduit **55** also connects the vapor space of the truck tank to the vapor space of the storage tank **5** to receive the expelled vapors from the tank **5** when it is being filled with gasoline. In the event that the pressure in tank **5** becomes excessive, the excess vapor can flow through the conduit **2** in a reverse direction and into the conduit **4** for liquefaction similar to the liquefaction of the vapors from the tank **14**.

The present process may be more fully understood by a description of the operation thereof. As the tank **14** is being filled with gasoline from the storage tank **5**, the vapor is expelled by its displacement with the liquid gasoline. The expelled vapors are collected in the conduit **2** and below a predetermined pressure as controlled by the controller **32** the vapor is returned to the storage tank **5**. Above the predetermined pressure, the vapor flows through the venturi **25** and most of the hydrocarbons in the vapor are liquefied by condensation and/or absorption due to contact with cold hydrocarbon or gasoline dispensed through the nozzle **26**. The liquefied vapor and circulating gasoline are collected in the chamber **18** for later use in liquefaction or return to the tank **5** for later consumption. Preferably, the chamber **18** is operated at a pressure above atmospheric say, for example, 0.1 psi above atmospheric as is controlled by the pressure relief valve **23**. The positive pressure in the chamber **18** and thereby the positive pressure in the various conduits prevents leakage of air and moisture into the system.

Typically the gasoline in the chamber **18** would be maintained at a temperature of approximately −20°F while gasoline is supplied to the dispensing unit **10** at a temperature of approximately 80°F. Typically, more vapor is produced during the warmer months of the year and the liquefaction means would be sized to handle the highest expected amount of vapor. Normally, it would be expected that approximately 10% or less of the vapor expelled from the tank **14** would flow to the liquefaction means **3** by regulation of flow through the conduit **4** by the pressure controller **32**. At a predetermined time interval during a time cycle, the valve **38** would open and liquid in the chamber **18** will displace the liquid in heat exchanger **45** with the displaced liquid being returned to the storage tank **5**. Preferably, the heat exchanger **45** is sized to hold approximately the same volume of liquid as is dumped from the chamber **18**. The valve **38** will remain open until a predetermined level, as sensed by the liquid level controller **37**, is reached at which time the valve will be closed by the liquid level controller **37**.

The cold gasoline contained in the heat exchanger **45** cools gasoline pumped by the pump means **8** through the conduits **9** and **41** to the heat exchanger **45** for cooling gasoline before it is charged into the chamber **18** via the conduit **42**. Only a portion of the vapors from the tank **14** need to be liquefied by the liquefaction means **3** as a portion of the vapors can be returned to the storage tank **5** to occupy the volume vacated by pumping the gasoline into the tank **14** from the storage tank **5**. This is also true of the excess vapor that is available due to filling of the storage tank **5** by the truck **53**.

As described above, water is collected in the trap **47** and can be periodically discharged through the discharge **48**. Because of the differences of specific gravities of the hydrocarbon and the water, they are separated, with water being collected in the trap **47** for eventual discharge. The amount of refrigerated or cooled gasoline from the chamber **18** dispensed through the nozzle **26** is controlled by the pressure controller **32** wherein excess gasoline is returned through the valve **34** and conduit **35** rather than flowing through the nozzle **26** whereby only the needed amount of gasoline is dispensed through the nozzle **26** as is determined by the amount of vapor flowing through the conduit **4**. Excess pressure is relieved through the vent **22** as determined by the pressure relief valve **23**. Normally, this excess pressure is caused by air or other gases which are not liquefied by contact with the gasoline dispensed through the nozzle **26**.

WATER POLLUTION CONTROL

Oil Transfer Operations

This part (9) applies to the transfer of oil to or from any vessel with a capacity of 250 or more barrels for that oil on the navigable waters or contiguous zone of the U.S., except the transfer of lubricating oil for use on board the vessel; and non-petroleum-based oil that is transferred to or from a vessel other than a tank vessel.

1) Waivers. The USCG Captain of the Port may waive, in whole or in part, compliance with any requirement in this part if:
   a. Application for the waiver is submitted to the Captain of the Port 30 days before operations under the waiver are proposed unless a shorter time is authorized by the Captain in the Port.
   b. The Captain of the Port finds that an equivalent level of protection of the navigable waters and contiguous zone from pollution by oil will be provided by the alternative procedures, methods or equipment standards to be used.

2) Person in Charge—Limitations. No person may serve as the person in charge of oil transfer operations on more than one vessel at a time unless the vessels are immediately adjacent, there is a ready means of access between vessels, and the person in charge is not also the person in charge of the facility. Further, no person may serve as the person in charge of both the vessel and the facility during oil transfer operations except when the Captain of the Port authorizes such procedure.
3) Requirements for Oil Transfer. No person may transfer oil to or from a vessel unless:
   a. The vessel's moorings are strong enough to hold in all expected conditions of surge, current, and weather and are long enough to allow adjustment for changes in draft, drift, and tide during the transfer operation.
   b. Oil transfer hoses or loading arms are long enough to allow the vessel to move to the limits of its moorings without placing strain on the hose, loading arm, or transfer piping system.
   c. Each hose is supported in a manner that prevents strain on its coupling.
   d. Each part of the transfer system necessary to allow the flow of oil is lined up for the transfer.
   e. Each part of the transfer system not necessary for the transfer operation is securely blanked or shut off.
   f. The transfer system is connected to a fixed piping system on the receiving vessel or facility, except that when receiving fuel for the vessel, an automatic back pressure shutoff nozzle may be used.
   g. Except when used to receive or discharge ballast, each overboard discharge or sea suction valve that is connected to the vessel's oil transfer, ballast, or cargo tank system is sealed, lashed, or locked in the closed position.
   h. Each oil transfer hose has no loose covers, kinks, bulges, soft spots, and no gouges, cuts, or slashes that penetrate the hose reinforcement.
   i. Each coupling meets the requirement of 33 CFR 156.130.
   j. The discharge containment required by 33 CFR 154.530, 155.310, 155.320, as appropriate, is in place.
   k. Each scupper or drain in a discharge containment system is closed.
   l. Any continuing loss of oil from any transfer component is at a rate that will not exceed the capacity of the containment system.
   m. The communications required by 33 CFR 154.560 are operable for the transfer operation.
   n. The emergency means of shutdown required by 33 CFR 154.550 and 155.780, as appropriate, is in position and operable.
   o. The designated personnel are on duty to conduct the transfer operations in accordance with the facility operations manual and vessel oil transfer procedures that apply to the transfer operation.
   p. At least one person is present who fluently speaks the language spoken by each person in charge.

q. The person in charge of oil transfer operations on the transferring vessel or facility and the person in charge of oil transfer operations on the receiving vessel or facility have held a conference to assure that each person in charge understands the identity of the product to be transferred; the sequence of transfer operations; the transfer rate; the name or title and location of each person participating in the transfer operation; particulars of the transferring and receiving systems; critical stages of the transfer operation; federal, state, and local rules that apply to the transfer of oil, emergency procedures, discharge containment procedures, discharge reporting procedures, watch or shift arrangement, and transfer shutdown procedures.

r. The person in charge of oil transfer operations on the transferring vessel or facility and the person in charge of oil transfer operations on the receiving vessel or facility agree to begin the transfer operation.

s. Each person in charge required by this part is present.

t. Between sunset and sunrise the lighting required by 33 CFG 154.570 and 155.790 is provided.

u. For vessel to vessel transfer operations involving a tank barge between sunset and sunrise, lighting of the intensity specified in 33 CFR 155.790 is provided on the barge.

4) Connections. Each person who makes a connection for oil transfer operations shall:

a. Use suitable material in joints and couplings to make a tight seal.

b. Use a bolt in at least every other hole and in no case less than four bolts in each temporary connection utilizing an American National Standards Institute (ANSI) standard flange coupling.

c. Use a bolt in each hole of couplings other than ANSI standard flange couplings.

d. Use a bolt in each hole of each permanently connected flange coupling.

e. Use bolts of the same size in each bolted coupling.

f. Tighten each bolt and nut uniformly to distribute the load.

No person who makes a connection for oil transfer operations may use any bolt that shows signs of strain or is elongated or deteriorated. No person may use a connection for oil transfer operations unless it is a bolted or full-threaded connection; a quick-connect coupling approved by the Commandant, USCG; or an automatic back pressure shutoff nozzle used to fuel the vessel.

5) Declaration of Inspection. No person may transfer oil to or from a vessel unless each person in charge, designated under 33 CFR 154.710 and 33 CFR 155.700, has signed the prescribed declaration of inspection form. No person in charge may sign the declaration of inspection unless he or the other person in charge has determined by inspection that the facility and vessel meet the requirements in 33 CFR 156.120. The declaration of inspection required to be signed may be in any form but must contain at least:

a. The name or other identification of the transferring vessel or facility and the receiving vessel or facility.
b. The address of the facility or location of the transfer operation if not at a facility.
c. The date the transfer operation is started.
d. A list of the requirements in 33 CFR 156.120 with spaces on the form following each requirement for the person in charge of the vessel or facility to indicate whether the requirement is met for the transfer operations.
e. A space for the date, time of signing, signature, and title of each person in charge during oil transfer operations on the transferring vessel or facility and a space for the date, time of signing, signature, and title of each person in charge during the oil transfer operations on the receiving facility or vessel.

The form for the required declaration of inspection may incorporate the declaration-of-inspection requirements of 46 CFR 35.35-30. The operator of each facility shall retain for at least one month from the date of signature, a signed copy of each declaration of inspection required for that facility.

6) Supervision by Person in Charge. No person may connect, top off, disconnect, or engage in any other critical oil transfer operation unless the person in charge, designated under 33 CFR 154.710 or 33 CFR 155.700, personally supervises the operation. No person may start the flow of oil to or from a vessel unless instructed to do so by the person in charge. No person may transfer oil to or from a vessel unless the person in charge is in the immediate vicinity of the transfer operation and immediately available to the oil transfer personnel.
7) Equipment Tests and Inspections. No person may use any item of equipment in oil transfer operations unless the operator of the vessel or facility has tested and inspected it annually, and found the equipment to meet the following conditions:
   a. Each nonmetallic oil transfer hose that is larger than three inches inside diameter must have no loose covers, kinks, bulges, soft spots, and no gouges, cuts, or slashes that penetrate the hose reinforcement; have no external and, to the extent internal inspection is possible with both ends of the hose open, no internal deterioration; and have no burst, bulge, leak, or abnormal distortion under static liquid pressure at least as great as the pressure of the relief valve setting (or maximum pump pressure when no relief valve is installed) plus any static head pressure of the system in which the hose is used.
   b. Each transfer system relief valve must open at or below the pressure at which it is set to open; each pressure gage must show pressure within 10% of the actual pressure; each loading arm and each oil transfer piping system, including each metallic hose, must not leak under static liquid pressure at least as great as the pressure of the relief valve setting (or maximum pump pressure when no relief valve is installed) plus any static head

pressure in the system; and each item of remote operating or indicating equipment, such as a remotely operated valve, tank level alarm, or emergency shutdown device, must perform its intended function.

During any test or inspection required by this section, a hose must be in a straight and horizontal position and the entire external surface must be accessible. The only exception to the above conditions is hose used in underwater transfer. However, no person may use any hose in underwater service for oil transfer operations unless the operator of the vessel or facility has tested and inspected in biennially in accordance with paragraph 3.2.2.4 (7), as applicable.

A proprietary system developed by *S.J. Memoli; U.S. Patent 3,958,521; May 25, 1976* includes a plurality of deep sea tanks which provide a containment barrier surrounding one or more towers which function as a docking facility for oil tankers. The deep sea tanks are connected to a cable surrounding the tower, with an opening at either end of the tank arrangement to permit access by the oil tankers to the tower. The tower employs features which ensure a stabilized platform, while allowing easy access to the tankers. The tower is equipped with living and operational spaces for the crew, and a pipeline is provided to allow transfer of oil to the shore. The present system is advantageously employed around oil drilling rigs and in other environments in which it is desired to contain oil spills or other pollutants floating on the water.

Such a system is shown in rough schematic form in Figure 28. The deep sea tank and seaport arrangement includes a plurality of interconnected deep sea tanks **11** which float on the surface of a body of water such as the ocean and which encircle or otherwise surround one or more stationary towers **12**. In the schematic view, only a representative number of tanks **11** are shown, but it should be understood that the tanks **11** will normally extend around the entire periphery of the containment area in order to provide an uninterrupted barrier.

A cable hook-up bracket **19** is provided on the bottom of each tank **11** to allow the tanks **11** to be anchored to the bottom of the sea using means such as a conventional anchor **20** and chain **21**, for example. Generally, the main tanks **11** at each end of the barrier, to be described hereinafter, are provided with such anchoring means and alternating tanks **11** around the barrier are also provided with the anchoring means. Each tank **11** is linked to the adjacent tanks **11** by a cable **22** which extends in a generally circular path around the towers **12**, passing through a pair of receiving brackets attached to the inner vertical surface of each tank **11**.

In order to prevent oil from passing between the tanks **11**, a skirt member **26** is attached between adjacent tanks **11** around the perimeter of the containment barrier. The skirt member is in the form of an accordion-shaped construction **26**. The skirt member **26** is attached along its length on each side to the tank **11** on either side thereof so as to close off the space between tanks **11**. The accordion-shaped configuration **26** may be constructed of fiberglass or plastic material of sturdy construction and is easily attached to tanks **11** constructed of fiberglass or plastic material by the use of, for example, a plastic weld. Other conventional attachment means are employed for other materials. The material for the skirt member **26** is preferably somewhat pliable, in order to allow the skirt member **26** to adjust to variations in the distance between tanks **11**.

FIGURE 28: DEEP SEA TANK AND SEAPORT SYSTEM FOR SPILL CONTAINMENT IN TANKER OPERATIONS

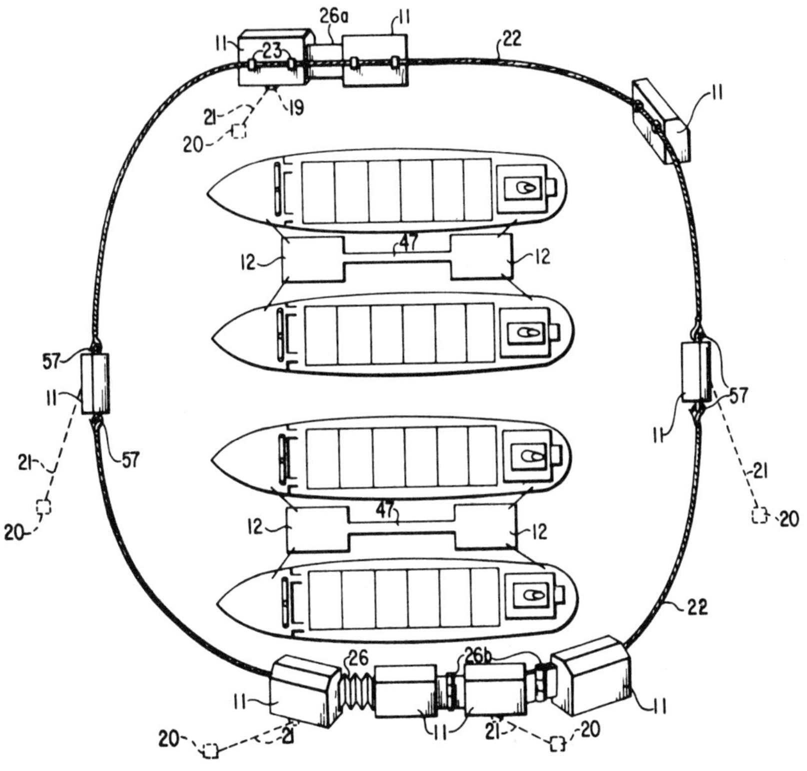

Source: U.S. Patent 3,958,521

A skirt member **26a** may be employed in place of member **26**, this member **26a** including a close wire mesh screen, of stainless steel or similar material, to which is attached a vertical metal bar on each side thereof. The bar is attached by means such as screws or welding to the tank **11** on either side.

The embodiment of the skirt member **26b** includes a pair of L-shaped flanges **30** of general angle iron configuration with inner ring-shaped hinge portions which mesh in a conventional manner and are connected by a hinge pin. The two panels which form L-shaped flanges are joined at an angle of about 90°. The outer portions of each flange are attached to the tanks **11** on each side by means such as screws or welding.

An entrance to the area occupied by the towers **12** is provided at each end of the containment barrier by a main tank **11** having a cable attachment ring **57** mounted on either side thereof, as shown. The cable **22** which links the tanks **11** is in two sections, one section for each of the two sides of the barrier, and each of the two section of cable **22** is connected at each end to the attachment ring **57** of the respective main tank **11** by suitable connecting means. Such connecting means is preferably easily disconnected from the ring **57** to allow quick and easy opening of the barrier for passage of ships and may include means such as a swivel link and senhouse slip attached to the end of the cable **22** for releasable connection to the ring **57**.

As shown, the seaport includes one or more double A-frame towers **12**, with the two A-frame members in each tower unit **12** being joined by a catwalk **47**. The towers **12** are constructed of steel I-beam and steel plate materials and with a top deck having a generally rectangular shape.

Thus the present system is provided with many oil spill containment precautions, including a containment barrier in the form of a series of deep sea tanks **11** which are especially constructed to contain oil spills from such sources as oil tankers and broken oil lines. An oil spill would be contained within the link of tanks **11**, then by means of a skimmer or other collecting device (not shown), a deep sea tug would be able to pick up the oil and transmit it to settling tanks located within compartments on the towers **12**, from which the oil could be reclaimed. Depending upon the sea conditions, the tanks **11** which serve as a protection from oil spills may be raised or lowered by pumping water in or out, so as to modify the buoyancy and oil-retention characteristics of the barrier.

Oil tankers or ships would normally be towed by tugs into the port through the barrier openings in the vicinity of the main tanks **11** at each end of the barrier. The tankers should be equipped with both bow and stern anchors so that the tankers could be stationary and yet need not totally depend upon the ships' securing lines to maintain position. In addition to towing the tankers in and out, tugs may be used to open and close the links of tanks **11** surrounding the towers **12**.

As an example of specifications which may be employed in the present apparatus, ships having a length of about 1,000 feet and with approximately 90-foot draft are advantageously employed with the present configuration. The deep sea tanks **11** could be, for example, approximately 40 feet in length, having a width of about 15 feet and with an overall height of about 20 feet. The barrier created by the tanks **11** may encircle an area of approximately 1½ miles in diameter. The towers **12** would be located in the center of this barrier and a pipeline provided to stretch from the towers **12** to the shore, advantageously a distance of 10 miles or less, depending upon the depth of the water and the draft of the particular tanker.

Tank Cleaning

Petroleum fuels and good water are two of the modern world's most important necessities, with the demand for both continually increasing. One of the most economical methods of transporting liquid petroleum fuels, such as crude oil, fuel oil, heavy diesel oil and lubricating oil, has been by maritime carriers and, unfortunately, this has led to considerable pollution problems on the high seas and on waterways. This pollution problem has been intensified by the development of very large crude carriers, which formerly were in the size class of 20,000-30,000 dwt, and now range up to 326,000 dwt. It is expected that the pollution problem will be further aggravated by the use of even larger carriers, now being built.

While water pollution occurs through accidental oil spills, an equally serious source of pollution is the petroleum fuel that is intentionally discharged by carriers during the washing of the emptied compartment tanks, and also with the ballast water. The cargo compartments contain considerable amounts of residual fuel oil after they have been emptied and they must be cleaned to eliminate a fire hazard. A recent study of three 70,000 dwt carriers by Cities Service Tankers Corporation over a 3-year period indicated that approximately 0.25-0.30% of the oil remained in the emptied cargo tank. Projecting this percentage to a 326,000 dwt carrier, would indicate a residual oil weight of about 1,000 tons.

The ballast water is taken on by the emptied carrier during its trip back to the originating port terminal, in order to properly immerse the propeller and rudder for controllability, as well as immerse the ship's hull to reduce structural stress. Efforts to control this source of pollution have not been completely successful even though international laws relating to the problem are continually being strengthened. For example, regulations of the International Convention for Prevention of Pollution of the Seas by Oil (1954, amended 1962, and 1969) specified (1) that the instantaneous rate of oil discharge should not exceed 10 liters per mile, (2) that the total quantity discharged should not exceed 1/15,000 of the total cargo-carrying capacity, and (3) that the tanker should not be less than 50 miles from the nearest land at the time of discharge. These regulations do not apply, however, if the carrier's compartments have been cleaned and the subsequent discharged ballast water does not produce visible traces of oil on the water surface.

At the 1970 NATO meeting in Brussels, international support was given for the eventual termination of all intentional discharge of oil from ships into the oceans. These are strong stringent requirements that are obviously necessary to prevent the continuing pollution buildup. Unfortunately, these regulations are difficult to enforce, and surveillance techniques for detection and identification of discharges are not completely developed.

There are some alternative solutions for eliminating the oily ballast discharge. For example, the carriers could be constructed with sufficient clean ballast space so that the water would not have to be taken into the oily cargo compartments. Obviously, this would be an expensive modification in the carrier design and would reduce the ship's carrying capacity. Another suggestion is to have ballast water discharge and treatment facilities at the terminal ports, to receive the oily ballast before loading the fresh cargo. Still another suggestion is to clean the ship at a special cleaning station after the cargo has been discharged. These procedures would entail considerable time delay in port as well as the construction of expensive special port facilities. Thus, it would appear that the elimination of water pollution due to the cleaning of cargo compartments or the discharge of oily ballast, requires expensive and time consuming corrective measures.

There is another problem that has to be considered in the washing of the cargo tanks after the oil has been discharged. Serious explosions may occur during the washing procedure, or later on passage. Earlier investigations had indicated that forced ventilation of the cargo tanks before and during the cleaning would reduce the danger of explosions. Recently, however, explosions have occurred in very large crude carriers even though the atmosphere in the cargo tanks was kept below the lower explosive limit. The washing techniques in very large crude carriers involve the use of high velocity rotating jets of cold, clean, unrecirculated sea water, usually at flow rates of approximately 180 tons per hour at 140 psig. The disintegration of the water jet on the tank walls has been shown to give rise to a cloud of charged water droplets, and it is thought that this electrostatic condition is responsible for the ignition of the explosive atmosphere.

Obviously it would be very desirable to reduce these electrostatic hazards. Crude oil, however, is an impure product containing insoluble solids and sludge, and the heavy deposits formed on the tank surfaces necessitate stringent cleaning methods. One possible solution would be to use very low pressure water containing chemical detergents; however, the toxicity of the chemical detergents on marine life would have to be considered. Reports in 1968, after the Torrey Canyon Oil Spill, indicated that the chemical detergent that was used, did more biological damage than the oil itself. Present indications are that the use of chemical detergents would add to the pollution problem, unless the cleaning operation was carried out at a shore facility having regulated disposal procedures.

Thus, it would be very desirable to develop other economical methods of cleaning cargo tanks without concurrently increasing the hazard of explosion or the danger of water pollution.

A process developed by *D. Gutnick and E. Rosenberg; U.S. Patent 3,941,692; March 2, 1976* is one in which oil is removed from sea water, for example, in the cleaning of tanks of oil transport ships, by utilizing bacteria or cell-free solutions from such bacteria in a confined space with the addition of a source of nitrogen and a source of phosphorus, under aerated conditions. The resultant microbial fermentation converts the oil to protein-containing by-products making it possible to discharge the contents of the tanks without oil contamination and, if desired, to recover useful by-products.

The following is a specific example of the operation of this process. The starboard slop tank (Tank A) and the port slop tank (Tank B) of a 120,000 ton oil carrier were carefully cleaned prior to taking on a cargo of Agajari crude oil at Kargh Island in the Persian Gulf. These slop tanks measured about 12 meters by 5 meters by 25 meters (depth). After cleaning, and prior to taking on the oil cargo, Tank A was fitted with an aeration system which consisted of a polyethylene pipe (32 mm diameter) which ran from an air compressor on the deck down into the tank, where it was connected to branched polyethylene sections on the tank bottom. The branched bottom polyethylene sections had a total of 50 air holes (2 mm diameter) drilled in the piping to provide uniform distribution of air. The compressor provided air to this system at the rate of

1-3 cubic meters per minute. Tank B was not fitted with an aeration system.

The tanker discharged its cargo of Agajari crude oil at Eilat in the normal manner, and Tanks A and B were not cleaned. The carrier left Eilat, and about 7 hours later, sea water was added to both tanks. The total volume of liquid in Tank A was 107 cubic meters and that in Tank B 121 cubic meters. Urea (20 kilograms) and K_2HPO_4 (1 kilogram) were dissolved in sea water and added to Tank A, and similar amounts to Tank B. Air was then introduced into Tank A at the rate of 1-3 cubic meters per minute. Tank A was then inoculated with a suspension of bacteria containing a total of 1×10^{12} bacterial cells. The value of bacteria at 0 hour was 10^4 per ml which comprises both the inoculum and bacteria present in the sea water and residual oil.

The tanks were sampled at the start of the test, and every day thereafter. No increase of bacteria occurred in Tank A during the first day; however, there was an increase of over a thousandfold from the first day to the fourth day, when the bacterial count increased from 10^4 to over 10^7 per ml. In contrast, Tank B, which was essentially identical to Tank A except that it was not aerated or inoculated, showed a smaller bacterial increase, from 10^4 to 10^5 per ml.

Oily ballast water in untreated tanks on the carrier had a bacterial count of approximately 5×10^3 per ml, and sea water had a count of approximately 0.5×10^3 per ml.

When the test was started, a thick layer of oil could be seen floating on the surface of the solutions in Tanks A and B. The amount of oil present was estimated to be in the range of 2-5%. The oil layer in Tank B did not change in appearance throughout the run. The oil in Tank A, however, started to coagulate at about the 96th hour and streaks could be seen on the solution surface. The oil became "mushy" with the consistency of pudding at about the 100th hour, and as the test proceeded, more and more of the oil dispersed into the water phase.

At the 156th hour of the test, the solution in Tank A was discharged into the sea, and no oily material could be seen in the wake of the carrier. In contrast, when Tank B was discharged in a similar manner, a thick black expelled mixture was immediately observed, followed by a yellow oil slick in the wake of the carrier. Both tanks were then vented and washed for a few minutes with sea water using a hose at low pressure. In contrast to the normal oily appearance of emptied tanks, Tank A was relatively clean; there was no crude oil sludge and only small amounts of oil visible where the aerated culture solution reacted on the tank walls. It was evident that the tank could be used for clean ballast with only a low pressure water rinse. Tank B exhibited oil sludge throughout the compartment.

A technique developed by *R.R. Goodrich and E.R. Corino; U.S. Patent 3,948,770; April 6, 1976; assigned to Exxon Research and Engineering Co.* is one in which mixtures of finely dispersed oil droplets in sea water which are often present in oil tanker compartments may be effectively separated by a chemical flocculating agent comprising a dry powdered mixture of an anionic polyelectrolyte and a sodium or calcium montmorillonite clay.

Several methods are available by which the dry powder mixture may be applied, including the following:

1) Using an eductor system operated by water pressure which disperses the powder mixture into the slop tank, preferentially into the oily water layer.
2) Recirculating the slop tank while simultaneously injecting the dry powder mixture near the pump suction.
3) Transferring the oil-water mixture to another tank for settling and, during the transfer, injecting the dry powder flocculant into the transfer pump section.
4) Injecting the flocculant at a controlled rate directly into the tank washings as they are being transferred into the slop tank during the washing of oil cargo tanks.

The performance of the dry powdered system is especially effective in speeding the separation of oil and water under difficult conditions aboard ship. It also has the capacity of making a superior separation compared to that which has been heretofore available and further improving the operation of existing Load on Top oil-handling systems whereby tank washing is carried out while the ship is underway and only clean ballast water is pumped over the side with residual oil being recovered by loading new cargo on top of the residual oil. Not only does the flocculating cause a reduction in amount of pollution caused by oil entrained by water pumped over the ship's side, but, in addition, an economic advantage is obtained in that additional oil is recovered by settling, which may be later picked up by new cargo. Because the flocculant is effective at low dosages, the amount of solids introduced into the crude is insignificant. Thus the flocculant agent obtains a unique dual advantage.

Ballast Disposal

Tankers which are used to ship intermediate and final products generally arrive at the refinery in ballast (approximately 30% of the cargo capacity is generally required to maintain vessel stability) (8).

The ballast waters discharged by product tankers are contaminated with product materials which are the crude feedstock in use at the refinery, ranging from water-soluble alcohol to residual fuels. In addition to the oil products contamination, brackish water and sediments are present, contributing high COD, and dissolved solids to the refinery wastewater. These wastewaters are generally discharged to either a ballast water tank or holding ponds at the refinery. In many cases, the ballast water is discharged directly to the wastewater treatment system, and constitutes a shock load on the system.

As the size of tankers and refineries increases, the amount of ballast waters discharged to the refinery wastewater system will also increase. The discharge of ballast water to the sea or estuary without treatment, as had been the previous practice by many tankers, is no longer a practical alternative for disposal of ballast water. Consequently, the ballast water will require treatment for the removal of pollutants prior to discharge. The use of larger ballast water storage tanks or ponds, for control of flow into the wastewater treatment system, should increase as ballast water flow increases.

A system developed by *J. Di Perna; U.S. Patent 3,957,009; May 18, 1976* is a system installed within a tanker vessel for collecting all residue oil within a single tank of the vessel so that ballast water within other tanks can be readily discharged into the sea, without oil pollution thereof. The system consists of a floating oil collector in each tank that collects all oil above a water surface therewithin, the collected oil being transported through a pipe to the single collection tank. The vessel already incorporates systems for pumping sea water into the tanks to serve as ballast, and for pumping the ballast water back into the sea thereafter.

Figure 29 shows a vessel incorporating such a system in plan and elevation. Referring now to the drawing in detail, the reference numeral **10** represents a ship ballast, water and oil separation system which is installed within an oil tanker vessel **11** having a series of transverse rows of tanks including amid-ship tanks **12**, starboard wing tanks **13** and port wing tanks **14**, all of which are for carrying oil **15**.

The vessel already includes an existing system **16** of pipes **17** running along the bottom of the tanks and having downward discharge outlets **18** in each tank located near the tank bottom wall **19** for the purpose of admitting sea water **20** into the tanks and which is delivered from a pump **21** connected to a supply sea chest **22**. Appropriate valves, not shown in the drawing are located so to control water movement into selected ones of the tanks.

One of the end amid-ship tanks **12a** comprises a collection tank. A downward facing intake **23** located near the tank bottom serves to pump the ballast water out of the collection tank by being connected to outlet pipe **24** connected to pump **25** communicating with discharge port **26**.

The oil and ship ballast water separation system **10** includes a main pipe **30** extending longitudinally through the amid-ship tanks **12** and which is connected by cross-tees to a series of branch pipes **32** extending sideways therefrom and leading into the wing tanks **13** and **14**.

In each one of the tanks **12**, except in the collection tank **12a**, a tee along the main line, and in each one of the tanks **13** and **14**, an elbow on the branch pipe end are connected to an upward extending flexible, accordion pipe or duct which at its upper end is connected to a floating oil collection unit **36** which includes a buoyant collar around a downwardly converging funnel that is open on top. The buoyancy of the collar is such that the flotation of the unit is with the upper edge of the funnel being at water level surface, so that only oil above the water level flows down into the funnel. The main pipe **30** extends at one end into the collection tank **12a** where it is fitted with an upwardly turned elbow **39**, so that the oil from all the tanks is thus collected in the collection tank. Valves **40** manually controlled from the deck selectively operate the oil collection units **36**.

In operative use, it is now evident that prior to returning ballast water back into the sea, the residue oil is thus removed from the tanks on a day when the ship is not excessively rolling or pitching so that the oil is steady on top thereof. After the oil is removed, the ballast water is returned to the sea prior to entry into port without contaminating the sea.

The collected oil in the collection tank can be either pumped out afterwards for purification or can be delivered at an oil receiving port on a next voyage. A further feature is that the oil and water mixture which is transferred

FIGURE 29: PLAN AND ELEVATION OF TANKER INCORPORATING BALLAST OIL POLLUTION CONTROL SYSTEM

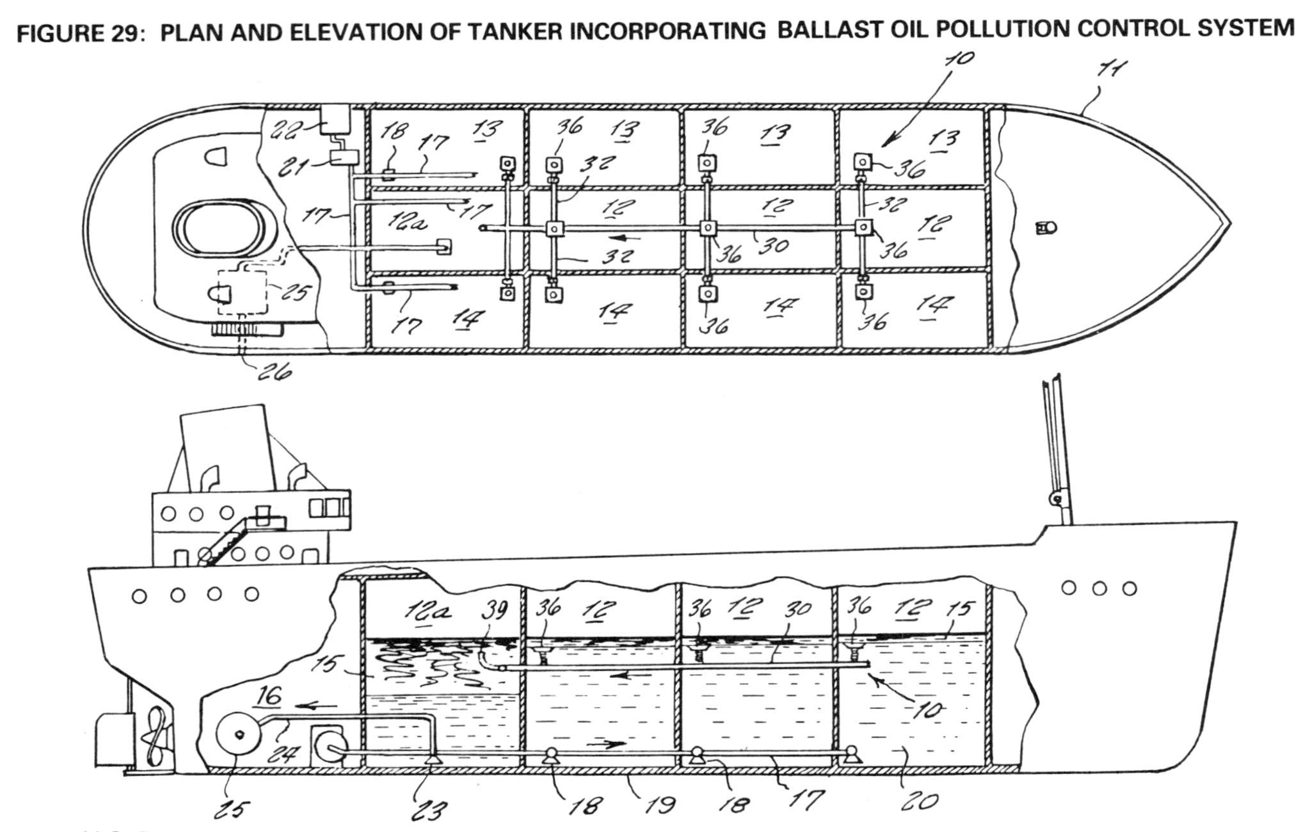

Source: U.S. Patent 3,957,009

to the collection tank need never flow through any pumps since the difference between liquid levels in the tanks causes the liquid to transfer due to gravity. This is an important improvement over systems which pump oil and water mixtures into a slop tank or to onshore processing stations, because the pump emulsifies the mixture making separation much more difficult or almost impossible.

A system developed by *D.C. Garber; U.S. Patent 3,965,004; June 22, 1976; assigned to Sun Shipbuilding & Drydock Co.* is a water decontamination system which is particularly applicable to oil tankers. In tankers running under ballast, the ballast water becomes contaminated with oil. Also, the engine room bilge water is oily, and the liquid cleaning of the cargo tanks results in the production of slop, which is oily. These various oily waters also commonly contain particulate matter such as silt, rust, waxes, asphalts, etc. Before this ballast, bilge water, or slop is discharged overboard from the tanker, it is essential, or a least highly desirable, that substantially all of the oil be removed therefrom, as well as the particulate matter. The main reason for this is to avoid pollution of the rivers and oceans. By way of example, it may be desired to reduce the oil content of the water to be discharged overboard to less than about six parts per million.

In this system, the oily water is first passed through an oil-water separator such as a gravity-type which provides a rough separation of the oil and water. The effluent from this unit is pumped through a ceramic dewaxer containing porous ceramic pellets. Waxes, asphalts, gums, and similar materials in the separator effluent (influent for the dewaxer) impinge upon and are retained by these pellets, which may be regenerated when necessary. Effluent from the dewaxer is passed through means to remove particulate matter such as a sand filter and then through a coalescer, which coalesces the remaining oil droplets to complete the oil-water separation. Such a system is shown in Figure 30.

The oily water influent for the separation system, which may comprise the ship's ballast, or slop or bilge water stored in tanks aboard the ship, is pumped through a line **1** toward and into a gravity feed tank (storage tank) **2**, by means of a pump (not shown; usually a centrifugal pump) installed on the ship. A level controller **3** (which may be of the conventional float type), installed on tank **2**, controls the valve operator **4** of a valve **5** in line **1** in such a way as to normally prevent overflow of tank **2**. An emergency overflow pipe **6** leads from the top of tank **2** to the bilge or slop tank or tanks.

The contaminated water flows by gravity from tank **2** through a line **7** to the first processing unit or component **8**. This unit is a gravity-type oil-water separator operating on the density difference between oil and water, and providing a rough separation of the oil and water. The separator **8** may, for example, be a "Model DE-5 Separator," manufactured by Butterworth System, Inc. of Bayonne, New Jersey. It operates at ambient temperatures and requires only gravity flow throughout. Under favorable operating conditions, the nominal separatory refinement capability of unit **8** is approximately 100 ppm oil, which is to say that the oil content of the effluent from this unit (in the effluent line **9**) may be reduced by the unit to approximately 100 ppm. The capability mentioned varies somewhat with the type and amount of oil in the influent (line **7**) to this unit.

FIGURE 30: BALLAST WATER DECONTAMINATION SYSTEM FOR SHIPBOARD USE

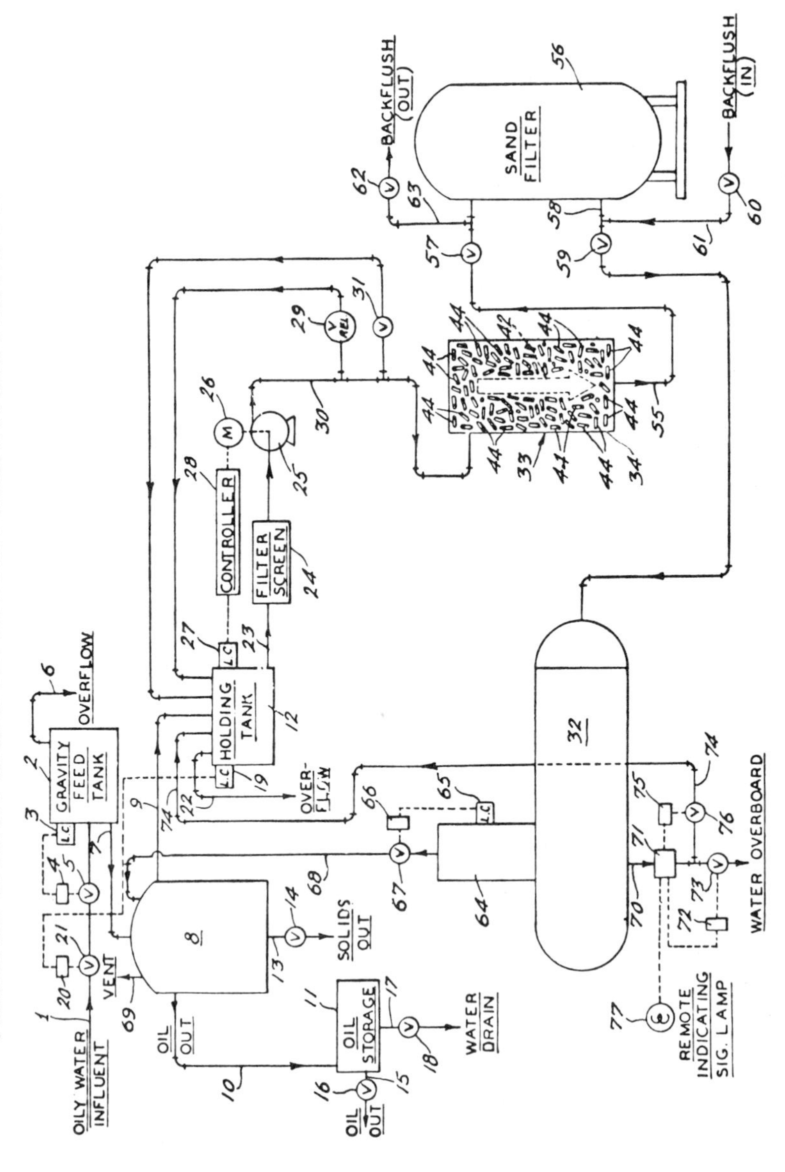

Source: U.S. Patent 3,965,004

In the interest of completeness, a brief description of the structure and mode of operation of separator **8** will now be given. This unit is a three-stage gravity-type unit utilizing fins or vanes to promote the coalescing of the larger oil droplets which then rise to the top, agglomerate and overflow weirs leading to the oil outlet pipe **10**. It may be noted here that this oil outlet pipe leads to an oil storage tank or reservoir **11**. The deoiled watery effluent overflows a central conically topped standpipe and then exits through the water outlet pipe **9** to a pump holding tank **12**. The oil overflow weirs are set hydraulically (vertically) higher than the overflow edges of the water standpipe, in order to pass the oil but retain the water.

The only servicing required by the separator **8** is periodic cleaning. A line **13** and valve **14** are provided for draining liquid or for the removal of collected solids from the separator **8**. For effective operation, the flow through separator **8** should not exceed its rated capacity.

A particularly useful characteristic of the separator **8** is its capability of completely bypassing to the oil outlet **10** an influent stream (to this unit) comprising 100% oil. Thus accidental or random "flooding" of downstream units of the separation system is avoided.

A line **15** and valve **16** are provided for removing oil from the reservoir **11**. In addition, a bottom drain line **17** and valve **18** are provided for draining out water which may accumulate in reservoir **11**.

The first processing unit (separator device) **8**, operating at ambient temperatures, will provide an effluent (at **9**) consisting of water plus the following:

a. Neutrally buoyant particles large enough to be screenable, and that might interfere with effective pump operation if not filtered out.
b. Neutrally buoyant particles (e.g., silt) too small to be screenable.
c. Neutrally buoyant globules consisting of fine metal particles encased in oil.
d. Neutrally buoyant waxes, asphalts, and similar materials that quickly clog porous filtration media such as fine hole size coalescers and porous ceramic filtration media.
e. Microscopic oil globules in the micron and submicron size ranges, whose settling-out rates due to density differences would result in prohibitively long residence times in the settling-out chamber **8**.

The pump holding tank **12** is provided with a level controller **19** (holding tank overfill float control) which controls the valve operator **20** of a valve **21** in line **1**, upstream from valve **5**. An emergency overflow pipe **22** leads from the top of tank **12** to the bilge or slop tank or tanks.

The deoiled water in the holding tank **12** is fed by means of a line **23** through a filter screen **24** to the intake side of a pump **25**. The filter screen **24** separates out item *"a"* above, thus protecting pump **25**.

Pump **25** (illustrated as of centrifugal type, though it may be reciprocating) functions to pressurize the oily water stream (deoiled effluent from the separator **8**) exiting from the holding tank **12**, in order to force this stream through the subsequent components. Pump **25** is driven by a motor **26** adapted to obtain its power from a steam, a compressed air, an alternating current, or a direct current source. Pump **25** is activated by a float control sensor (level controller) **27** in the holding tank **12**, which controls a motor controller **28** for motor **26**. It may be noted that the float of controller **27** is set lower than the float of controller **19**, so that pump **25** will be activated before valve **21** is operated.

The system is protected against overpressurization by a pressure relief valve (safety valve) **29** which is connected between the discharge side of pump **25** (line **30**) and the upper portion of tank **12**.

The desired (designed) net flow for the system is obtained by means of a flow control bypass valve **31**, connected between line **30** and the upper portion of tank **12**.

A coalescer **32**, comprises the final unit of the separation system of the process; this unit operates on the microscopic oil globules described in item *"e"* above. However, as stated in item *"d"* above, waxes, asphalts, gums, and similar materials quickly clog fine hole size coalescers, as well as porous ceramic filtration media. If these contaminants are also neutrally buoyant, they cannot be removed from the (coalescer) influent by the separator **8**. Therefore, a ceramic dewaxer, the cartridge of which is denoted by numeral **33**, is inserted upstream or ahead of the coalescer **32**. In this connection, it is desired to be pointed out that the name "ceramic" dewaxer indicates that a ceramic filtration medium is used in the dewaxer. The ceramic dewaxer basically functions to separate out all of item *"d"* above, thus preventing this material from reaching and clogging the coalescer. The ceramic dewaxer also separates out some of items *"b"* and *"c"* above.

For simplicity, only the cartridge of the ceramic dewaxer is illustrated at **33**. The ceramic dewaxer cartridge comprises a stainless steel tube, open at both ends, with a stainless steel wire cloth screen covering one end and fixed to it by welding. The other end of the tube has a set of female threads which mate with the threads on a removable stainless steel plug having a central bore the inner end of which is covered with a stainless steel wire cloth screen fixed to the plug by welding. A round stainless steel bar is welded across the open outer end of the hollow plug, and one end of a positioning rod, which extends outwardly, in the axial direction of the plug, is welded to the bar.

The tube **34** has lengthwise arrows **42** on it, indicating the direction of normal fluid flow during operation, these arrows being formed by drilling small holes through the tube wall. These holes also serve to vent the gas formed during the wax burn-out operation (this will be described later). These holes (about ⅛-inch in diameter) and the openings in the wire cloth screens **35** and **39** are small enough to ensure that the ceramic pellets **44** (which are packed into tube **34**) are always retained in the tube and between the wire cloth screens.

To load the tube **34**, the threaded plug **37** is removed, the ceramic pellets **44** are randomly inserted to the proper level, and then the tube is vibrated to obtain maximum compaction, following which the plug **37** is tightly screwed home. This constitutes the ceramic cartridge **33**.

A preferred embodiment of the ceramic pellets **44** is a solid, cylindrical configuration, 0.205 inches in diameter by 1¼ inches long. Although a cylindrical shape is generally preferred, the pellets could be of various other shapes. The material is a special formulation of abrasive, similar to that used for abrasive deburring of metal parts by barrel tumbling a mixture of the abrasive pellets and the metal parts. It is a very porous material, with open pores. Typically the porous ceramic material will be a fired, but unglazed clay, but it will be understood that other porous clay-type materials are equivalent. A supplier of the preferred ceramic material is Almco-Queen Products Division of King-Seeley Thermos Co., located in Albert Lea, Minnesota.

As previously stated, the pressurized oily water output from the pump **25** flows through the ceramic dewaxer in the direction of the arrows **42** on the ceramic cartridge **33**. The ceramic dewaxer separates out the waxes, asphalts, gums, and similar materials (which quickly clog porous ceramic filtration media) by virtue of the fine pores within the bodies of the ceramic pellets or cylinders **44** becoming clogged therewith. Specifically, the oily water stream, in passing through the perforated (at its ends) stainless steel ceramic dewaxer cartridge **33**, is constrained to course back and forth as it flows through the constricted passageways formed by the tightly compacted cylindrical ceramic pellets. In so doing, the waxy, asphaltic, and gummy type materials described in item *"d"* above impinge upon and are retained in the open pores that are peculiar to the abrasive ceramic structure. Some of the other materials, also, are retained on this structure. It may be noted here that pellets of other shapes than cylindrical will provide similar constricted passageways, through which the oily water may course back and forth.

During operation, prior to the ceramic cartridge's becoming fully "loaded up" with the above-described contaminants the cartridge **33** is replaced with a "clean" one, and the "dirty" cartridge is processed through one or more of the following regeneration steps.

1) The "waxy" material is burned off the pellets by placing the ceramic cartridge inside a boiler, near the bottom of the combustion zone. This may be effected, for example, by utilizing a stainless steel insertion tube which extends into the boiler combustion zone and which is adapted to receive and retain the ceramic cartridge while allowing the boiler heat to "bake out" the waxy materials which have been collected by the ceramic pellets **44**. As previously stated, the holes **43** which form the arrows **42** serve to vent the gas formed during the wax burn-out operation.
2) Alternatively, or in addition to the baking out, the ceramic cartridge **33** may be placed in a high pressure, high temperature pipe line through a bolted flange, and then cleaned with superheated steam which flows through the cartridge in a direction opposite to that indicated by the "operational flow" arrows **42**.
3) Alternatively, or in addition to the above-mentioned cartridge regeneration processes, the ceramic cartridge can be processed by a solvent cleaning operation, utilizing an agent such as trichlorethylene, or other solvents, or caustic.
4) Alternatively, or in addition to the above-mentioned cartridge regeneration processes, the ceramic pellets **44** can be removed from the cartridge **33** and cleaned in acid, by dumping them into an acid vat in order to remove metal particles.

5) Alternatively, or in addition to the previously mentioned cartridge regeneration processes, the pellets may be wet with water and then exposed to microwave energy, for example in a microwave oven. In this case, the microwave energy flashes the water in the pores into steam, which then pushes the dirt out of the pores.
6) Alternatively, the "spent" ceramic pellets **44** are discarded and the cartridge repacked with new pellets at low cost. However, by virtue of all of the previously mentioned steps, the mean time between replacement of materials is considered to be very long.

Although the previous description has referred to a single ceramic cartridge **33** inside a pipe, to provide for increased rates of flow multiple cartridges (i.e., a plurality of cartridges) may be utilized within a single pressure vessel. In this case, a manifolded flow arrangement may be employed, wherein all of the cartridges are paralleled insofar as the flow is concerned.

Another contaminant (in addition to the waxes, asphalts, gums, and similar materials, which are removed from the stream by the ceramic dewaxer) inimical to the continued operation of the coalescer **32** is waterborne silt. Some of this material passes through the separator **8** and the ceramic dewaxer and, absent any means to eliminate it from the coalescer influent, would be retained by and thus clog the coalescer.

In order to remove from the stream and retain the silt particles of larger size (e.g., those larger than about 7-10 microns), which larger particles would clog the pores of the coalescer (pore size about 25 microns), a sand filter **56** is placed upstream or ahead of the coalescer **32**. It may be noted at this juncture that smaller size silt (particles smaller than about 7-10 microns, not retained by the filter **56**) will readily pass through the pores of the coalescer **32**.

The sand filter **56** is preferably a "Baker Hi-Rate" filter, manufactured by Baker Filtration Company of Huntington Beach, California. The influent for filter **56** is the dewaxed effluent in line **55**, and this is fed to the inlet connection of filter **56** by way of valve **57**. The desilted effluent leaves filter **56** by way of a line **58** and is supplied through a valve **59** as the influent for the fine pore coalescer **32**.

Periodically, the sand filter **56** is backflushed for a period of 2½ to 3 minutes with clean sea or fresh water, to remove the silt. For this, valves **57** and **59** are closed, valve **60** in a backflush water inlet line **61** is opened, and valve **62** in a backflush water and silt outlet line **63** is opened. The backflushing flow rate is approximately 75% of the maximum forward flow filtration rate. At time intervals of between 6 months and several years, the filter sand in unit or component **56** is replaced, at low cost.

The finely dispersed oil in the watery, desilted effluent from filter **56** is removed in its entirety by the last component in the separation system, the coalescer **32**. The separator **8**, the ceramic dewaxer **33**, and the sand filter **56** all function to properly precondition the oily water influent before it is presented to the coalescer **32** for such final and complete separation of the finely dispersed oil in the water. While the separator unit **8** operates primarily on a quantitative basis the ceramic dewaxer **33** and sand filter **56** operate principally on a qualitative basis, selectively removing those contaminants that are inimical to the continued operation of the coalescer **32**.

The preferred embodiment of the coalescer **32** is one having a fine hole size. As mentioned hereinabove, this type of unit is quickly clogged by waxes, asphalts, gums, and similar materials; it is also clogged by silt. Therefore, it is a requirement that the ceramic dewaxer **33** and the sand filter **56** precede this coalescer, in order that a reasonable service life shall ensue.

The construction of the coalescer element may be achieved by precise lathe winding of a monofilament material such as Teflon, nylon, glass fiber, or other material on a perforated tube so as to create a large number of radially-oriented holes of small diameter and long length. Coalescers using this type of construction are available from Selas Flotronics Division of Selas Corporation of America, Spring House, Pennsylvania.

The pressurized, deoiled, dewaxed, desilted effluent issuing from the sand filter **56** (via pipe **58** and valve **59**) is fed to the inside of the perforated tube, and is forced by the pressure radially outwardly through the small diameter holes. (It may be noted here that a number of these coalescing elements or fiber-wound tubes or cartridges are mounted in the right-hand or inlet end of the shell or tank of coalescer **32**.) While traversing these holes, the finely dispersed, microscopic oil globules have their surfaces cleaned or polished as they brush past the monofilament material that lines the hole sides. The combination of surface cleanliness and physical intimacy enforced on the oil globules on their through transit causes them to coalesce into significantly larger globules, which quickly settle out due to density differences after exiting from the fine holes.

It is highly desirable that separator **8** remove the maximum amount of oil from the influent oily water mixture, in order to minimize the resultant "oil loading" of the coalescer **32**. This desirability stems from the fact that the finely dispersed microscopic oil droplets in the water have the dirt removed from their surfaces as they traverse the fine hole size pores of the coalescer cartridges. This dirt is retained within the body of the cartridges. As the dirt accumulates, the pressure drop through the cartridges for a given flow rate increases. When the cartridges eventually require a 100 psi pressure drop to maintain the flow rate, they are replaced if not cleanable. Thus, the life of the cartridges is a function of the amount of dirt retained within the cartridges, which in turn is a function of the amount of oil actually handled.

Periodically, on the order of 6 months, the coalescer cartridges (coalescing elements) are replaced, if not cleanable in solvents, at reasonable cost. As explained in the preceding paragraph, the "lifetimes" of these cartridges are sensitive to the rate of actual "dirt loading" that the cartridges experience.

A storage chamber **64** is provided on the top of the tank of coalescer **32**, in communication with this tank. A level controller **65** (oil-water interface float control) is located approximately at the midpoint of the length of chamber **64**. This controller controls the valve operator **66** of a valve **67** in an air/oil feedback line **68** connected to the top of chamber **64**. The line **68** extends to the separator **8**.

The coalesced oil droplets and any coalesced air or other gases rise to the top of the storage chamber **64**, and accumulate therein. As the oil/water interface moves down, the float control **65** opens the storage chamber valve **67**, which results in feeding back the coalesced oil and air to the separator **8**. The oil merges with that in the unit **8** and goes out to the oil storage tank **11**. The air is vented to the atmosphere, by way of a vent **69** provided on the top unit **8**. The clean water effluent exits from the coalescer, for going overboard, by way of an effluent line **70** connected to the bottom of the coalescer tank, in line with chamber **64**. This clean water effluent is sensed by an effluent oil monitor **71** which is coupled into line **70**. Monitor **71** is preferably a Bull and Roberts Model 240 Dual Beam oil-in-water detector, with integral sampling cell window self-cleaning capability using steam jets, manufactured by Bull and Roberts, Inc. of Murray Hill, New Jersey. One beam in this device uses ultraviolet radiation to sense the oil content, while the second beam uses visible radiation to sense suspended particulates and thus prevent signal ambiguity. This device can detect as little as 1 ppm oil content and can actuate relays and signal out remotely.

One of the relays actuated by device **71** controls the valve operator **72** for a cut-off valve **73** in line **70**. A bypass line **74**, for rejected effluent feedback, branches off from the effluent line **70** between the outlet side of monitor device **71** and the valve **73**. Line **74** extends from line **70** to the top of holding tank **12**. Another of the relays actuated by device **71** controls the valve operator **75** for a bypass valve **76** in line **74**. A remote indicating singnal lamp **77** is also actuated by device **71**.

By setting the oil-in-water monitor **71** to trip the relays associated therewith at from 6-10 ppm of oil, any substandard effluent from coalescer **32**, including that experienced during start-up or shutdown, will be automatically reprocessed or recycled (by closing valve **73** and opening valve **76**, thus feeding the water effluent from the coalescer back into holding tank **12**), and this operating condition will be remotely indicated by lamp **77**.

As previously stated, the separator **8** has the capability of completely bypassing to the oil outlet **10** and influent stream comprised 100% of oil. The coalescer **32**, independently, also has this capability. In addition, as just described, the effluent oil monitor **71**, which senses the oil content of the effluent water stream issuing from the coalescer **32**, will cut off the overboard water discharge if its oil content is too high, i.e., above 6-10 ppm of oil. Thus, the system of this invention is triply protected against random or accidental overboard oil discharges.

Although the ceramic dewaxer has been previously described as utilizing the ceramic pellets **44** in the tube **34** (a cartridge **33**), it may be more convenient, in some instances, to omit the cartridge structure and simply utilize the ceramic material, in bulk, in a pressure vessel. From a materials handling standpoint, this latter expedient might be more economical.

A system developed by *J.O. Moreau; U.S. Patent 3,985,020; October 12, 1976; assigned to Exxon Research and Engineering Co.* is a monitoring system for proving compliance with pollution regulations, which enables a tanker to prove that restrictions on total oil or rate of oil pollution have not been violated. The system comprises adsorbing an oil sample which is proportional to total oil discharged and rate of oil discharged on a continuously moving lipophilic belt and subsequently analyzing the belt after the voyage has been terminated when required to prove that the tanker has not exceeded ocean pollution requirements.

Figure 31 is a schematic diagram of such a monitoring system. An oily water sample is withdrawn in a conventional manner (typically through a sample tap) via tubular or conduit line **14** at a location **16** in the conduit **12** and is passed through a constant pressure regulator **18** and then through the tubular line of conduit **20** into a sample manifold **22** of the monitor **10**. The regulator **18** insures constant velocity of any portion of the sample of oily water which contacts the tape by maintaining a constant pressure drop across the orifice which restricts the sample flow. A typical regulator comprises a variable orifice flow restrictor.

FIGURE 31: OIL POLLUTION COMPLIANCE MONITOR SYSTEM FOR USE IN MONITORING TANKER DISCHARGES

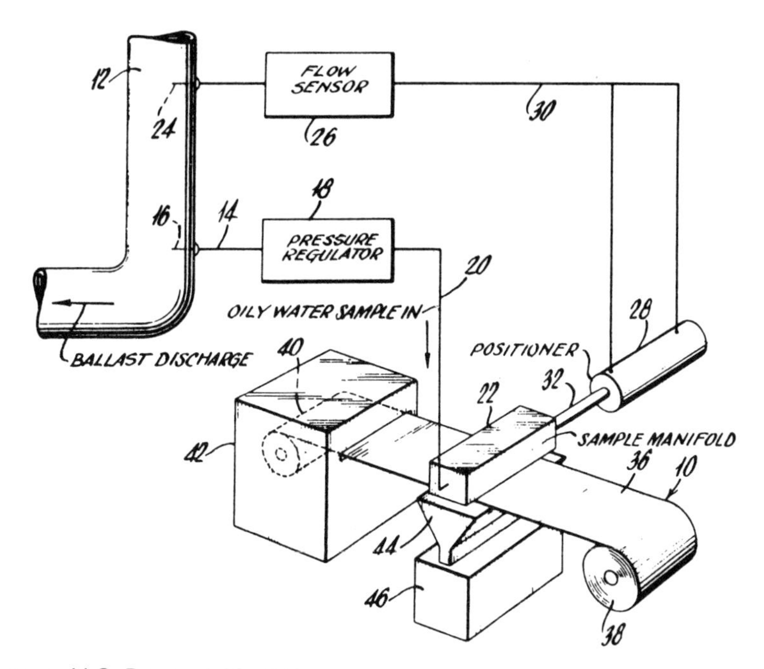

Source: U.S. Patent 3,985,020

Other devices which also are sufficient for this purpose include a constant-head tube with overflow or a pump with constant output pressure. At a point **24** upstream from the sample withdrawal point **16**, there is provided a flow sensor **26**, which typically comprises a conventional orifice and pneumatic amplifier, for sensing the flow rate of the ballast discharge and producing an output pneumatic signal of 3 to 15 psi which is proportional to the rate of ballast discharge flow. The signal is coupled to a pneumatic positioner **28** (e.g., a control valve actuator) via conduit **30**. The positioner **28** in turn is coupled through its output shaft **32** which is movable in an axial direction, to a movable vane which is positioned within the manifold **22**. The control positioner causes the vane to move axially such that the total oily water sample flow rate out of the manifold is proportional to the total ballast discharge flow rate.

A belt of lipophilic tape **36** is pulled at a constant speed from a supply reel **38** (which rotates counterclockwise) in a direction passing beneath the sample manifold **22** which is transversely located relative to the direction of tape movement, for depositing sample on the tape as discussed in detail hereinafter. The tape is collected and stored on a takeup reel **40** (also rotating counterclockwise) which in the preferred embodiment is located in a refrigerated container **42** which minimizes the deleterious effects of vaporization, oxidation and biodegradation of the oil on the tape. As the tape traverses from the supply reel **38** to the takeup reel **40**, it passes under the sample manifold **22**, which directs oily sample water through all or a portion of the tape width dependent on ballast discharge rate.

The percentage or portion of tape width contacted with the sample oily water is adjusted by means of a piston which is controlled by the pneumatic signal (from flow sensor **26**) proportional to the flow rate of the ballast water being discharged. As the sample passes down through the tape, it is received in a funnel or chamber **44** located directly on the opposite side of the tape **36** below the sample manifold **22** and is discarded into a suitable storage or chamber **46**.

Tanker Accidents

The matter of tanker accidents involves, as with most cases of oil pollution, the two aspects of (1) prevention and (2) removal if prevention fails.

The pollution resulting from accidents involving ships and especially petroleum tankers and the public concern over the ecological effects of such pollution have increased markedly in recent years. In 1959, the world's total loss of tankers was 350,000 deadweight tons and in 1969 this increased to almost 600,000 tons. This is the equivalent of thirty-three ships the size of the small tanker Arrow which ran aground off Nova Scotia, Canada in February 1970. This "small" accident alone fouled hundreds of miles of beaches, caused serious harm to marine and bird life and necessitated the expenditure of more than three million dollars in direct and indirect costs for salvage and clean-up operations.

When the relatively small size of this ship (18,000 deadweight tons) is considered in relation to the super tankers of 200,000 tons or more which are now in operation and those of up to 400,000 tons which are soon to come into operation, the enormity of the economic and ecological implications of an accident at sea involving one of these giant ships can be readily appreciated. The importance of salvaging oil from such vessels in the event of their loss at sea needs no elaboration. Methods and equipment presently available for this purpose are not adequate to meet the task even under "ideal" salvage conditions much less under severe weather and/or deep water conditions.

Present methods for salvage of oil from sunken ships require that divers work directly on the vessel to manually connect and disconnect the oil removal pipes to the sunken vessel. As a result, there are severe limitations on the depth and climatic conditions under which oil can be successfully removed from a sunken ship. This is particularly so in the case of removal of heavy oil such as Bunker "C" fuel oil from tankers under cold water conditions since such materials have a tendency to congeal at low temperatures thereby necessitating the use of heat in order to permit the pumping and transport of the oil through conduits to the surface.

A technique of *J. Rolleman; U.S. Patent 3,831,387; August 27, 1974; assigned to Salvage Oil Systems, Ltd., Canada* involves removal of oil from a sunken vessel by means of a remotely controlled submersible "pump-house" or salvage capsule which performs, without the services of divers, all the necessary functions to gain access to, and remove to the surface, oil contained in the sunken vessel. Instead of relying on manual operations performed by divers on the sunken vessel, the essential features of the present process include remotely controlled means on and within the salvage capsule for effecting the necessary functions including: means for securely but detachably fixing the capsule in the decking or hull of the vessel in proximity to the compartment from which the oil is to be salvaged; drill means for providing access to the compartment through one or more openings, extensible oil suction pipe means for insertion into the compartment through the opening, first pump means for removing oil from the ship compartment into a holding chamber within the capsule and second pump means for removing oil from the holding chamber to the surface where it may be held in suitable storage such as balloons or salvage tankers.

The essential elements of the present system are shown in the diagram in Figure 32. The oil salvage capsule **10** is shown being utilized with a recovery supply ship **11** to remove oil from a sunken tanker **12**. Service lines **13**, which connect the salvage capsule **10** with the ship **11**, lead into the interior of the capsule through a sealed port. The service lines include all conduits, cables, etc. (which are not individually shown), necessary to supply electricity, water, steam and compressed air to the salvage capsule **10** as well as an oil recovery line for pumping oil to the surface. Recovery cable **15** extends from the ship **11** and is attached to the lower side of the capsule. The recovery cables **15** are utilized not only for raising and lowering the capsule **10** but also for aiding in the vertical and horizontal positioning of the capsule in relation to the sunken vessel **12**. A tower **18** is secured to the upper side of capsule **10** and is adapted to support and selectively reciprocate a suction pipe **19**.

A technique developed by *E.C. Garcia; U.S. Patent 3,832,966; September 3, 1974* involves building a tanker with double sides defining side tanks extending from a top deck to hull bottom and adjacent to and associated with cargo tanks. The side tanks from hull bottom to waterline are made equal in volume to respective volumes of associated cargo tanks above the waterline.

It is old in the art to build tankers with double bottoms and double sides for an additional cost of some 30% increase in the price of the tanker. In case of wreck or other marine catastrophe, it is expected that at least the inner bottoms and sides will remain undamaged or liquid leakage proof.

It is an object of the present mode of construction to build a tanker that will provide security from spills at an additional cost of only 10% of the tanker's cost.

FIGURE 32: APPARATUS FOR SALVAGING OIL FROM SUNKEN VESSELS

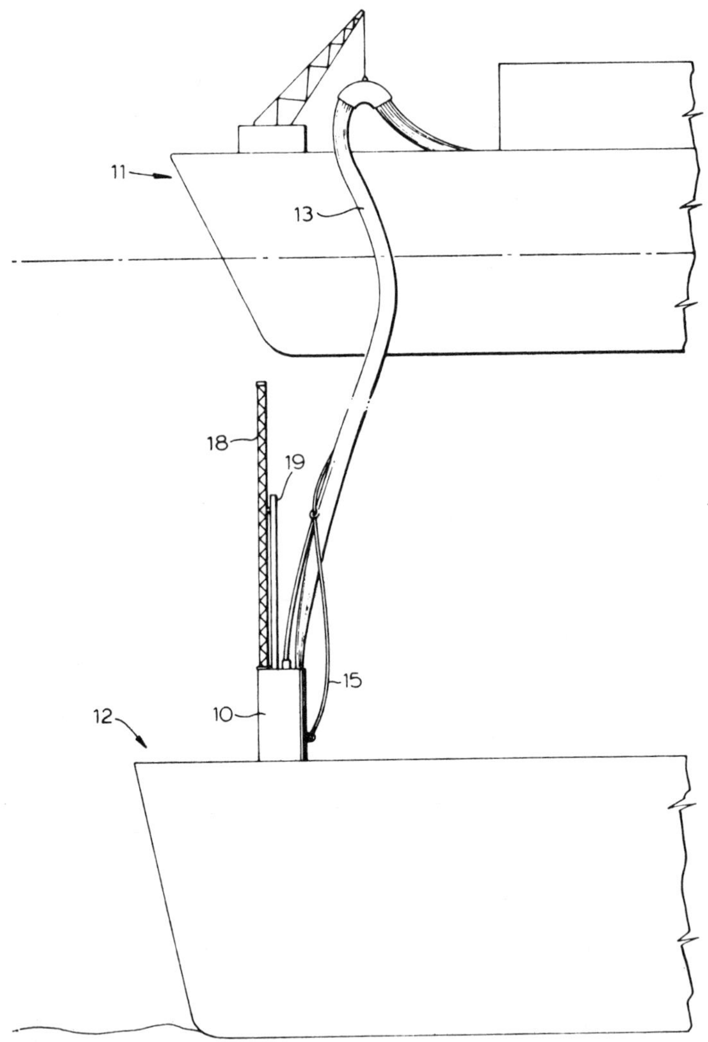

Source: U.S. Patent 3,831,387

Figure 33 shows the present method of construction. Referring to the drawing, tanker **10** has center cargo tanks **12** and **14**, extending longitudinally for any convenient distance, defined by liquid proof bulkheads **16, 18** and **20**. Valved openings **22** and **24** are defined as illustrated in bulkheads **16** and **20**, which also define side tanks **26** and **28** with the hull of tanker **10**. Tanks **26** and **28** are sized to equal, below the loaded waterline **29**, the respective volumes **30** and **32** above the loaded waterline of tanks **12** and **14**. Side tanks **26** and **28** have air vents **34** and **36**, larger than air vents **38** and **40** in cargo tanks **12** and **14** to ensure that the downflow of liquid from volumes **30** and **32** never exceed the flow of liquid from tanks **12** and **14** into side tanks **26** and **28**. Thus no liquid escapes from the ruptured hull bottom because the amount of downflow from volumes **30** and **32** is dependent on the relative sizes of vents **38** and **40** in comparison with vents **34** and **36**.

Of course if only side damage is suffered, spills are prevented by the double sides defining side tanks **26** and **28**.

FIGURE 33: METHOD OF BUILDING TANKERS FOR PREVENTING OIL SPILLS IN CASE OF WRECK

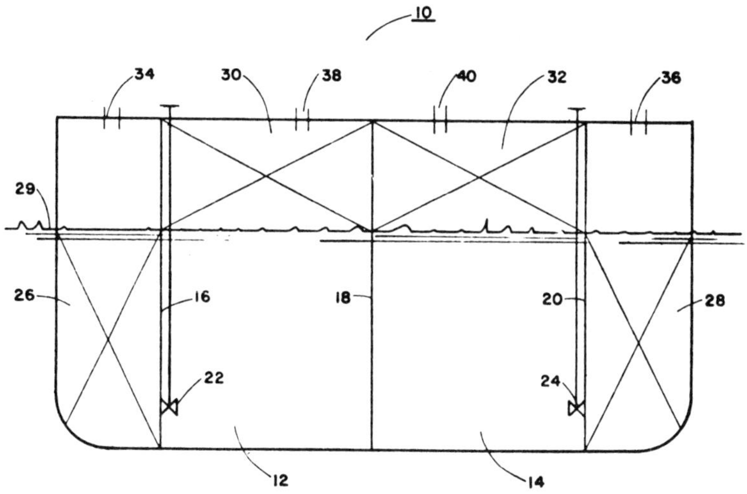

Source: U.S. Patent 3,832,966

A very interesting report in this connection is one (12) entitled "Being Prepared for Future Argo Merchants" and substantial sections of that report are reproduced here.

The grounding and break-up of the Argo Merchant in December 1976 and the subsequent oil spill brought to public attention just how inadequately prepared we are for dealing with offshore oil tanker accidents. This lack of preparedness applies both to cleaning up oil from the seas as well as to salvage of the ship and cargo before or during oil spillage.

Although each marine tanker accident is different and requires a somewhat different response, there are a number of relatively common features. Therefore, by knowing the important events of the Argo Merchant incident, the reader can achieve an understanding of what is needed for dealing with such events in the future. Therefore, these events are described here. Then, possible means for dealing with such events, and with similar events, in the future are described.

Chronology of the Event: At approximately 6 a.m. on Wednesday, December 15, 1976 the Argo Merchant ran aground on Fishing Rip, which is a shoal located about twenty-seven miles southeast of Nantucket Island, Massachusetts. The grounding damaged the vessel and flooding of the engine room soon began. This flooding disabled the ship's power-making machinery which resulted in power-driven machinery, such as ship's pumps, being made inoperative. Furthermore, steam could no longer be supplied to the heating coils in the ship's tanks so that the oil began to cool slowly. The Argo Merchant carried No. 6 oil which is so viscous at low temperatures that it is difficult to pump. Therefore, during shipment it is usually kept warm (90° to 120°F) so that it can be pumped off the ship with relative ease when the ship arrives at its destination. Once heating steam is lost from the heating coils in the tanks, the cooling begins. This cooling takes place relatively slowly and generally it would take several days for the temperature of the oil in the ship to reach that of the surrounding sea.

At 7 a.m. on Wednesday the U.S. Coast Guard station in Woods Hole, Massachusetts received a Mayday message from the ship. During Wednesday, the Coast Guard delivered emergency water pumps to the ship and personnel from the Coast Guard cutters Sherman and Vigilant assisted with operating them.

During the day Wednesday, water was pumped out of the engine room by means of the pumps which the Coast Guard had brought aboard. Of course, water was leaking in at the same time because of the damage. The damage to the ship also resulted in some of the cargo of No. 6 oil leaking into the engine room. This oil was cooled off by the cold sea water and the resulting cold oil was so viscous that it fouled the pumps. The estimated sea water temperature in the engine room was about 10°C (outside in the sea the temperature was about 6°C). At 10°C the viscosity of the No. 6 oil carried by the Argo Merchant was about 35,000 centipoises. To give the reader an idea of just how viscous this is, it is remarked that the viscosity of water at room temperature is about 1 centipoise and that of a typical crude oil is about 100 centipoises. The cold No. 6 oil has a consistency not very unlike that of thin peanut butter.

By Wednesday evening, a Coast Guard helicopter had put strike team personnel and one ADAPTS (Air Deliverable Anti-Pollution Transfer System) pumping system aboard the vessel. Three strike teams, one on each of the Atlantic, Gulf, and Pacific coasts, are maintained by the Coast Guard for dealing with oil spills and potential oil spills. An ADAPTS pumping system is designed for offloading oil from a stricken vessel. It consists of a power source, an offloading pump, and hoses. The offloading pump is a submersible type, intended to be submerged in a ship's tank. The pump is driven by an integral hydraulic motor which operates from high pressure hydraulic fluid delivered to it through hoses by the power source. The power source is a self-contained diesel engine, hydraulic fluid pump, and associated machinery. The ADAPTS systems are designed to offload between 1,000 and 2,000 gallons of oil (having a low to moderate viscosity) per minute.

In the case of the Argo Merchant, the ADAPTS system was not used for offloading oil, but rather to pump water with some oil out of the flooding engine room, and into the sea. Because of moderately large winds and seas and the dark of night, more ADAPTS systems were not brought aboard during Wednesday evening. Even to get the one ADAPTS system aboard it was necessary for the Strike Force personnel to first cut loose all the ship's antenna wires that ran between the midships and afterhouses so that a helicopter could safely operate between the houses and lower the ADAPTS components to the ship on a cable. The ship was still the responsibility of the owners who hired the Murphy Pacific Salvage Company to carry out salvage operations, whose representative was brought aboard the ship by Coast Guard helicopter. During Wednesday evening and Thursday, while the single ADAPTS system was pumping water from the engine room, the Coast Guard personnel and the representative of Murphy Pacific studied the situation to determine the best course of action.

At this point, it is appropriate to describe some aspects of the situation in which these people found themselves. They were aboard a damaged and grounded ship with a flooding engine room, and with a heel angle sufficient for the starboard side of the main deck to be nearly awash. Furthermore, the ship had taken on an abnormal trim with the stern lower than normal and the bow higher. With the equipment and facilities that were available, it was impossible to determine the precise nature of the damage. Although the basic design parameters of the ship and even curves of ship stability were available on board, determination of the exact nature of the flooding in various parts of a ship from this information as well as heel and trim is impossible with a vessel which is simultaneously grounded and flooded. If only one of these two situations had existed, grounding or flooding, the basic loads on the vessel could be determined from measurements of the heel and trim and use of the available information. This would yield knowledge about the amount and location of flooding in the case of a damaged vessel or the magnitude and location of the load on the bottom of a grounded vessel. With the simultaneous existence of grounding and flooding, this determination could not be made.

Therefore, measurements of the height of the liquid in a number of tanks were made as well as could be done under the circumstances, with the intent of comparing these with later measurements to then obtain information about the amount of flooding and its distribution. The flooding information about a stricken vessel is generally needed in order to be able to determine the best way to use whatever pumping capacity is available; in other words, to be able to decide which compartments to pump out. In the particular instance of the Argo Merchant, with only a single adequate pump available and with the observed engine room flooding it was obvious that the pump should be used to empty water out of the engine room.

During Wednesday, the Coast Guard had requested that the nearest available empty barge and tug come to the scene to assist in oil offloading operations. A major problem in the rapid use of a barge for offloading a stricken tanker in the presence of large waves is that of providing fendering between the barge and stricken vessel so damage is not inflicted by waves smashing the two vessels together. The vessel owners did not have large fenders available to them and although the Coast Guard had four, three of these were hundreds of miles away. The Coast Guard made arrangements for two fenders and two more ADAPTS pumping systems to be transported to the Coast Guard Air Station on Cape Cod. The pumping systems arrived at the Air Station later on Wednesday and the fenders arrived there by Thursday.

By Thursday morning, the fifteen knot winds and six to eight foot seas that had existed in the vicinity of the Argo Merchant during Wednesday had diminished. Also, the people aboard the vessel reported that the initial engine room flooding which had reached a height of twenty-two feet, had been reduced to fifteen feet by the single ADAPTS pump. By 8 a.m. on Thursday morning, the two additional ADAPTS systems were aboard the U.S. Coast Guard buoy tender Bittersweet in Woods Hole, Massachusetts along with additional strike team members. Shortly thereafter, the Bittersweet left Woods Hole for the scene of the Argo Merchant.

During early Thursday afternoon, the personnel aboard the Argo Merchant found that the water in the engine room was again rising. These people aboard knew that additional ADAPTS pumping systems would soon be aboard, and there must have been some question in their minds as to how they should be used. Should the additional pumps be used in the engine room to try to lower the water level there? On the other hand, some of the tanks of the vessel (the vessel contained thirty cargo tanks) appeared to be flooding. Since the vessel was heeled to starboard, should the additional ADAPTS pumps be used to pump water out of some of the apparently flooding starboard tanks in order to help right the vessel and possibly float her free of the shoal?

If any tank had a large hole in its bottom, no good could be accomplished by pumping out that tank inasmuch as the tank would flood to the waterline level no matter how much water was pumped out of it. If the hole in a flooding tank were so small that water entered through the hole more slowly than water was pumped out of the tank, then the water level could be lowered and additional buoyancy thereby provided. If it could be ascertained that there were some undamaged cargo tanks, additional buoyancy could be provided to the vessel by pumping oil from such tanks overboard. Although this may have been the most appropriate thing to do if undamaged tanks could have been located, the responsibility for pumping oil overboard could not have been assumed by anybody involved, so nobody would make a decision to do this.

Here, another diversion is in order. Under circumstances that existed aboard the Argo Merchant at this time, or which have and will exist on other ships under similar conditions, there is no decision an individual can take which can do him very much good. On the other hand there are many possible decisions which can do one much personal harm. Thus, situations like this tend to inhibit decision making when effective response actually requires firm and decisive decision making. The technical matters are difficult enough. Somehow, the social and institutional pressures must be eliminated in order to encourage decision makers to take the best possible action.

The Bittersweet arrived alongside the Argo Merchant at about 3 p.m. on Thursday. The two ADAPTS systems and additional strike force personnel were offloaded and the Bittersweet left the scene. Prior to this time, the responsibility for the Argo Merchant was that of the owners and this was being exercised by the representative of the Murphy Pacific Salvage Company. He had decided to use one of the additional ADAPTS systems for pumping water out of one of the starboard cargo tanks which was supposed to have contained no oil when the ship left Venezuela, but which appeared to be flooding. By this time, the heel of the vessel had increased, the sinkage towards the stern was larger, and the sea state was increasing with waves beginning to break onto the deck. This resulted in considerable time being required to set up the ADAPTS system and associated hoses. The reader should try to appreciate how difficult it is for men to handle large, heavy six-inch diameter hoses covered with slippery oil on a tilted deck covered with slippery oil with waves and spray coming down upon them.

It was dark by the time the pumping system had been set up on Thursday. When pumping began, it was not water which came out of the tank, but oil. This further increased the uncertainty of the situation. Could it have been that that tank was not empty when the ship began the voyage even though the crew reported that it was empty? Could a bulkhead between that tank and another tank have been damaged in the grounding, resulting in a leak so that oil from an adjacent tank poured into a previously empty tank? Could the condition of the ship before the voyage have been so bad that there was leakage between one tank and another so that an initially empty tank slowly filled up? These questions and others must have been going through people's minds and no answers to them were available.

During Thursday afternoon, the Coast Guard assumed command of the salvage operation under authority of the 1974 Federal Intervention on the High Seas Act. No doubt this was done because the owners of the vessel had not accomplished any positive steps toward salvage. All pumping up until that time had been done by Coast Guard personnel with Coast Guard equipment. The owners had not made any plans for rapid delivery of barges, fenders, or pumps for offloading cargo, nor had they made any arrangements for cleaning up oil that had spilled or might spill later. The strike team aboard the vessel was informed of the Coast Guard intervention by radio.

At about 7 p.m. Thursday evening, a tug, the Sheila Moran, arrived at the scene and it appears that her assistance had been requested by the Murphy Pacific Company. Since there was nothing she could do to help at that time, she stood by on scene.

During Thursday evening, the wind strength, which was now from the northwest, increased as did the size of the waves and the amount of wave breaking onto the main deck of the vessel, whose low (starboard) side was towards the waves. Some buckling of the main deck on the aft portion of the ship had been observed, and leaking of oil from a cargo tank into the engine room around bolts or rivets in a bulkhead could be seen. In the region of this bulkhead, strange sounds were emanating from the ship structure as a result of the loads caused by the seas and the bottom against the grounded vessel. Only the one ADAPTS pump taking water out of the engine room was being used. No one aboard knew how long the ship would last. No one could know. Even though the Coast Guard had assumed command of the salvage operation, both the Coast Guard personnel and the representative of Murphy Pacific were trying to figure out the best thing to do. It was clear that at that time there was very little that could be done immediately.

Many of the ship tanks were inspected by opening cover plates on them and a considerable number of the tanks exhibited much agitation and sloshing of the surface of the oil. This indicated that it was quite possible that the bottom of the ship was torn open, in which case pumping oil out of the tanks and overboard would most likely not have aided in ship salvage. Water was rising in the engine room and it seemed doubtful that pumping more water from the engine room with additional ADAPTS pumps would "stem the tide." Furthermore, it was deemed by everyone aboard to be extremely dangerous to work in the vicinity of the engine room because the behavior of the deck and the bulkhead between the aftermost pump room which was full of oil and the engine room, together with the sounds the structure was making, indicated that the vessel might break there at any time.

As a result, a decision was reached to take the people off the vessel. This was accomplished by Coast Guard helicopters late Thursday evening. When the lights from the helicopters illuminated the scene, it became clear that a substantial rate of oil leakage into the sea had begun. It was uncertain as to how much of this oil was coming out of deck openings and how much was coming out of the damaged bottom at this time.

At about 4.30 Friday morning, the 140,000 barrel barge Nepco 140, towed by the tug Marjorie D. McAllister, finally arrived about forty-seven hours after the grounding. However, seas were about four to six feet high with worse weather predicted and the Coast Guard chose to concentrate on other tasks rather than deliver fenders to the ship and bring the barge alongside the Argo Merchant. There was substantial oil pollution at this time and the pollution rate was estimated to have been approximately 40,000 gallons per hour.

Personnel of the Coast Guard had realized the pollution threat from the beginning and had brought their high seas oil booms and skimmers to the air station at Cape Cod. Also, the Coast Guard had contracted the Murphy Pacific Company to supervise their salvage effort.

Following the grounding of the Argo Merchant, the heading of the vessel changed from time to time. It is not certain how much of this heading change was due to wave forces and how much was due to forces of the currents. The currents at the location of the grounding are somewhat unique in the sense that they are rotary, not reciprocating. Whereas in most locations, tidal currents go in one direction and then switch and go in the opposite direction, on Nantucket Shoals the current direction rotates through all headings.

The salvage plans being generated by Murphy Pacific were to begin with stopping the heading changes of the ship by putting out two bow anchors. Then, plans were to locate a group of heavy moorings, each with a mooring buoy, around the ship to which barges could be tied. A work vessel was to be brought alongside the Argo Merchant with fendering to be provided by the two large Coast Guard fenders which had arrived at Cape Cod. This work vessel was to contain a steam heater which could be used to pump steam through a portable coil which could be put in one Argo Merchant tank after the other to heat the oil again to a temperature at which it could be pumped.

Conditions on Friday, December 17th, were somewhat rough and work was limited to inspection of the ship since all of the planned equipment for salvage was not yet available. Saturday, December 18th, was even rougher. Wind strength increased to over 40 knots and seas were nine to twelve feet high with almost every wave breaking on the shoals. Although the amount of heeling of the vessel seemed to change as the tide changed, the stern of the vessel was definitely getting lower and lower.

By Sunday morning, December 19th, the wind and seas had abated and conditions were nearly calm. In a combination of effort by Coast Guard Strike Force personnel and the tug Sheila Moran and her crew, one of the bow anchors of the Argo Merchant was put out. In the calmer conditions, the oil leakage rate appeared to be somewhat less than before. Wind and sea conditions were also moderate on Monday, December 20th, but during the night conditions worsened. By the morning of Tuesday, December 21st, strong northwest winds and large seas were again present. At 8:30 a.m. the Argo Merchant split in two and a great deal of oil escaped. By Wednesday morning, the wind strength had reached 45 knots and the seas were about twelve feet high. At about 9 a.m. on Wednesday, a section of the bow which had been afloat broke in two and most of the remaining oil escaped.

From the standpoint of pollution damage, we were all very lucky in the case of the Argo Merchant accident. Although at the time of the grounding the wind was from the southwest, during all of the time that oil escaped from the vessel, all strong winds were from the north or the northwest. This resulted in the oil being driven away from the shore, but to the south of George's Bank. Following the spilling of the oil, only for one short period did the wind blow towards land and although some oil came to within fifteen miles of Nantucket Island, before it got closer the wind direction again changed to the northwest and the oil was blown out to sea.

We will not always be so lucky. Statistics about wind direction indicate that such good fortune can be expected most of the time during winter in the location of the Nantucket shore. However, most of the time does not mean all of the time. In addition, winds toward shore are more prevalent there in the summer. There are, of course, many locations in the United States where the situation is reversed and the most frequent winds blow towards the shore.

If the Argo Merchant oil had been blown onto shore, we would be dealing with a coastal disaster of major proportions. Any region which has a large quantity of oil blown onto its shores will have such a disaster. The preceding description of events provides a useful framework for considering optimum equipment, personnel training and planning for diminishing the magnitude of such disasters.

Preparedness for Response to Stricken Vessels: The technology of the salvage of vessels which are grounded offshore has shown no essential advancements during the past thirty years. A technological advance in this field can be made now if funding and the attention of competent engineers are "focused" on the problem. The subject of regulations to decrease the likelihood of tanker accidents is not a subject of this report (it is being given much attention elsewhere). However, the subject of regulations intended to make tankers easier to salvage if they run aground is indeed a subject of this report and will be considered here.

The Argo Merchant lasted just slightly longer than six days after it grounded. During part of this time, the weather was quite rough. In 1970, the tanker Arrow ran aground in Chedabucto Bay, Nova Scotia. The Arrow lasted four days before breaking up. The Torrey Canyon lasted about one week.

The fact that grounded tankers generally seem to last several days after grounding indicates that the mechanism of breakup is not that of any particular instantaneous load exceeding that which the ship can initially withstand, but rather the process is one of fatigue, whereby reciprocating loads deteriorate either the macrostructure (frames, joints, etc.), the microstructure (metallurgical properties of the steel), or both over a period of several days. A feature of fatigue failure is that for specified loading conditions, a small increase in strength will often greatly extend the number of cycles a structure can withstand before ultimate failure.

Since grounded tankers generally seem to last several days before breaking up, it seems quite possible that a relatively small increase in structural strength could result in grounded tankers generally lasting several weeks. Studies to determine whether this would be the case are within the capabilities of present day ship structures experts and such studies should certainly take place. If the expected lifetime of a grounded vessel could be materially increased, many salvage operations could take place which are not possible with an expected lifetime of only a few days. Therefore, if studies indicated that a modest structural strength increase would markedly increase the expected lifetime of a grounded vessel, regulations upgrading the structural standards for tankers entering U.S. waters would be appropriate.

Much has been said and written about the advisability of requiring tankers to have double bottoms to minimize the pollution threat if a tanker should run aground. In the present context, the use of double bottoms could be quite helpful lengthening the expected longevity of grounded vessels. When a vessel runs aground, the bottom of the vessel is usually damaged. The cross-sectional shapes of large ships of today are such that the beam substantially exceeds the depth. This has been caused by the need to increase cargo holding capacity without increasing ship draft, which would limit the areas the ship could use because of limited water depth.

With cross-sectional shapes which are relatively wide and shallow, the ability of the structure to withstand sidewise bending is far greater than the ability to withstand vertical bending. In vertical bending, the maximum loads are carried in the ship's bottom and the ship's dock. If the bottom is damaged upon grounding, one of the primary structural members (the bottom) for withstanding vertical bending is either less efficient or unable to contribute at all to bending restraint. On the other hand, if a vessel had a double bottom and the outer bottom were ruptured, the inner bottom could still contribute a significant amount to the provision of vertical bending restraint.

It is useful to understand the enormous amount of buoyancy which can be required to re-float a grounded vessel. The Argo Merchant was a relatively small tanker by today's standards. It could carry approximately twenty-seven thousand tons of oil. The weight of the ship itself, exclusive of cargo, was about 18,000 tons. Suppose, for example, that the degree of damage to the ship was such that an external buoyancy equal to half of the weight of the steel of the ship had to be provided, this would be four thousand tons. One person once asked why the ship could not have been lifted up high enough to get it off the shoal with helicopters. The helicopters having the largest lifting capacity known (Sikorski Skycranes) can lift about 12 tons. Seven hundred fifty such helicopters simultaneously lifting would be required to lift the weight of half the steel of the Argo Merchant. Obviously, that would not be a practical solution. There are more practical possibilities.

By far, the most practical of these would be an arrangement whereby a stricken ship could float itself. It was impossible to do this with the Argo Merchant with the equipment that was aboard. However, it is feasible to require that all tankers entering waters be capable of sealing all deck openings in a time of one hour or less. Retrofitting existing vessels to meet such a requirement would be entirely practical. Suppose it had been possible to completely seal all deck openings on the Argo Merchant. If this could have been done, then if air were pumped into the tanks above the cargo while removing as much cargo as was displaced by the air, the four thousand tons of buoyancy could have been provided by depressing the liquid level in the ship's tanks three feet. This would have been effective if the bottom were ruptured or not. Fittings to accept air lines could be required on the top of each tank. Emergency salvage equipment could include compressors and hoses to supply the air.

Let us consider what this would have involved, had it been possible, in the case of the Argo Merchant. First of all, the deck opening seals and the deck structure itself would have to have been strong enough to withstand an internal pressure of approximately two pounds per square inch above atmospheric pressure (a practical requirement). Furthermore, approximately one seventh of the ship's cargo would have to have been discharged to allow a space for the air. Under the conditions of the grounding, had this even been possible, the only practical way to discharge this cargo (which would amount to about one million gallons) would have been to discharge it overboard.

Authorizing the discharge of one million gallons of oil into the sea is a responsibility which an individual simply cannot take under existing political conditions. In the case of the Argo Merchant, the impact of the oil, while severe on certain forms of life, especially sea birds, was the least possible since the oil moved offshore and apparently without damaging Georges Bank. In spite of that, all actions of the Coast Guard were mercilessly and unjustifiably attacked by the Lieutenant Governor and the Secretary for Environmental Affairs of the Commonwealth of Massachusetts. It is not hard to imagine what the nature of these attacks would have been like if the Coast Guard officers had been in a position to save the ship and most of the cargo by discharging one million gallons. It is certain that they would have been even worse than they were in the Argo Merchant case, since the public officials could link all the pollution that would then exist to direct actions of the Coast Guard.

A workable procedure for exercising human responsibility must be prearranged in a special way if optimum response to stricken tankers is to occur in the future. For each vulnerable region of the U.S. coastline, the most appropriate individual should be designated in advance as the one who will have the ultimate responsibility for making decisions regarding stricken vessels in his or her area. Arrangements must be made so that these individuals know in advance that they will not be held accountable for any unpleasant results resulting from well-founded decisions. For example, suppose as in the case of the Argo Merchant a northwest wind would blow the oil safely offshore and that a northwest wind was forecast for at least five days. Further suppose that the deck openings could be sealed and air pumped into the tanks above the oil. The best immediate decision under these circumstances could very well be to pump one million gallons of oil overboard to re-float the ship and tow her free and to safety. Now suppose that one day after pumping one million gallons overboard the wind unexpectedly shifted and the oil were blown ashore. Under no circumstances should the person who made the decision to pump the oil overboard or the weather forecaster be held accountable for this occurrence, and this fact should be a law.

As mentioned in the beginning of this report, every oil tanker accident is different. If a grounded vessel is equipped to be able to close deck openings and withstand internal air pressure, salvage by the means described above might be appropriate in some instances. An example of such an incident is the set of conditions that surrounded the Argo Merchant accident. Since the oil was No. 6 and since cargo heating had been lost, the oil could only be moved with relative ease for a few days. One day after the accident the wind began coming from the northwest with the weather forecast being for northwesterly winds of increasing strength for several days. If the tanker had been equipped to close deck openings, discharge some cargo, and fill the resulting spaces above the cargo with air, this action, with the pumping of oil overboard, would have been appropriate for the environmental conditions that existed for many days starting on Thursday morning. In other instances, such a course of action might not be appropriate. Such instances would include those where it would be more practical to offload the oil into barges and those when the prevailing wind would be certain to blow discharged oil ashore.

How might a grounded tanker be salvaged under such circumstances? The first thing which would have to be known is the extent of damage. If the tanker had deck openings which could be sealed and deck structures which could stand internal pressure, much could be learned about the condition of the ship structure by measuring the pressure in each tank resulting from air being pumped into it. If there were no openings between the tank and the outside environment, and if a sudden increase in air pressure above the cargo were applied, the pressure would not subsequently slowly drop. If there was a path from the tank to the outside environment, the rise in pressure would slowly diminish after air were suddenly pumped in. Were there broken bulkheads between otherwise intact tanks in the Argo Merchant? We will never know for sure. On the other hand, if the deck openings could have been sealed, we could have found out. By first measuring the size of the air space above cargo and then measuring the relationship between the amount of air pumped into a tank and the resulting rise in pressure, it would be possible to determine if the internal bulkheads were intact or broken. We are a long way from being able to do this now. Not only are tankers built without the provision for complete sealing of all deck openings above tanks, but the necessary equipment for rapidly making the aforementioned measurements has not been developed.

If the state of damage of a vessel were known and if some time for salvage operations could be expected to be available, the most appropriate steps could be planned and taken. It is useless to try to pump liquid cargo from a tank having a large hole to the sea. Water will enter as fast as it is pumped out. With large pumping capacity, some flotation can be provided by pumping liquid cargo from a tank having only a small hole to the sea. The best way to provide flotation by offloading cargo is to remove the cargo from intact tanks so the resulting air space will not flood, and hence be able to provide buoyancy. If this were to be done without pumping the cargo overboard, the most rapid technique would be to tie barges alongside with fenders between the barges and the stricken vessel. Doing this rapidly requires the availability not only of barges, but of lightweight rapidly deployable fendering systems. Such systems do not exist now. They could be developed.

In the case of the Argo Merchant, it took about forty-seven hours for a barge to reach the scene of the incident, which was only twenty-six miles from land and 90 miles from Providence, the nearest large commercial port. Such response is too slow by at least a factor of 5. What is required for faster response by barges? The answer is that it is necessary for state or federal governments to have contracts with barge operators all around the coastline of the nation to be able to provide a prearranged amount of barge capacity on very short notice. The most appropriate contractual arrangements would appear to be those which provided barge capacity according to a certain schedule. A small amount of capacity would have to be available on very short notice. More capacity would have to be available on somewhat longer notice, and still more capacity would have to be available on still longer notice. Contracts for barge capacity on a "best effort" basis would not be sufficient. It would be necessary for barge operators to continuously demonstrate their ability to meet such contracts by means of "surprise tests" called by the contracting agency.

The individuals who would actually carry out the tasks of salvage operations would necessarily have to be highly trained. Presumably they would be groups something like the existing U.S. Coast Guard Strike Forces and quite logically could be the Strike Forces themselves. How could adequate training be assured? Again, surprise tests would have to be carried out on a frequent basis. The tests would really have to test how well the people could do. For example, at random and unannounced intervals a derelict ship filled with a non hazardous dye could be towed up on a shoal and the strike forces called. The performance of the forces could be measured by examining how much of the dye escaped into the sea before either the ship was floated free and taken to a prearranged location or the entire cargo of dye was offloaded into barges.

One technique of marine salvage of damaged and grounded vessels is that of supplying external buoyancy to the stricken vessel by means of special flotation tanks taken to the grounded vessel which are subsequently flooded, attached to the vessel, and then pumped out to provide buoyancy. This has never been done on a scale which is appropriate for salvaging as large a ship as a modern tanker. It seems appropriate to study the possibility of developing a technology which could apply this technique on a large scale. Many changes from the smaller scale operations would be needed. For example, the only way that a flotation chamber of sufficient magnitude could be attached to a stricken vessel with sufficient strength would be to weld the chamber to the vessel. Can the technology to do this in the presence of substantial seas be developed? The answer to this question is not now known; however, it could be determined by means of a relatively straightforward feasibility study.

After the grounding of a large tanker, one possibility which always seems to come to the minds of many people is that of burning the oil. Usually this cannot be done. However, it seems entirely feasible to design, develop, and construct special burners which could be placed aboard a stricken vessel for the purpose of burning the cargo. The questions which must be answered first are: How much air pollution would this cause and how long would the burning take? Engineering studies to answer these questions are in order.

Preparedness for Offshore Spill Cleanup: Now we are in a position to consider what is necessary to achieve a state of preparedness which is adequate to collect large quantities of oil from offshore spills and to provide significant protection to land areas from oil which is so viscous that it is not practical to pump it. First, we must consider the means of transportation of equipment and people to the scene of an oil spill. Since maneuvering vessels are going to be needed in any case, waterborne transportation is the method of choice. It should be noted that this is different from some aspects of tanker salvage operations wherein optimum transportation of people and at least lightweight equipment is by helicopter. Waterborne transportation can be expected to have a speed of about 14 knots in moderate weather. The number of locations at which equipment and recovery vessels must be stockpiled depends on the distance from shore that we wish to be able to work at, as well as the response time. It is sufficient to discuss these matters in approximate terms. Roughly speaking, a response time of five hours for distances up to 25 miles offshore would seem acceptable. This would certainly be much better than what is being done now. To protect a stretch of coastline under such conditions would require stations of equipment to be located approximately 130 miles apart.

In consideration of the fact that it is impractical to protect the entire U.S. coastline to this extent, the approximate number of stockpiling stations would have to be about 20. If Great Lakes protection were also to be provided as well, about thirty stockpiling locations would be needed. The items needed at each station shortly upon notification of a spill are barriers with built-in skimmers or separate skimmers, storage vessels, tow vessels, and trained personnel. Each of these will be described in turn.

Barriers are the most highly developed of all of the required items for a complete spill cleanup system. Barriers capable of working on the high seas in breaking waves up to 8 feet high and in very large nonbreaking waves exist, although in inadequate numbers. Some of these barriers are available in lengths of about 600 feet packaged in containers with the total loaded package weight of about 15,000 pounds. The U.S. Coast Guard has developed special sleds on which the packages of barrier can be towed at speeds up to 20 knots. These sleds are expected to be available in the very near future so that the problems of an adequate supply and ability to transport barriers are only those of stockpiling and routine maintenance.

The next element needed for a high seas cleanup system is the provision of skimmers that can work effectively together with barriers. At least three different skimmer types have been developed for high seas use. One is designed to be built directly into a high seas barrier so that once a barrier is deployed, the skimmer is automatically also deployed. None of the skimmer types is appropriate for use in a large spill of cold No. 6 oil because of the difficulty of pumping this material, but most oil transported is not No. 6 oil and initially it would seem advisable to forego the possibility of collecting No. 6. None of the seemingly useful types which have been developed for large offshore spills have been thoroughly tested offshore with large quantities of oil. Each has been tested in testing tanks. Offshore tests of the most appropriate skimmer types with oil are an absolute necessity so that confirmation of their capabilities will be available in order to aid in planning for cleaning up oil spills.

Packaged barriers and skimmers can be carried to the scene of a spill on high speed planing sleds towed by high speed vessels. For example, the barriers and skimmers whose use is presently being contemplated by the U.S. Coast Guard can all be towed at speeds in excess of fourteen knots by the Coast Guard 82-foot cutters as well as some of the larger Coast Guard vessels. The packaged barriers are especially designed for rapid deployment (it takes about 20 minutes to deploy a 600-foot long barrier from a container and barriers can be connected together to give a longer barrier), so that rapid availability of skimmers and barriers is possible.

Rapid availability of collection vessels which can be used to store collected oil does not exist. Several attempts have been made to develop large lightweight rubber bags suitable for this purpose, but tests of these bags have resulted in their structural failure. One element which must be developed in order to fulfill the needs of a total spill cleanup system is that of storage capability. A consideration of this subject indicates that two different types of storage capability are needed. The first is that of special newly designed small barges. These would be vessels having an overall length of about 75 feet, which would be stored near stockpiles of barriers and skimmers. These vessels would be designed as lightly as possible in order that they could be towed by the same type of vessels that could be used to tow sleds containing boxes of barriers and skimmers.

In effect, the capacity of such a collection barge would be that of the largest barge, which vessels such as Coast Guard 82-foot cutters could tow at speeds of 14 knots or more when empty. Preliminary calculations indicate that vessels of this type could have a storage capacity of about 100,000 gallons of oil which would represent about three hours of oil collection from the barrier skimmer combination that was collecting oil at a rate of approximately 600 gallons per minute. These storage barges should also be designed to achieve gravity separation of oil and water since some water is collected with all skimmers and separation would allow discharge of the water so more oil could be collected.

The second aspect of the storage system is that of the availability of commercial barges. Some organization or agency must take the responsibility for generating contracts with a great many barge operations around the coasts of the U.S. These contracts must be arranged so that empty barge capacity is available on a few hours notice. The required capacity could be on a scheduled basis where an initial capacity in each location is provided within a relatively short time (for example, 8 hours) and more capacity is provided over a longer time. By this means, the immediate storage capacity response to an oil spill could be provided by the special small barges which could be towed at high speed, with arrangements made for these barges to be able to offload collected oil into the commercial barges at a later time.

The next needed item for total spill cleanup system is that of maneuvering vessels. As has been explained previously, it will generally be impractical to moor oil spill cleanup system in the vicinity of a tanker grounding. Effective cleanup requires towing of barrier-skimmer combinations at a speed of one knot or less. Vessels which are capable of towing at speeds of one knot or less and still maintaining steerage control in waves of substantial size are few in number. The power requirements for this towing are particularly small; a few hundred horsepower is more than sufficient.

The two problems that do exist are the ability to tow continuously at a slow enough speed and the ability to have adequate steering control to properly handle barrier-skimmer combinations at such low speeds. It is quite within our capabilities to retrofit a large number of existing vessels to provide them with this added capability. To date, hardly any such retrofitting has been done. Since several barrier-skimmer-barge combinations can be expected to be needed at the scene of an offshore oil spill, numerous tow vessels must be available at each location. It would be entirely feasible to add the low speed towing capability to the same vessels which were to be used for high speed response with equipment and personnel.

This brings us to the matter of trained personnel. Cleanup of an oil spill offshore is a difficult task and requires personnel who are thoroughly trained in the job that they are to do. Just as is the case with training and practice for salvaging stricken vessels and their cargoes, training and practice of cleanup personnel are needed as well. Because of the availability of air transportation for people, trained personnel need not be stationed at every location where equipment is stockpiled. The present U.S. Coast Guard Strike Force concept could be used if the number of strike forces were increased in number and if thorough and regular training took place.

This completes the description of what is needed for a total spill cleanup system: barriers, skimmers, storage vessels, tow vessels, and trained personnel. If any one of these elements is absent, essentially no oil cleanup can take place even if all of the remaining items are provided.

Dock Operations

A device developed by *P. Preus; U.S. Patent 3,786,773; January 22, 1974* is one placed on the floor of a drydock to encompass the damaged area of the hull of a ship drydocked therein which intercepts the flow of water from the area of the hull during drydocking to filter out hydrocarbons emanating from the damaged hull to preclude contamination of the waters surrounding the drydock and the drydock itself.

The device as shown in Figure 34, incorporates longitudinal barriers in the drydock floor **14** and **16** and transverse barriers (not shown in the figure) which create a containment "box" on the floor of the drydock.

FIGURE 34: HYDROCARBON RETAINER DEVICE FOR DRYDOCKS

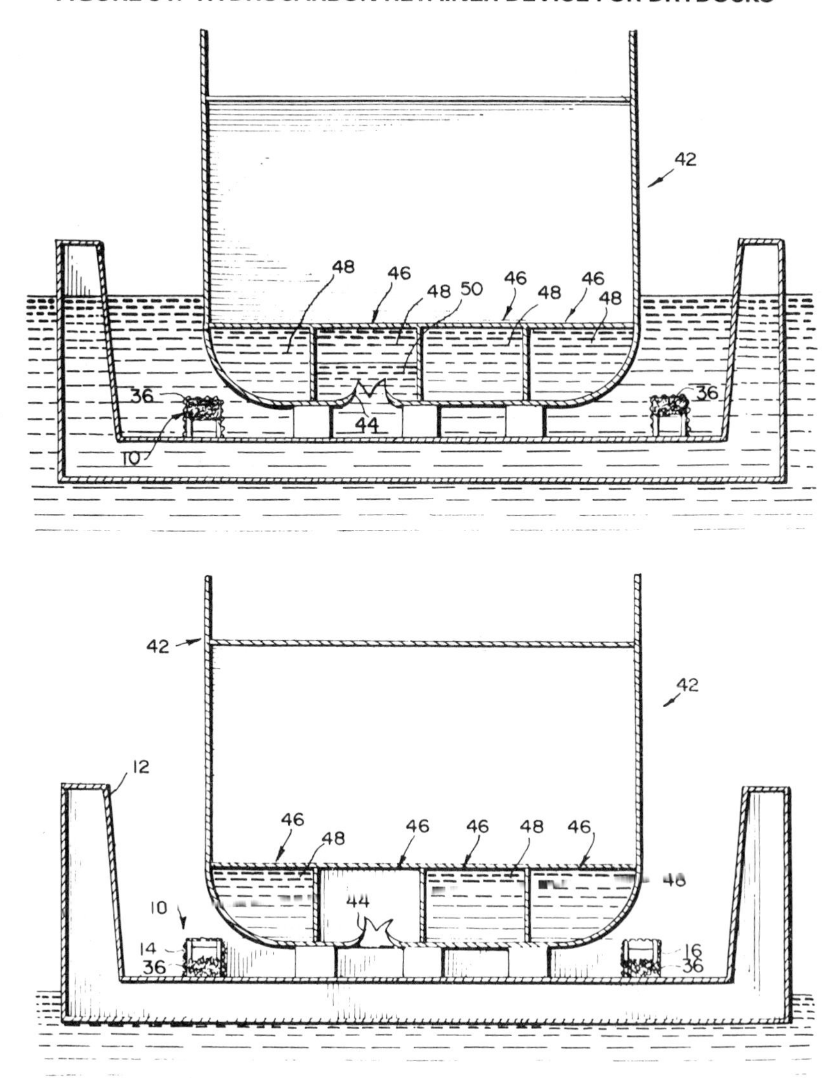

Source: U.S. Patent 3,786,773

In the upper view in the figure, double bottom tanks **46** containing the ship's fuel oil or "bunkers" **48**, are disposed in the bottom of the hull of the ship **42**. The damaged portion **44** has ruptured one of these tanks **46** and a portion of the bunkers **48** has been displaced and replaced by sea water **50**. Since oil is less dense than water, the bunkers **48** in the ruptured tank **46** are pressed against the tank top and are retained in the upper portion of the tank. The oil absorbent material **36**, because of its bouyancy and water repellent characteristics, floats to the top of the device **10** as the drydock is submerged and remains there, pressed against the top mesh of the device, so long as the drydock is submerged.

In the lower view in the figure, the drydock **12** has been refloated bringing the ship **42** out of the water for repair. The water in the drydock is discharged as the drydock is raised, through the open end thereof. As this water is discharged, the oil or bunkers **48** in the damaged tank **46** drains from that tank and flows, with the water level beneath the tank, toward the open ends of the drydock. As the water level drops in the drydock, the oil absorbent material **36** also follows the water level thereby remaining in a blocking position relative to the path the oil being discharged from the holed tank **46** must take to leave the area of the hull damage. As the oil and water reach the blocking position assumed by the oil absorbent material **36**, the oil is absorbed at this point and effectively "filtered" from the water flowing through the device **10** thereby retaining all of the oil in the absorbent material and/or within the confines of the device **10** while allowing all of the water to drain from the drydock without pollution of surrounding waters.

In actual use of such a device, the M/V Singapore Trader was successfully drydocked without any overboard discharge of bunker fuel although a considerable amount of fuel was discharged onto the drydock floor within the confines of the device.

Although the device **10** is illustrated having a water pervious construction throughout the entire perimeter thereof, it should be understood that portions thereof may be formed as a water tight "wall" with water pervious "box" portions interspersed along the length thereof as required. The primary feature that must be considered in such design is that the oil absorbent material **36** be present in sufficient quantity to absorb the full amount of oil expected to be discharged from the ship. Absorbent C pertoleum absorbent material has an absorbtion capability of about 0.55 gallons of bunker C per pound of absorbent material and, utilizing the probable capacity of the damaged tanks, the quantity of material needed can readily be determined beforehand.

A device developed by *E.T. Tedeschi, Jr.; U.S. Patent 3,906,732; September 23, 1975; assigned to The B.F. Goodrich Co.* is a pier-mounted shipside oil barrier seal structure floating on the water surface for use between a generally fixed connection point floating on the water surface and a ship having a side surface. The structure comprises a floating shipside oil seal member vertically extended above and below the water level, having at its ship-contacting end a flexible elastic rubber-like vertically extended sealing element for contacting the ship side surface. It is mounted on the pier by a pair of transversely spaced supports slideably supporting it for transverse movement, the supports being arranged for free vertical movement of the floating oil seal member with the water level as it changes with respect to the pier and the ship side. Springs are connected between the oil seal member and the supports for urging the oil seal element into contact with the ship's side. It may also have a bendably collapsible portion collapsible upon horizontal movement of the ship side toward the fixed connection point.

A barrier (or barrage) developed by *A. Grihangne; U.S. Patent 3,922,861; December 2, 1975* is particularly adapted to dockside operations as shown in Figure 35.

FIGURE 35: FLOATABLE-SINKABLE FENCE INSTALLATION FOR SPILL CONTROL AT DOCKSIDE

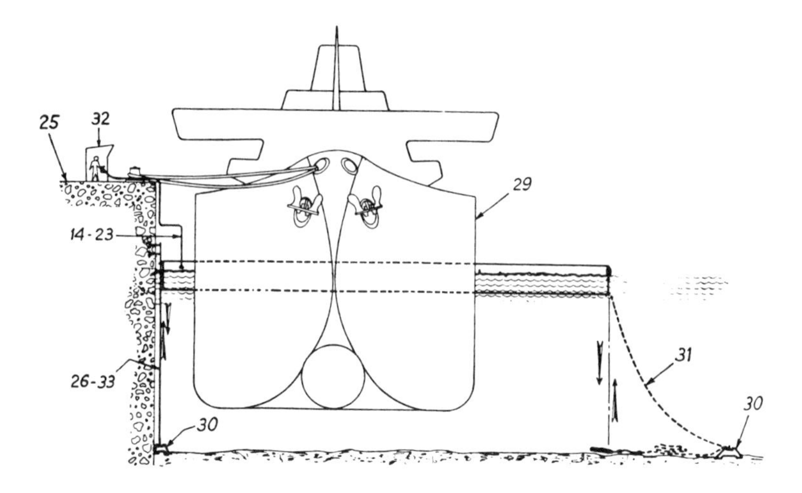

Source: U.S. Patent 3,922,861

A tanker **29** moored at its loading or unloading station is surrounded by the barrage which is anchored by its two ends to the quay **25** and retained at the bottom by deadweights **30** by means of chains **31**. The two flexible hoses **14** and **23**, which respectively control inflation of the barrage and its emersion on the one hand, and its immersion on the other hand, extend from a handling station **32**. A cable **33**, having one end hooked to a ballast resting on the bottom or to a suspended deadweight **30** and the other end secured to a tensioning system secured to the upper portion of the pipe **26**, acts as an extension of the vertical guiding function of the pipe **26** down to the sea or river bed.

Sailing of the tanker necessitates removal of the barrage. Reduction of the pressure in the flexible hose **23** enables the air in the buoyancy chamber to escape. Gradually as the barrage sinks under the effect of the chains or weighted belts the air is expelled to atmosphere by the increasing hydrostatic pressure exerted on the envelope of the buoyancy chamber. Under the effect of the chains or ballasted belts the barrage will lie flat on the bottom, totally evacuated, and be unaffected by any possible current of ebb or flood tide or the wake of the screws of passing vessels.

The scheme described permits remote control of the barrage between the two extreme situations namely (a) lying flat on the bottom of the sea, or (b) floating in operative state.

A device developed by *S.C. Light, Jr.; U.S. Patent 3,984,987; October 12, 1976; assigned to Sun Shipbuilding and Dry Dock Co.* is an apparatus for a marine facility which prevents silt from entering the facility and also provides spill containment. A floating barrier provides the containment and a flexible curtain is attached to the barrier. Having devices anchoring the curtain to the floor of the water prevents silt from entering the facility. The barrier rises and falls with the tides and openings in the curtain permit the tidal flow to enter and leave the facility. The apparatus is attached to the fixed structure in a suitable manner. The barrier and curtain can be moved to permit passage to a ship.

LAND SPILL PREVENTION AND CONTROL

The procedure selected to contain spills on land will vary with the amount and type of oil spilled, the type of soil and the terrain. Less viscous oil and more porous soil will allow greater and more rapid penetration and lateral migration in the soil. Where feasible, absorbent materials should be applied as soon as possible. Larger spills may require containment techniques such as digging interceptor trenches or collecting pools from which the oil may be pumped as shown in Figure 36.

FIGURE 36: INTERCEPTOR TRENCH FOR LAND SPILL CONTROL

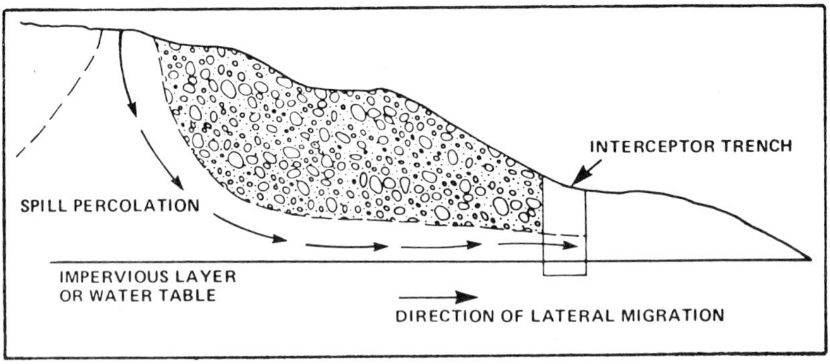

Source: Reference (9)

Spills of petroleum products in land areas rank with beach areas as the most difficult, time consuming and expensive spills to effectively clean up. The primary concern in handling oil spills on land is the leaching of the product to a ground water supply, as illustrated in Figure 37. Spreading sorbents over the spill with tamping to insure oil saturation of the sorbent is one good procedure for preventing ground leaching. Sorbent booms are an effec-

FIGURE 37: LAND SPILL PROCEDURES

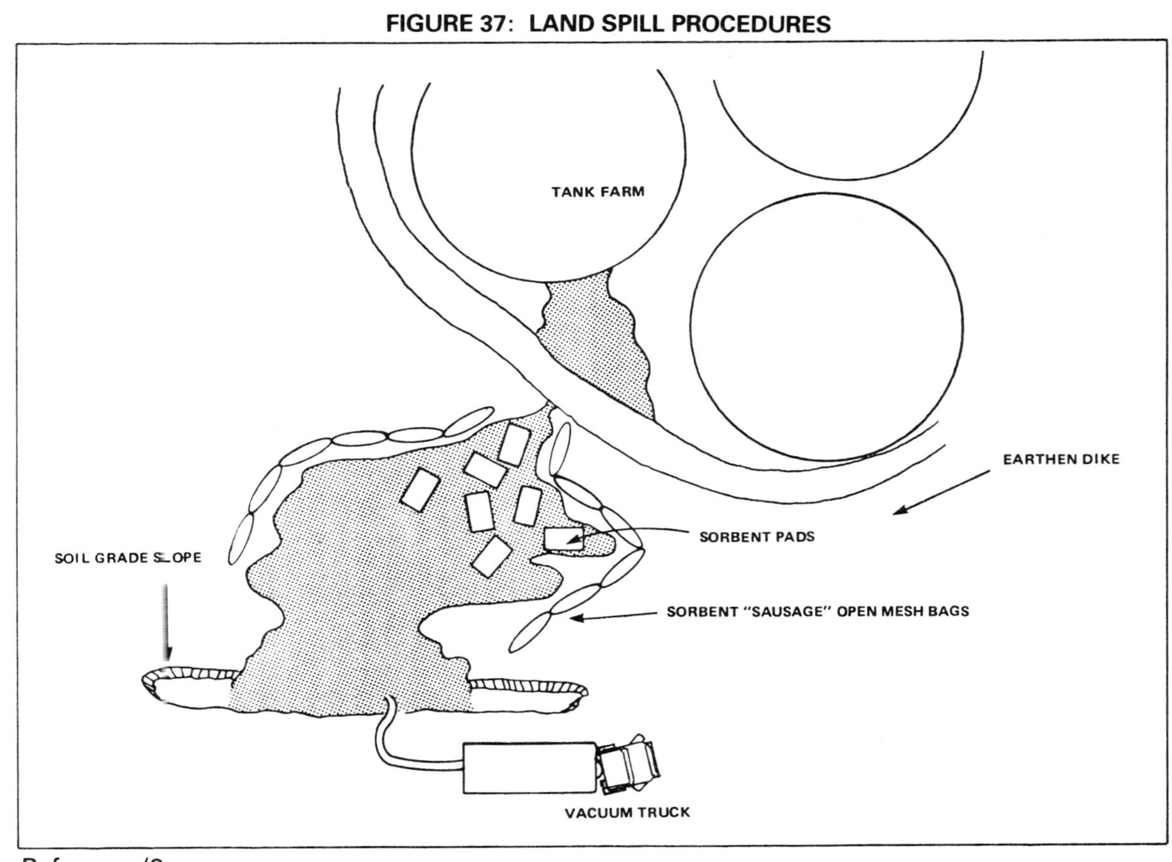

Source: Reference (9

tive way of minimizing the extent of the ground area affected by the spill. A trench can be constructed to allow containment of oil draining off the spill area. Oil collecting in the trench can then be pumped out with diaphragm pumps or vacuum trucks. In some soils it is possible to drive pipes equipped with sand points, as used in the drilling of water wells, into the ground and pump water into the soil to create a locally high "water table" and percolate the oil up out of the ground and into collection trenches. Personnel with knowledge of the local soil and water table conditions should be contacted for guidance. Another alternative is to obtain and use earthmoving scraper equipment to remove the heavily saturated top soil for disposal in an approved landfill.

Pipeline Operations

A device developed by *G.F. Wittgenstein; U.S. Patent 3,802,456; April 9, 1974* is a safety installation for protecting the environment from leakages in pipelines. It comprises a fluid-tight jacket which surrounds the pipeline and forms therewith an intermediate space containing inserts which are separated by ducts sealed with a liquid or gas. The inserts are joined together by cables or wires and there is provided at least one tank towards which the leakage liquid flows through the ducts. A liquid leakage detector thereupon acts to remotely control the pumps and valves of the pipeline. The jacket is formed from short rigid runs connected by flexible joints; breaks between the runs contain separation bands and are covered by a flexible fluid-tight structure resting on the inserts and projecting onto the ends of the runs.

Storage Tanks

A system developed by *E.N. Fisher; U.S. Patent 3,863,694; February 4, 1975* is one for containing oil spills or leaks from small oil storage tanks. The system comprises a diaphragm of polyvinyl chloride, synthetic rubber or the like, formed in a circular pattern of radius approximately equal to the radius of the storage tank plus its height. The circular diaphragm is placed under the tank and centered with respect thereto. The edge of the diaphragm is pleated and fastened to a tension ring which surrounds the upper part of the tank and is held in place by friction or other suitable means. When a leak occurs, the tension ring releases and slides down the tank as the diaphragm fills with the oil escaping from the tank. In the event of tank failure, the diaphragm is capable of containing the entire contents of the tank in its fully deployed position. A constraint dike may be used, although its use is entirely optional, not to structurally reinforce the diaphragm since such reinforcement is not needed, but rather to protect the tank and undeployed diaphragm from mechanical damage.

Oil Spills On Water

In the disaster where the tanker Torrey Canyon ran aground and leaked 117,000 tons of crude oil into the English Channel, a panic to clean up the spill ensued as miles of beaches and valuable scenery were polluted and blighted, not to mention the biological damage to sea life. The British utilized nonionic detergents, trying with some success to emulsify the oil, breaking it up into small particles which with evaporation greatly reduced the masses of oil reaching shore. Even so, over 60 miles of English shoreline was contaminated.

The Torrey Canyon spill also drifted in large patches across the English Channel to the French shore. At sea, the French used powdered chalk spread on the slick and churned it up with small but powerful boats. This caused the oil to break up and sink into the sea. They also tried this same technique along with sawdust with some success. Both methods still did not remove all of the oil from the water and a large amount of it still floated ashore. One French ship tried to "pile up the slick" against the side of the ship and then tried to, in effect, vacuum it with a hose and a broad nozzle. This worked, but the amount that could be gathered was relatively small.

Straw, hay and sawdust have been spread on the waters along the shore to soak up the oily mess caused by the oil slick. While this effort produced some results, the scars cannot be removed entirely for years to come. Detergents and other cleaners were used on the rocks and other shore items, but these too left their mark.

In addition to tanker accidents, there are of course offshore production well blowouts such as that which occurred in 1977 in the North Sea.

The reader is also referred to a number of reviews (4)-(7) as well as a compendium of available commercial devices (11) cited in the bibliography.

When oil is spilled, it triggers a series of actions that are common to all spills and which have been categorized into the following general operational phases:

1) Discovery and notification.
2) Evaluation and initiation of action.
3) Containment and countermeasures.
4) Recovery, mitigation and disposal.
5) Cleaning and repositioning equipment.
6) Documentation and cost recovery.

Spill phases do not necessarily follow in sequence, but may and generally do overlap. Figure 38 shows this overlap and summarizes some of the actions in each phase of an oil spill.

Equipment and procedures for the effective control and cleanup of oil spills depend upon the type of oil spilled, the wind, waves, and currents in the area. All of these conditions can vary widely for different spills and even during the course of a given spill. It is the responsibility of the On-Scene Coordinator to use knowledge of local conditions and equipment resources to minimize the environmental damage and the attendant cost of an oil spill incident.

Following the discovery of an oil spill the following actions should be initiated immediately and concurrently:

a. Identify and Secure Spill Source. If the source cannot be immediately secured, booms must be placed around the source to minimize the area affected.
b. Identify Spilled Product. For personnel safety it is necessary to know what type of product was spilled and its explosive or toxic vapor properties. It is recommended that MOGAS, high-octane AVGAS, and JP-4 fuels be cautiously contained in open water areas away from confined spaces between ships or under piers and allowed to evaporate naturally. If cleanup of any residual amounts, after evaporation of volatile fractions, is deemed necessary, no attempt should be made until the area has been declared gas-free by a safety officer using a combustible gas indicator of the type used by marine chemists and supplied with the medium and large skimmer systems. For all other oil products (JP-5, Navy Distillate, Marine Diesel, Navy Special Fuel Oil, Bunker C, lube oils) cleanup can and must be initiated immediately.
c. Contain the Spill. An excellent first response, particularly if the spill is in open waters, is to surround the slick with piston film herding agent. In any case, booms should be deployed as soon as possible to minimize the spill area.
d. Determine Spill Size and Predict Spill Movement. To position men and equipment effectively, it is necessary to know how the spill will move with time. Knowledge of the local wave, tide, current, and wind conditions are essential and should be established in the Navy Area Contingency Plan prior to a spill. Oil slicks move at the speed of the local surface current and at approximately 3% of the wind velocity. In open water areas where waves reach two feet or more, oil slicks will take the form of windrows in wave troughs. Figure 39 illustrates the procedure of determining slick speed and direction by the addition of current and wind vectors. Constant field reports should be obtained to determine how the slick is actually moving and to update predictions for effective positioning of men and equipment.

FIGURE 38: OPERATIONAL PHASES IN AN OIL SPILL

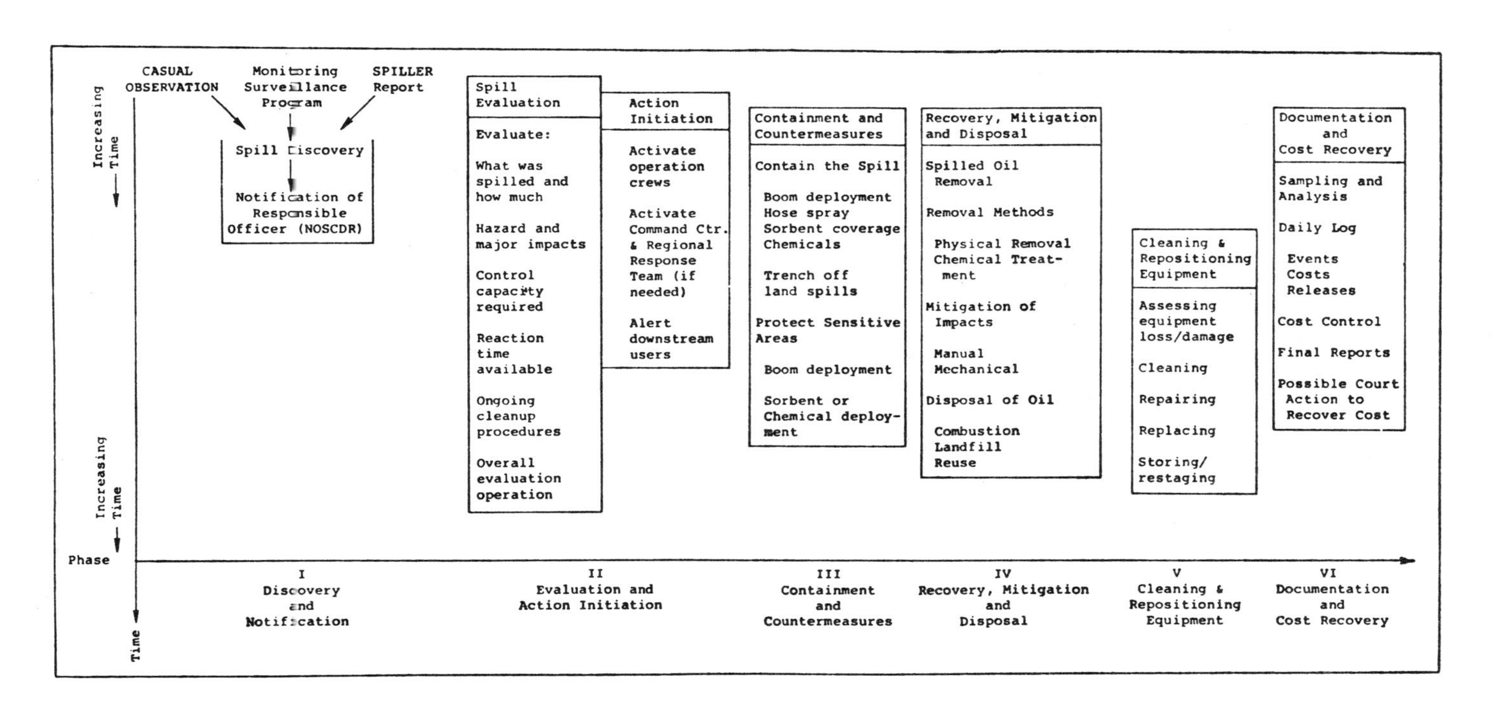

Source: Reference (9)

FIGURE 39: PREDICTION OF SLICK MOVEMENT

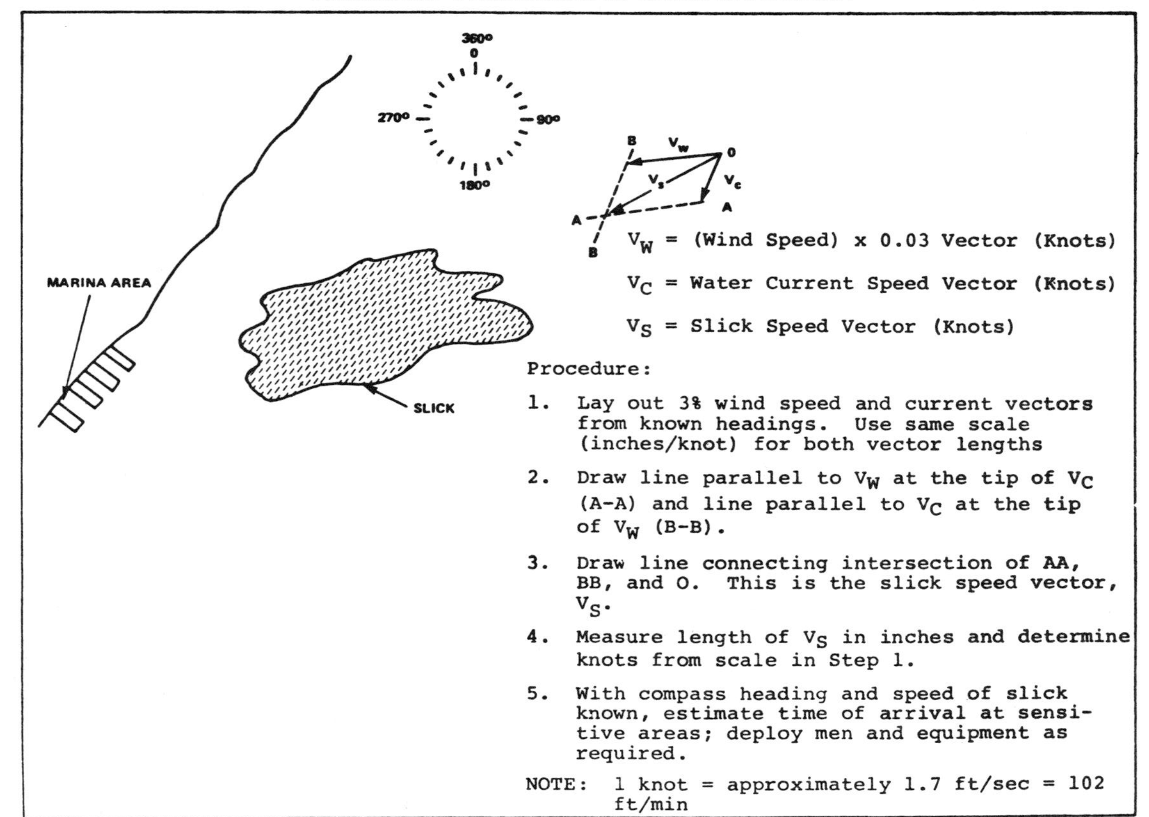

V_W = (Wind Speed) x 0.03 Vector (Knots)

V_C = Water Current Speed Vector (Knots)

V_S = Slick Speed Vector (Knots)

Procedure:

1. Lay out 3% wind speed and current vectors from known headings. Use same scale (inches/knot) for both vector lengths
2. Draw line parallel to V_W at the tip of V_C (A-A) and line parallel to V_C at the tip of V_W (B-B).
3. Draw line connecting intersection of AA, BB, and O. This is the slick speed vector, V_S.
4. Measure length of V_S in inches and determine knots from scale in Step 1.
5. With compass heading and speed of slick known, estimate time of arrival at sensitive areas; deploy men and equipment as required.

NOTE: 1 knot = approximately 1.7 ft/sec = 102 ft/min

Source: Reference (9)

REGULATORY CONSTRAINTS

Table 18 summarizes the laws and regulations pertaining to oil pollution as outlined in a U.S. Navy publication (9).

TABLE 18: SUMMARY OF LAWS AND REGULATIONS PERTAINING TO OIL POLLUTION

Public Law/Executive Order/Regulation	Major Effect
(1) Federal Water Pollution Control Act Amendments	Prohibits the discharge of oil or hazardous substances from any vessel, from any onshore or offshore facility, into or upon the navigable waters of the U.S., adjoining shorelines, or the waters of the contiguous zone.
	Prohibits a discharge that would cause a visible sheen upon the surface of the water or adjoining shorelines or cause a sludge or emulsion to be deposited beneath surface of the water.
(2) Oil Pollution Act, as Amended	Provides that oil or oily wastes shall not be discharged from any Navy activity or ship within any "prohibited zone."
	Prohibited zones are prescribed in the Oil Pollution Act and 33 CFR. It is noted that the prohibited zone for the U.S. is designated as waters within 50 miles of the U.S. coastline. In some cases, for other nations, the distance is greater than 50 miles.
	Any reduction or extension of the zones described under the terms of the International Convention for the Prevention of Pollution of the Sea by Oil, 1954, as amended, will be published in *Notices to Mariners* as issued by U.S. Coast Guard and U.S. Navy.
(3) National Environmental Policy Act	Establishes requirement for CEQ.
	Directs that all agencies of the Federal government shall include in every recommendation or report on proposals for legislation and other major Federal actions significantly affecting the quality of the human environment, a detailed statement on the environmental impact of the proposed action.
(4) Executive Order (E.O) 11752	Requires compliance by Federal facilities with Federal, state, interstate, and local substantive standards and substantive limitations to the same extent that any person is subject to such standards and limitations. However, excludes Federal facilities from compliance with state or local administrative procedures. The term "facility" is used in the broadest sense and includes aircraft, ships and vehicles.
(5) E.O. 11472	Establishes Council on Environmental Quality.
	Constitutes primary advisory body to the President with respect to environmental quality matters.
(6) E.O. 11514	Provides for responsibilities of executive agencies of the Federal government in furtherance of NEPA.
(7) E.O. 11735	Assigns functions among Federal agencies, particularly EPA and USCG, relative to oil pollution prevention under the terms of the Federal Water Pollution Control Act, as amended.

(continued)

TABLE 18: (continued)

Public Law/Executive Order/Regulation	Major Effect
(8) OMB Circular No. A-106	Sets forth the requirements for the OMB Report. Provides for the basic "tool" for budget requests to correct environmental deficiencies.
(9) National Oil and Hazardous Substances Pollution Contingency Plan (CEQ, 40 CFR 1510)	The plan applies to all Federal agencies. Its objectives are to provide for efficient, coordinated and effective action to minimize damage from oil and hazardous substance discharges, including containment, dispersal and removal.
(10) Pollution Prevention; Vessels and Oil Transfer Facilities (USCG, 33 CFR 151-156)	Provides regulation for each onshore and offshore facility on the transfer of oil in bulk to or from any vessel that has a capacity of 250 or more barrels of oil. Prescribes rules for the operation of vessels on the navigable waters and contiguous zone in order to prevent the discharge of oil. Provides regulations on the transfer of oil to or from any vessel with a capacity of 250 or more barrels, on the navigable waters or contiguous zone of the U.S.
(11) Oil Pollution Prevention; Nontransportation-Related Onshore and Offshore Facilities (EPA, 40 CFR 112)	Establishes the procedures, methods and equipment to prevent the discharge of oil from nontransportation-related onshore and offshore facilities into or upon the navigable waters of the U.S. or adjoining shorelines. The Navy is subject to these regulations to the same extent as any other agency or person. Requires the preparation and implementation of Spill Prevention Control and Countermeasure Plans.

Source: Reference (9)

DETECTION

The existence of oil spills in the oceans creates an unpleasant sight and odor as well as actual harm to marine life. The need for an early warning system to detect the violator and to initiate clean-up procedures is brought out by the cost of clean-up in some recent oil spill cases wherein millions of dollars have been expended by major oil companies to clean up various coastlines.

However, the largest portion of oil which is spilled does not result from the large spectacular events publicized, such as the leak from or destruction of an oil tanker or leak from an offshore oil operation. The major portion of the oil pollution results from the smaller, more common spills of between 100 and 1,000 barrels of oil which inundate the inland waterways. Accordingly, there is a need for collecting sufficient data, on a continuous real-time basis, to be able to identify the party responsible for the detected effluent and quantitatively determine the amount and type of discharge.

An apparatus developed by *K. McCormack; U.S. Patent 3,783,284; Jan-*

uary 1, 1974; assigned to Texas Instruments Inc. is one for indicating the presence or absence of petroleum products in a water area by utilizing an active infrared source which illuminates the water area which may contain a petroleum product. The reflected infrared radiation is filtered by two filters at two different wavelengths, λ_1 and λ_2. Two infrared detectors produce signals which are proportional to the detected reflected radiation at the wavelengths λ_1 and λ_2. A processing channel is connected to each detector, the processing channels each including a log amplifier, the output of which is coupled to a a differencing circuit which produces an output signature signal:

$$\ln [V(\lambda_1)/V(\lambda_2)]$$

which indicates either the presence or absence of the petroleum products in the water area. Such a device is shown in Figure 40.

FIGURE 40: INFRARED DEVICE FOR DETECTION OF OIL SPILLS ON WATER

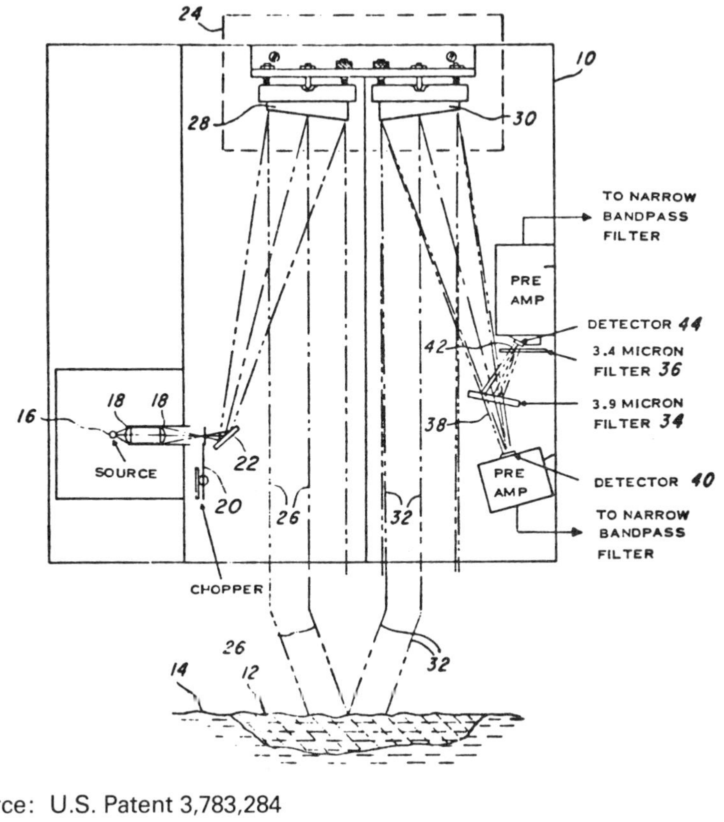

Source: U.S. Patent 3,783,284

This system **10** is mounted on a structure (such as a tower or loading dock) to detect the presence or absence of petroleum products **12**, such as oil, in or on a water area **14**. For the system **10** to work during day or night and be independent of meteorological conditions, it is desirable to use an active energy source **16**, which in the preferred embodiment emits radiation in the infrared region to illuminate the water area **14**. Active energy source **16** may be, for example, a laser or a discharge tube (e.g., xenon). Relay lens system **18** is utilized for efficient collection of the radiation from source **16** and refocused for chopping. Chopping or modulation of the radiation from source **16** may be performed by a tuning fork **20** whose resonance is at a selected frequency, such as 400 hertz. The tuning fork **20** is driven by electronic circuitry (not shown). The modulated infrared radiation is reflected by a folding mirror **22** into optical system **24**. The optical system **24** transmits the radiation **26** onto water surface **14** which may have present a petroleum product, such as oil slick **12**, on the surface thereof.

Optical system **24** is shown as comprising transmitting and receiving paraboloid mirrors **28** and **30**, respectively. Accordingly, the optical system **24** results in the system **10** being a staring system, i.e., one which has a single resolution element. However, system **10** could scan an area, if required, by the use of additional mirrors associated with optical system **24**. A suitable scanning optical system is shown in U.S. Patent 3,211,046.

The transmitted radiation **26** is reflected from the water surface area **14** (or oil slick **12**) and impinges upon receiving paraboloid mirror **30**. This reflected modulated infrared radiation **32** is in turn reflected from said mirror **30** onto two narrow-band interference filters **34** and **36**. Filter **34** operates as a beam-splitter to allow the reflected radiation **32** to pass in a very narrow bandwidth of radiation **38** centered at a wavelength of λ_2. This radiation **38** centered at a wavelength of λ_2 impinges upon detector **40**. All of the remaining radiation is reflected from filter **34** and passes through filter **36** which in turn passes only radiation in a very narrow bandwidth of radiation **42** centered at a second wavelength λ_1. The radiation **42** centered at λ_1 impinges on detector **44**. Detectors **40** and **44** may be PbSe IR detectors.

A device developed by *D.A. Leonard and C.H. Chang; U.S. Patent 3,806,727; April 23, 1974; assigned to Avco Everett Research Laboratory Inc.* is one in which the oil pollution content of water is continuously monitored by measuring the oil fluorescence spectrum producted by an ultraviolet light source and by comparing it with the Raman spectrum of water.

This device is particularly useful in measuring trace levels of oil in water since the technique is highly sensitive to oil concentrations in the range of 10 to 50 ppm. The sensitivity decreases substantially in the concentration range of 50 to 1,000 ppm, and this device is relatively insensitive to concentrations of more than 1,000 ppm. The essential features of the device are illustrated in Figure 41.

As shown, a conventional ultraviolet light source **10** is used to generate appropriate light energy. For example, the source **10** may be a conventional mercury arc lamp energized from a 60 hertz source. In that case the light emissions will have strong lines centered near 3660 A. Alternatively, the ultraviolet light source may be a laser having strong emissions centered at 3371 A. In any case, the light emissions are filtered by band-pass filter **16** having a band-pass equal to or less than the width of the Raman spectrum for water and is focused by means of lenses **18** and **20** into a pipe **22** through which oil-polluted water is flowing. The pipe **22** is provided with an entry window **24** for admitting the filtered light emissions and with an exit window **26** positioned at an angle of 90 degrees with respect to the window **24**.

FIGURE 41: OPTICAL DETECTOR FOR OIL SPILLS ON WATER USING UV SOURCE

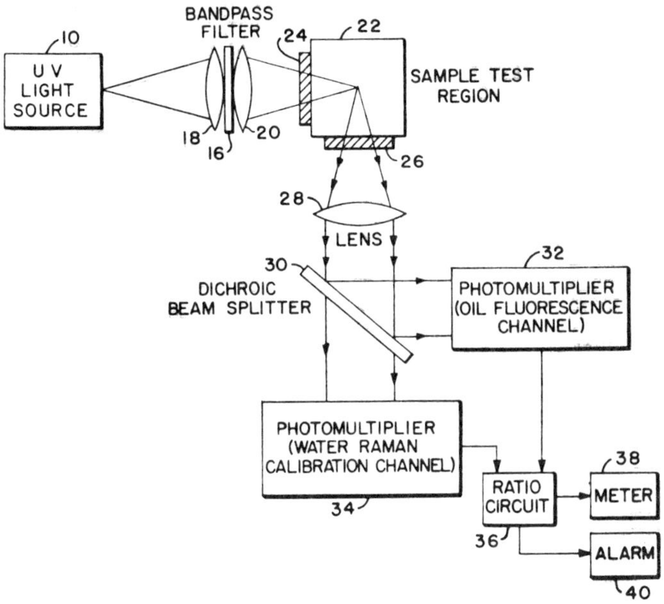

Source: U.S. Patent 3,806,727

The scattered and re-emitted light is collected through the 90 degree exit window **26** so that the oil-polluted water flowing through the pipe **22** is the only element which is common to both the transmitter and receiver optics. The re-emitted light spectrum from the window **26** is focused through a lens **28** onto a dichroic beam splitter **30**. The beam splitter **30** is coated so that it divides the re-emitted light into two wavelength channels, one channel for oil fluorescence spectrum, and one channel for water Raman spectrum. The oil fluorescence spectrum is applied to a photomultiplier **32**, the output voltage of which is a measure of the intensity of the energy of the oil fluorescence spectrum.

The water Raman spectrum is applied to a photomultiplier **34**, the output voltage of which is a measure of the intensity of the energy of the water Raman. Both output voltages are applied to a ratio circuit **36** having two output circuits. The first output circuit is applied to a meter **38** calibrated to indicate the ratio of the intensity of oil fluorescence to the intensity of water Raman. The other output from ratio circuit **36** is to an alarm **40** which serves to provide an appropriate warning whenever the ratio of the energy of the oil fluorescence to the water Raman exceeds a predetermined level.

If the light source is a mercury lamp, it has an output band centered at approximately 3660 A, and its power level is determined by the accuracy required for the system. The wavelength of the water Raman transmitted through the dichroic beam splitter **30** is centered at approximately 4170 A, while the wavelength of the oil fluorescence transmitted through the dichroic beam splitter **30** is centered at approximately 3800 A. While the above specified wavelengths may be suitable for a particular application, it will be recognized that other wavelengths may be preferable under different circumstances and for different types of oil or other pollutants. Where the light source is a laser, the strong lines are centered at 3371 A while the wavelength of the water Raman and oil fluorescence are centered at 3800 A and 4000 A, respectively.

The particular spectra were selected by measuring the water Raman and oil fluorescence for different contents of oil and by noting that for most types of oils found on board ship (for example, light crude medium detergent lubricating oil, diesel oil and bunker C fuel oil) there is a linear relationship between the oil content and the ratio of the oil fluorescence to the water Raman until the level of oil exceeds 50 ppm. Thereafter the ratio does not increase at a linear rate and the system becomes practically insensitive to changes above 1,000 ppm.

An apparatus developed by *M.D. Gregory, J.E. Stolhand and M.E. Yost; U.S. Patent 3,842,270; October 15, 1974; assigned to Continental Oil Co.* is for monitoring the presence in an aqueous medium of an oil which fluoresces when irradiated by ultraviolet light wherein the fluid stream being monitored does not come in direct contact with the light source or detection means. Such an apparatus is shown in sectional elevation in Figure 42.

The apparatus, generally denoted by reference **10**, comprises means for supplying a stream of fluid to be monitored such as valve **12** through which the composition to be monitored enters and preferably passes through inlet tube **14**. The main body of the composition passes through flow controller **16** and tubing **18** into rotameter **20** which measures the rate of flow.

From rotameter **20**, the main body of composition passes through tubing **22** into the top of housing **24**. The composition then falls through housing **24** as stream **26** collects in the bottom of housing **24** and forms fluid column **28**. The composition drains from the bottom of housing **24** and is withdrawn from the apparatus via tubing **30**, valve **32**, and valve **36**. It is preferred that a small portion of the composition be withdrawn from tubing **14** via tubing **38** and valve **40**, passed through constant temperature jacket **42** and then through tubing **44** which discharges into tubing **34**.

Housing **24** is conveniently made in upper section **46** and lower section **48** connected by flange **50** and sealing mechanism **52** such as an O-ring. Housing **24** is provided with first aperture **54**, covered by a transparent window which may be made of quartz, glass or other solid transparent to light. A sealing mechanism, such as an O-ring, is provided between housing **24** and the transparent window. The transparent window is held in position by a locking mechanism, such as a locking nut. A source of ultraviolet radiation is positioned outside housing **24** and located so as to pass a beam of ultraviolet radiation through the transparent window and through composition stream **26**.

FIGURE 42: PRESSURIZED OIL-IN-WATER MONITOR

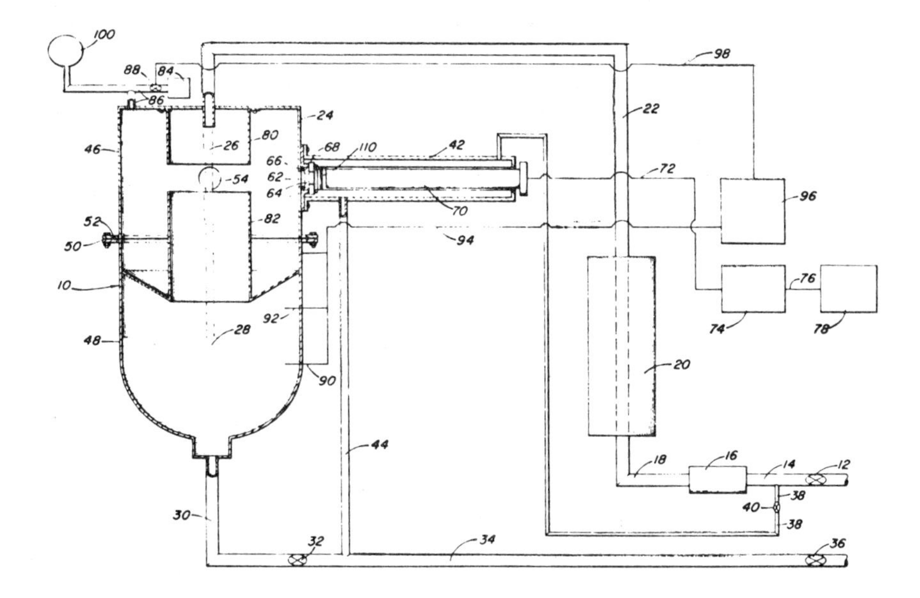

Source: U.S. Patent 3,842,270

Housing **24** is also provided with second aperture **62** positioned at the same height as first aperture **54** and at any position around the circumference of housing **24**, preferably at right angles to first aperture **54**.

This universal positioning of aperture **62** with respect to aperture **54** is made possible by a filter **110** in the photocell detector **70** that keeps the incident ultraviolet light from activating the photocell. The only light that may enter the photocell detector **70** is long wavelength fluorescent light the intensity of which is proportional to the contaminant concentration in the stream to be monitored. Thus, photocell detector **70** detects the fluorescence radiation created by ultraviolet light striking composition stream **26**. This is generally visible light.

Second aperture **62** is covered by transparent window **64**. Sealing mechanism **66**, such as an O-ring, is provided between housing **24** and transparent window **64**. Transparent window **64** is held in position by locking mechanism **68**, such as a locking nut. Visible light detecting means, **70** such as a photocell, is positioned outside housing **24** behind transparent window **64**. Visible violet light detecting means **70** is surrounded by constant temperature jacket **42** which decreases fogging of transparent window **64** which can occur during operation of the apparatus in the absence of constant temperature jacket **42**. The fluorescent light transmitted from stream **26** is picked up by visible detecting means **70** and the results observed visually or recorded on a suitable recording means, such as wires **72** which connect visible detecting means **70** to amplifier **74** which in turn is electrically connected by wires **76** to recorder **78**.

It is preferred that housing **24** be provided with splash guard means **80** and **82** positioned between the lateral sidewall of housing **24** and stream **26**. Upper splash guard means **80** is preferably a hollow cylinder attached to the top of housing **24** and provides a shroud around the upper portion of stream **26**. The bottom of upper splash guard **80** is just above apertures **54** and **62**. Lower splash guard **82** is preferably a hollow cylinder attached to the lateral sidewall of housing **24** at a point below apertures **54** and **62**. The top of lower splash guard **82** extends to just below apertures **54** and **62**. The function of splash guards **80** and **82** is to reduce the amount of liquid from stream **26** and fluid column **28** that splashes onto the transparent windows during operation of the apparatus.

It was experienced during operation of the apparatus as described above that falling stream **26** dissolved and entrained a portion of the air in the housing and carried this air with it into fluid column **28** and then on out of housing **24**. Thus, as the quantity of air in housing **24** decreased, the height of fluid column **28** would gradually rise in housing **24** until it reached the level of the transparent windows and covered the same, disrupting the desired observations. To prevent housing **24** from filling with liquid in this manner, it is necessary to inject gas into housing **24**. This can be done by constantly bleeding gas from gas supply **84** which is at a pressure above the pressure in housing **24** into housing **24** via tubing **86** and control valve **88**. Preferably gas is intermittently injected into housing **24** before the height of fluid column **28** gets dangerously close to the transparent windows.

The intermittent injection can be achieved by a means for detecting the height of fluid column **28** in housing **24** which controls means for injecting gas into housing **24**. The means for detecting the height of fluid column **28** in housing **24** can be lower level conductance probe **90** and upper level conductance probe **92**. These level probes extend through the side of housing **24**. Lower level probe **90** is positioned near the bottom of housing **24**. Upper level probe **92** is vertically positioned therefrom at a point below the transparent windows. Lower level probe **90** and upper level probe **92** are electrically connected by wire **94** to level controller **96**. Level controller **96** is electrically connected by wire **98** to solenoid valve **88** which controls flow of gas from gas supply **84** into housing **24**.

Pressure gauge **100** indicates the pressure in the housing. The pressure inside housing **24** should be maintained about 10 psi above the pressure downstream of monitor **10**. When the height of fluid column **28** rises in housing **24** and touches upper level probe **92**, level controller **96** opens solenoid valve **88** allowing gas to flow into housing **24**, thus depressing the height of fluid column **28**. When the height of fluid column **18** falls in housing **24** below lower level probe **90**, the level controller **96** closes the solenoid valve stopping flow of gas into housing **24**.

A device developed by *A.R. Kriebel; U.S. Patent 3,885,418; May 27, 1975* is an oil slick detector comprising a spinner located in a shroud surrounding the spinner and providing a radial gap between the spinner and the shroud, the spinner being driven at a constant speed by a motor, any changes in torque on the motor resulting in changes in the drive current of the motor, the free end of the spinner and shroud being positioned in the water surface such that the water and any oil on the water will enter the radial gap between the spinner and shroud, the torque on the motor being higher in value if oil is present than that torque on the motor if only water is present, said difference in torque resulting in a measurable change in the motor drive current.

In a preferred embodiment, the spinner is provided with a spiral groove in its outer surface to encourage the flow of oil and water in the radial gap. By running the spinner first in a forward direction to move oil and water up the spinner and thereafter running the spinner in a reverse direction to move the oil and water off of the spinner, and by measuring the average motor current during the forward and reverse runs, a measurement related to the thickness of the oil slick is obtained. Such a device is shown in Figure 43.

FIGURE 43: FLOATING OIL FILM DETECTION USING TORQUE MEASUREMENT ON ROTATING SENSOR TO DETECT THE PRESENCE OF OIL

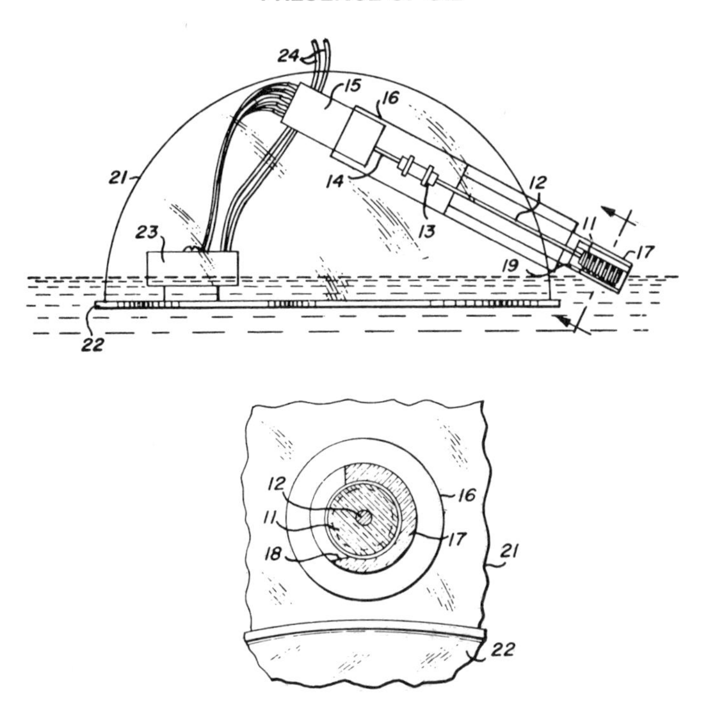

Source: U.S. Patent 3,885,418

The device comprises a cylindrical shaped spinner **11** having an integral drive shaft **12** affixed along the axis of the spinner, the drive shaft **12** being coupled by a flexible coupler **13** to the drive shaft **14** of a drive motor **15**. The drive motor **15** is fixedly mounted in the upper end of a first hollow cylindrical sleeve **16**, a second hollow cylindrical sleeve or shroud **17** being fixedly mounted in the other end of the sleeve **16**. The shroud **17** extends down over the spinner **11** with a small radial gap between the outer surface of the spinner **11** and the inner wall of the shroud **17**. The lower open end of shroud **17** terminates just beyond the lower end of the spinner **11**. The spinner **11** and shroud **17** may be made of suitable materials resistant to corrosion in water such as acrylic plastic or stainless steel. One wall section at the lower end of the shroud **17** is cut away to form an opening **18** in the sleeve **17** to provide access for the water and oil onto the surface of the spinner **11**. A small opening **19** is provided in the wall of the sleeve **17** at the location of the upper end of the spinner **11** to provide a spill-over opening for the fluid from within the sleeve **17**.

In one example of a spinner and shroud assembly, the spinner **11** is 0.790 inches long and has a diameter of 0.540 inches and the shroud **17** is about 2.093 inches long with an inner diameter of 0.550 inches and an outer diameter of 0.625 inches.

The preferable type of motor drive is the brushless dc motor and control circuit manufactured and sold by Siemens Co. under Model Type 1AD3001 which has good speed regulation, high efficiency, long lifetime, and good sensitivity. This motor operates off of a 10-15 volt source and will operate at practically constant speed until stalled under load.

The motor draws about 50 mA under torque, the current increasing almost in linear fashion with increasing torque until, at about a torque of 0.420 and a current of about 270 mA, the motor stalls. Since the motor is brushless, it has a long unattended life, for example 10,000 hours. Also, since the brushless motor does not produce electrical sparks, it is safe for using in hazardous environments.

This oil detector assembly is fixedly mounted on a float comprising a hollow hemisphere **21** sealed on a base plate **22**, the detector being mounted within the hollow hemisphere with the spinner end thereof extending out through an opening in the wall of the hemisphere. The detector assembly is mounted with its longitudinal axis at about 45° relative to the plane of the base plate **22**, the detector extending out from the float to a point where about one-half to three-quarters of the end surface of the spinner **11** is submerged below the water surface. The interior of the float is large enough to accommodate the control circuit **23** for the motor drive, electrical leads **24** extending from the float to provide dc current input to the control circuit which is monitored as the motor current related "oil detected" signal output. The motor drives the spinner at a constant speed of about 750 rpm and the surface of the spinner **11** within the shroud becomes covered with water when the water surface is free of an oil slick, and covered with water and oil when an oil slick appears on the water surface.

In one method of operation of the detector, the spinner is operated continuously and at a constant speed, for example 750 rpm. So long as the spinner is submerged only in water, the torque on the motor, and thus the motor current, will be on the low side of the scale. When oil appears on the water surface, the torque on the motor, and thus the motor current, increases. When the motor current reaches a preselected value, for example 120 mA, a relay or other known form of trigger circuit associated with the control circuitry is operated to signal the presence of the oil. Given such a trigger signal, many forms of alerting or recording techniques may be employed to make an observer aware of the oil slick condition.

A device developed by *E. Hakansson; U.S. Patent 3,908,443; September 30, 1975; assigned to Gotaverkin Oresundsvarvet AB, Sweden* is one in which contaminants in process water are detected by continuously dividing off a fraction of the water, subjecting this fraction to a settling action, preferably in at least two steps collecting a batch of residue so obtained in a receiver and transferring this batch first dropwise and finally suddenly to a heating means enclosed in a chamber, which is connected to a gas detector. This will not react to pure steam, so if the batch contains water only nothing will happen. If there is for instance oil mixed with the water the detector will react. The point during this last transfer where the gas detector reacts, or the intensity of the reaction is a measure of the content of contaminants.

A place where the risk of oil fouling of water is likely to occur, is the cargo heating systems for tank ships. The different cargo tanks are provided with heating coils sunk into the liquid cargo and heated by means of steam. This will condense in the coil and the condensate will be returned to the boiler to be transferred to steam once again.

Even an unsignificant crack in a coil will permit oil from the cargo to enter the heating system. This oil finally reaches the boiler, where it will form encrustations upon the water tubes and disturb the circulation in general.

Figure 44 shows the basic components of a device as used for monitoring the condensed water supply of the cargo heating system in a ship.

The condensate flowing from the cargo heating coils is by way of a conduit **10** conveyed back to a steam boiler (not shown). A fraction of the condensate is continuously divided off through a branch conduit **11**, and is by way of a pump **12** transferred to a tank **13**. This is divided into two compartments, **14** and **15**, by means of a partition, the upper edge of which forms a weir between the two compartments.

An outlet conduit **17** from compartment **14** is arranged in such a manner that a constant water level **18** is maintained in the compartment as long as the inlet and the outlet are open. The outlet conduit is connected to a filter tank **19**, and from the latter the water is transferred back to the process.

A two-way valve **20** is fitted between pump **12** and tank **13**, and in the outlet conduit **17** there is a further valve **21**.

After a predetermined time valve **21** is closed and the level in compartment **14** is momentarily raised until it reaches the weir **16**. During a further period of time the pump will supply water to compartment **14**, whereby the top layer therein will flow over to compartment **15**. Valve **21** is then opened and the level within compartment **14** will sink to level **18**.

The quantity of water divided off through conduit **11** is known, and experience tells how long a time possible oil contaminants in the water will require to rise to the surface. It therefore is comparatively easy to calculate the closing and opening times for valve **21**.

After a certain number of such overflow operations the level within compartment **15** has risen to value **22**. Valve **20** will then be shifted, so pump **12** during a short period will deliver water directly to compartment **15**.

The level therein will then rise, and a portion of the liquid therein will flow over a weir **23**, whereafter valve **20** is returned to its normal position and the pump again delivers water to compartment **14**. Compartment **15** will simultaneously be emptied through a conduit **43** provided with valve **44**, so the unseparated water used for transferring of the residue contained in compartment **14** will be emptied to filter tank **19**. The draining of compartment **15** may be included in the monitoring program, but level **22** may also be sensed by a switching device **24**.

FIGURE 44: DEVICE FOR DETECTING OIL IN WATER BY SEPARATION FOLLOWED BY HEATING AND SMOKE OR VAPOR DETECTION

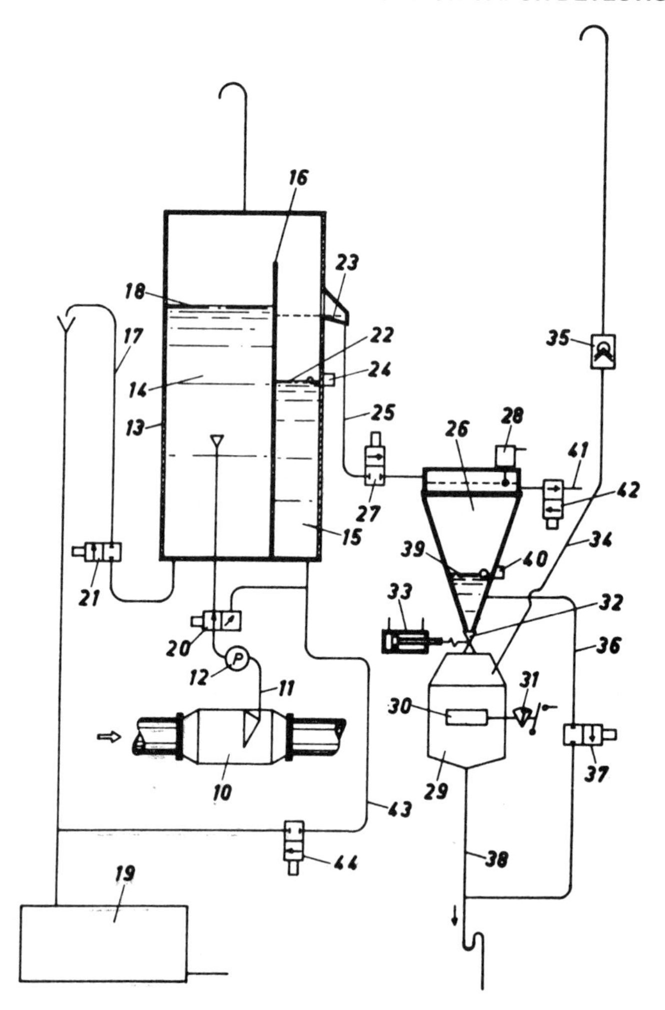

Source: U.S. Patent 3,908,443

From weir **23** the liquid is transferred by way of a conduit **25** to a funnel shaped receiver **26**. A valve **27** is provided in this conduit, and is normally maintained in closed position. It will however, receive an impulse to open simultaneously with valve **20** being shifted for delivery to compartment **15**. A further impulse will reset it simultaneously with the last mentioned valve.

Receiver **26** has sufficient volume to be able to receive a number of transfers from compartment **15**, and will serve as a last settling step. The highest level within the same is sensed by means of a device **28**, which, when activated, starts processing of the liquid contained within the receiver.

Below the latter there is a chamber **29** provided with heating means **30**. This heating means includes an obliquely located plate, which by means of a thermostat **31** is maintained at a temperature of between 250° and 350°C. A connection between the receiver and the chamber is governed by means of valve **32**, which by a servo motor **33** may be brought into different positions in a manner to be described herebelow. Chamber **29** is further, by means of a ventilating pipe **34**, connected to a gas detector **35** of arbitrary known type designed to react when fog or smoke appears in pipe **34**.

An outlet conduit **36** is connected to the lower end of the receiver, and is provided with a valve **37**. This conduit is connected to a further outlet conduit **38** from chamber **29**.

It is presupposed that valve **32** normally is closed. After a number of transfers from compartment **15** the level within the receiver will reach a value to activate the level sensing device **28**. This will open valve **37** and block valve **27**, so no further transfer from compartment **15** will occur during the time necessary to evacuate receiver **26**. By means of conduit **36** the level within the receiver **26** is reduced to value **39**, whereafter valve **37** is closed. The latter level is sensed by a device **40**, but the time necessary for reducing the level down to this value may be included in the program.

A compressed air conduit **41** is connected to the upper end of receiver **26**. This conduit is provided with a valve **42**, which on the one hand is a common shutoff, deaerating valve, and on the other hand, governed by an oscillator when brought into a certain position, will issure a series of short pressure air pulses within the receiver.

When the level within the receiver has been reduced to value **39** only a small volume of liquid will be retained therein, which will include at least a major portion of possible oil contaminants in the water divided off through conduit **11**. Valve **32** is now, by means of servo motor **33**, brought into a first position in which it opens a very restricted, narrow slot-formed passage between receiver **26** and chamber **29**.

Also within the small residue now present in the receptacle the oil will rise to the top. Due to the different surface tension characteristics of oil and water, respectively, water will start to dribble onto plate **30**, where it will immediately be transferred into steam. This steam will condense within the chamber, or in conduit **34**, and the water will flow out through conduit **38**, without causing any activity at the gas detector **35**. The compressed air pulses from valve **42** will aid in this drop-wise transfer.

When a small quantity remains in receiver **26** valve **32** is suddenly opened fully and the remaining fluid gushes over the plate, and a rapid rise of steam occurs. If this fluid still is water only the gas detector will not be activated. The actual quantity of fluid may be about 10 ml.

If a major oil leak has occurred it is possible that generation of smoke will occur already during the time when valve **32** is open in its first position, i.e., when the liquid still dribbles down. The time between the opening valve **32** and an activity at the gas detector **35** will then be a measure of the quantity of oil in the water. If the gas detector reacts just when the valve is shifted to its second position it is possible to estimate the quantity of oil by measuring the intensity of the reaction.

During normal use, pure water will be treated only, and as soon as the divided-off residue has passed valve **32** the apparatus is brought into action again, and receiver **26** is ready to take a new batch.

An oil spill detection system developed by *G.H. Miller and E.O. Renick, Jr.; U.S. Patent 3,916,674; November 4, 1975; assigned to Texaco, Inc.* is one which employs a rotating disc that is partially submerged in the body of water to be monitored. There is a doctor blade for removing adhering liquids from the surface of the disc, which liquids are accumulated in a settling vessel. Overflow is directed to apparatus for detecting the presence of oil. The sensitivity of the instrument is good without being too delicate.

Figure 45 shows the system which comprises an instrument **11** which is designed to float on the surface of a body of water **12**. The instrument includes a disc **15** which is mounted for rotation about its axis. The axis is located at the center of a supporting shaft **16** to which it is attached in a fixed manner for rotation therewith at all times. The shaft **16** is driven in rotation by a motor **17**.

FIGURE 45: OIL SPILL MONITOR USING CAPACITANCE TO DETECT THE PRESENCE OF OIL ON WATER

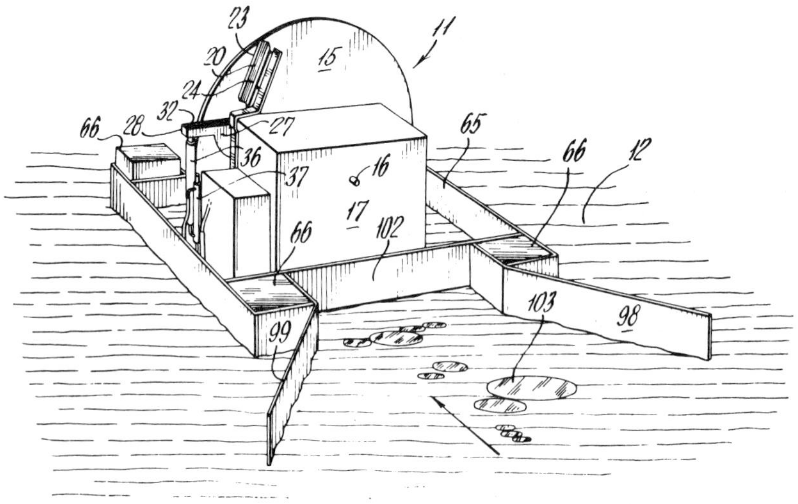

Source: U.S. Patent 3,916,674

The disc may be constructed of different materials. However, such materials should preferably be one of those which has at least partial affinity for oil.

Thus, the disc **15** might be metallic, and if desired, could be coated with any appropriate plastic material such as that known by the trade name Teflon. However, it has been found that a stainless-steel disc operates very satisfactorily by providing sufficient adherence of any oil floating on the surface of the body of water **12**.

It will be noted that the disc **15** is rotated clockwise. Consequently, as it rotates, the outer periphery sector which is submerged in the body of water **12**, will move into a contact line formed by a doctor blade **20**. The doctor blade acts to scrape off and thus remove a substantial portion of the adherents on the surface of the disc **15** over an area formed by contact with the doctor blade **20**. Such adherents will include any oil which may have been on the surface of the body of water **12** where it came into contact with the face of the disc **15** as it was rotated from beneath the surface out of the water. These liquids will then run down by gravity flow over a flexible edge **23** which is part of the blade **20**. They will continue down inside of a channel member **24** to drop off and fall into a vessel **27**. Vessel **27** has an arcuate shape to provide a deep bottom portion where any sediment that is included with the adherent liquids may settle out. The top of the vessel **27** is open but covered by a screen **32**. There is a radially extending portion **28** of the vessel **27** that has a port in the bottom near the extremity thereof. The adherent liquids may flow through the port, and it will be observed that this acts as an overflow from the contents of the vessel **27**.

The adherent liquids which are being removed from the surface of the disc **15**, after flowing over and off the doctor blade **20** (with its flexible edge **23**) and along the channel **24**, will have dropped into the vessel **27** through the screen **32**. After the vessel **27** has been filled to the level of the extension **28**, these liquids will flow outward along the extension **28** and through the port to a connecting conduit that directs them into the top of a tubular container **36**.

The level of the liquids in container **36** is maintained at a desired height above the capacitor, to be described below. This is accomplished by having the vertical location of an adjacent turbulence chamber **37** set for determining the level of overflow from it. Chamber **37** is connected to the bottom of the container **36** via a tube. The purpose of the chamber **37** is to provide turbulence in the liquids flowing through the system so that fine solids will not clog the fluid flow.

At an appropriate vertical distance beneath the level in the tubular container **36**, there is a pair of conductive-material plates or electrodes, which are situated diametrically opposite each other. These form an electrical capacitor the dielectric of which is made up of the fluid flowing through the container **36**.

The framework which makes up a supporting structure for the entire system might take various forms. As illustrated, it includes an outer rectangular frame **65** which has at each of the four corners an arrangement to support floatation members **66**. These are constructed of appropriate material having adequate buoyancy to create floatation for the entire device. It will be appreciated that the floatation effects may be adjusted by vertically positioning these members **66** in the rectangular frames which surround them. This would be done in order to cause the entire framework **65** and the rest of the system to float in a level position.

It will be understood that such a monitoring instrument would be anchored or otherwise tethered at a desired location on a body of water to be monitored for oil spills. The motor **17** would be energized and, consequently, the disc **15** would be rotated in the direction indicated by the arrows. As the submerged portion of the disc rises out of the water, it carries surface water and any oil or the like which clings to the surface of the disc, up until they contact the doctor blade **20** with its flexible edge **23**. This removes such adherents, and they run down the channel **24** to fall onto the screen **32** and through, into the vessel **27**.

As soon as the vessel **27** is full, the overflow goes out into the radial extension **28** and then through the port into the tubular container **36**. Here the container fills up until the overflow level is reached, as determined by the vertical level of the tube of the turbulence chamber **37**.

When these conditions are reached, the electrical bridge would be balanced. The balance would be indicated by a zero or minimum reading of a meter. This would represent the absence of oil on the body of water being monitored.

Thereafter, whenever any oil is encountered, it would be picked up and end up in the container **36** between the plates, so that the capacitance would change and the bridge would be unbalanced. Such unbalance would produce a signal which would be indicated by the meter, and an alarm or control signal would also be developed, if desired.

It will be understood that the system has the ability to be adjusted for desired sensitivity so that inconsequential oil presence on the surface of the body of water need not be indicated. Also, the system may be continuously operated, with appropriate monitoring, so that whenever an oil spill occurs, it will be indicated, and appropriate action may be taken.

It should be noted that there is a tendency under calm conditions for surface flow of the water **12** to be diverted around the instrument and so to create a diversion of any oil floating on the surface around the framework of the instrument. In order to avoid such condition, the illustrated structure includes a pair of wings **98** and **99** that are attached to the framework **65** in an appropriate manner. These are situated so as to create a wide mouth into which surface flow will take place.

Some movement of surface liquid will tend to be created by the action of picking up of adherents onto the surface of the disc **15** with removal thereof by the doctor blade **20**. However, under some circumstances it might be found necessary to add some pumping (not shown) of fluid from the vicinity of the disc **15** within the framework **65** to a point downstream in order to insure that surface liquid will be drawn into the instrument.

It is to be understood that the frame **65** includes a cross brace **102** which is narrower than the sides of the frame **65**, so that it (brace **102**) clears the surface of the water **12** in order to permit any floating oil **103** to go into the inside and then come in contact with the disc **15**.

A device developed by *J.O. Moreau and R.A. Halko; U.S. Patent 3,924,449; December 9, 1975; assigned to Exxon Research and Engineering Co.* is an oil pollution totalizer for accumulating all of the oil from a sample at a rate that is directly proportional to the rate of oil being discharged from the stream. The totalizer essentially comprises means for controlling the sample flow rate in proportion to the stream flow rate, means for removing the oil from the sample stream, and means for storing the oil for subsequent analysis. Figure 46 is a schematic diagram of such a device.

FIGURE 46: OIL POLLUTION TOTALIZER

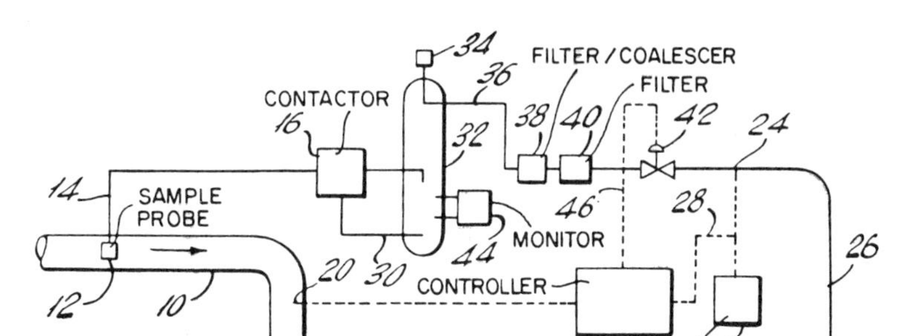

Source: U.S. Patent 3,924,449

The figure shows a typical discharge line **10** such as may be found on a tanker or as the output for a refinery effluent stream. In both situations there is a need to limit the rate and total amount of oil. As shown, the arrows illustrate the direction of flow of the discharge stream. Between the inlet (not shown) to the pipe **10** and its outlet (not shown), and preferably at a location such as a tanker pumproom or a turbulent portion of a refinery effluent stream, there is provided suitable means **12** for continuously withdrawing an oily water sample of a predertermined amount from the flow stream. For use on a refinery effluent stream, a pump may be required to propel the sample through the totalizer system. The amount of sample used then is conveyed via a conduit **14** into a solvent/sample contacting means **16**, the function and construction of which will be described in detail hereinafter.

There also is provided a flow controller **18** connected downstream of where the sample is taken. This controller typically may comprise a pneumatic ratio controller with recorder which operates to regulate the flow rate in the sample outlet line **26** at a predetermined fixed relation to the discharge flow rate. The controller is connected through a sensor **20** into the flowing discharge stream for sensing the rate of discharge. A flow recorder **22**, which may comprise a conventional pneumatic chart recorder, is connected to the outlet line **26** by sensor **24** for measuring the sample flow rate for comparison purposes with the sensed discharge flow rate by connection **28**.

The contactor **16** typically may comprise equipment such as a pump, a static mixer, an ultrasonic mixer, and the like, so long as it is capable of mixing and providing an intimate contact between the oily water sample taken from the discharge stream and the solvent (which may comprise a solvent compatible with the subsequent analysis for oil concentration, such as carbon tetrachloride with analysis by infrared adsorption) which flows through the line **30**. The solvent extracts the oil from the sample so that the water will form a separate phase. The mixture of solvent and sample passes from the contactor into a conventional settling chamber **32**, such as a section of cylindrical pipe. This chamber may be provided with an automatic air release **34** for elim-any entrapped air in the chamber which might be caused by air entrained in the discharge line **10** and otherwise result in transient surges in sample flow rate.

The chamber **32** separates the solvent with the oil from the combined solvent/sample mixture. The solvent with oil is then recycled to flow back through the line **30** to the contractor means **16** where the process is repeated. At the end of the chamber opposite from the solvent/oil recycle outlet, the clean water exits via the conduit **36** and passes through an optional means **38** for removing contaminants. This apparatus **38** may comprise a filter-coalescer for moving and retaining solid particulates, entrained solvent, etc. The stream then passes through a filter medium **40** which removes particulates and provides an additional filter medium to that provided by the filter-coalescer **38**. The contactor **16** provides the necessary mixing and intimate contacting between the recycled solvent and the sample for total extraction of the oil. By pumping this solvent/sample mixture into the settling chamber, the solvent containing the extracted oil separates from the water. If a heavier-than-water solvent is used as shown, the solvent plus oil settles to the bottom and is recycled, while the clean oil-free sample water flows out from the top of the chamber.

The clean water without solvent, oil or particulates then passes through the sample flow rate control valve **42** which is connected by line **46** to the controller **18**. The control valve **42** is continuously controlled as a function of flow rate in the discharge line **10**. Thus, valve **42** is variable in that it controls sample flow in line **26** in direct proportion to flow in line **10**. For example, a maximum flow in line **10** will result in a maximum sample flow in line **26** in accordance with a predetermined relationship as established by the controller **18**. The sample flow rate sensor **24** and the sample flow rate control valve **42** operate (i.e., they control) sample flow at a point of clean water on the sample after the solvent, oil and particulates have been removed. This technique enables accurate control of the low sample flow rate through the totalizer. Sensor **24** senses the sample flow rate in line **26**. Signals from sensors **20** and **24** are sent to controller **18**, which in turn controls throttling of valve **42**. The sensor **24** may be omitted and controller **18** set to throttle valve **42** in a predetermined pattern as a function of sensed flow in line **10**.

The total oil extracted from the sample is retained in the constant volume of recirculated solvent so that the oil concentration in the solvent is directly proportional to the total oil discharged in the ballast water or effluent stream. This occurs because the sample in the totalizer system is controlled by valve **42** to be directly proportional to the oily water discharged in line **10**. The total oil in the sample is therefore proportional to the total oil in the discharge line. Since all the oil from the sample stream is retained in the fixed volume of solvent, the oil concentration in the solvent is directly proportional to the total oil discharged through line **10**. Thus, the total oil discharged during a selected time interval, such as a tanker ballast voyage, could be determined by removing the solvent from the totalizer system and analyzing the oil concentration by standard laboratory techniques such as infrared absorption. If desired, a suitable device for continuously monitoring the oil concentration in the solvent may be provided as shown at **44**. This will provide information about the rate of oil pollution, since the time rate of change in the concentration of oil in the constant volume of solvent is directly proportional to the time rate of oil discharged through line **10**.

A typical monitoring device (not shown) may comprise an infrared absorption monitor equipped with a flow-through cell. The oil removed from the sample stream is retained in solution with the solvent. This technique not only facilitates analysis of oil concentrations, but even more important, it prevents changes in oil composition such as would be caused by evaporation, oxidation or biological degradation if the oily water were stored as such in a drum. Thus, this novel technique enables accurate measurements of the total oil discharged. By removing the oil from the water and storing it in a solvent phase, a separate extraction step is not required for subsequent analysis of the sample.

The arrangement also provides an excellent pollution control system, since the totalizer per se can be placed in a tamper-proof locked case so that the pollution control authority then can board a tanker in the port or go through a refinery and obtain detailed information about the total amount of oil pollution over any desired interval of time. The unit, when in a tamper-proof locked casing, could be an invaluable tool for the tanker or refinery operator by providing proof there was no violation of total oil limitations in the discharge. The totalizer system (excluding the concentration monitor **44**) could be so constructed as to use only pneumatically powered components, so that the entire system would be intrinsically safe. This capability would permit installation of the totalizer system in a hazardous location such as a tanker pump room, if such a location were desirable.

A device developed by *B.H. Stenstrom; U.S. Patent 3,964;295; June 22, 1976; assigned to Salen & Wicander AB, Sweden* is a device for determining or indicating the oil content of small amounts of oil in water. A defined amount of water is removed so that the oil is concentrated in the water, and thereafter the oil content in the concentrated oil-water mixture is measured. So as to concentrate the oil, the oily water can pass a filter material, the color change or dielectric constant of which is measured. As an alternative, the oily water is concentrated in a rotating cyclone or centrifuge so that substantially all oil is accumulated in the center together with a smaller portion of the original water volume. The clarity of the water is then determined and considered as a measure of the oil content.

A device developed by *R.M. deVial; U.S. Patent 3,965,920; June 29, 1976, assigned to Bailey Meters & Controls Ltd., England* is an apparatus for detecting the presence in a liquid (such as water) of a material (such as oil) that emits fluorescent radiation when stimulated by ultraviolet radiation. A chamber having an inlet and an outlet for passing a stream of liquid therethrough is provided with a source of ultraviolet radiation associated with said chamber with means for energizing same. Said radiation may pass to said stream and in turn to a receiver arranged to receive fluorescent radiation emitted by such a stream.

An oil-in water monitor developed by *M.G. Grant; U.S. Patent 3,966,603; June 29, 1976* is one for monitoring the effluent of an oil-water separator. A pressure vessel coalesces oil from the effluent and gathers the oil in a collection section which is automatically emptied after an hour. If more than a predetermined amount of oil is gathered in the hour, an oil-sensing probe sends a signal to activate a warning light valve which permit pipes to return the effluent to the separator.

If the effluent becomes too turbid, a turbidity sensor activates an alarm and the valves which permit pipes to return the effluent to the separator. If the coalescer unit clogs, an indication is given by a differential pressure gauge. Figure 47 shows such a monitor.

FIGURE 47: MONITOR-ALARM FOR CHECKING OIL CONTENT OF EFFLUENT FROM OIL/WATER SEPARATOR

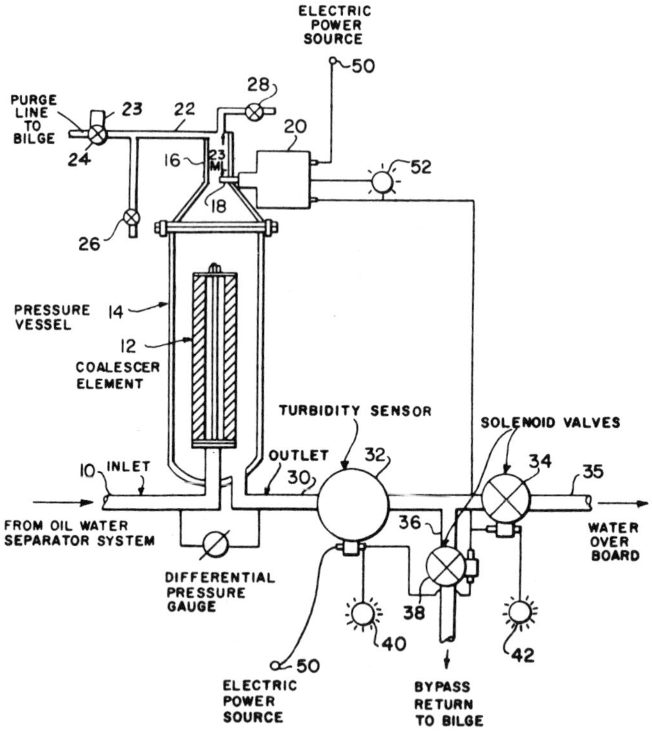

Source: U.S. Patent 3,966,603

An inlet line **10** from an oil-water separator system feeds into a filter/coalescer (or simply coalescer) element **12** located inside a pressure vessel **14**. The pressure vessel **14** has an oil-collection section **16** at the top for which the volume is easily calculated. The oil-collection section is convenient to have since it makes the volume from the top to any lower level easy to calculate if the shape is well-chosen as, for example, the cylindrical shape shown. However, an oil-collection section is not absolutely necessary. The oil would collect at the top of the pressure vessel anyhow and it is just a question of selecting the proper level for installation of the sensing probe **18** which must be at the bottom horizontal level of the desired oil volume.

An oil-sensing probe **18**, which is part of the interface sensing-probe assembly **20**, is placed in the oil-collection section **16** at a predetermined level which will indicate when the amount of oil in the pressure vessel (i.e., in the effluent from the separator system) is higher than desired.

A purge line **22** to the bilge is connected to the oil-collection section **16**. This bilge purge line can be opened by an electrically timed, bilge-purge-line valve **24** or by a manual bilge-purge-line valve **26**. The oil-collection section **16** also has an air-eliminator valve **28** at its top. This valve may be manual or automatic.

The pressure vessel **14** has an outlet line **30** at the bottom feeding into a turbidity sensor **32**, which may, for example, be of the photelectric cell type. The outlet line then goes through a solenoid shutoff valve **34** to an overboard vent line **35**. An overboard by-pass line **36** branches off the outlet line between the turbidity sensor **32** and the overboard-line shutoff valve **34** to return to the bilge. The by-pass line **36** also has a solenoid shutoff valve **38** connected to it which is activated by the turbidity sensor **32**. The latter also activates the turbidity alarm light **40**. The overboard-line solenoid valve **34** has an alarm indicator **42** associated with it which indicates that the valve has operated.

A differential pressure gauge is coupled across the coalescer element **12** (actually, coupled to the inlet line to and the outlet line from the coalescer element). If the coalescer element becomes clogged, the differential pressure across it increases and this is indicated by the gauge. The coalescer element can then be changed.

In operation, when the unit is turned on, liquid from the oil-water separator system flows through the coalescer element **12** and fills the pressure vessel **14**. The air eliminator valve **28** is opened until a few drops of liquid appear at its outlet. The valve is then closed since this indicates that no air remains in the vessel and that it is filled with liquid.

When the liquid goes through the coalescer **12**, any oil in the liquid is coalesced into droplets which rise to the top and accumulate in the oil-collection section **16**. The clear water in the bottom of the vessel **14** is vented overboard by the pressure in the system through the overboard valve **34** which is normally open.

The electrically timed valve **24** is set for a predetermined time, for example, one hour; the time depends on the purity desired in the water to be vented overboard. If insufficient oil has accumulated in the oil-collection section **16** in this period to activate the sensing probe **18**, the purge valve **24** is opened at the end of the timing period for a predetermined period of time, e.g., 3 seconds, which is sufficient to drain the oil from the oil-collection section.

If sufficient oil collects to be sensed by the probe **18** before the time period of the electrically timed valve expires, indicating that there is too much oil in the separator system liquid, the interfacing sensing-probe assembly **20** generates a signal which acts to close the overboard-line solenoid valve **34**, energize its associated alarm light **42** and open the by-pass return line valve **38** to the bilge.

The alarm light **42** informs operating personnel that the manual purge valve **26** to the bilge must be opened to drain the oil from the oil-collection section. When this is done, the signal from the probe assembly **20** ceases and the valves **34** and **38** return to their original conditions.

If the liquid contains a chemical emulsion of oil (e.g., oil, water and a surfactant) or if the amount of oil in the liquid is very greatly above the coalescing capacity of the coalescer element, the liquid coming through the outlet line **30** will be turbid. The turbidity sensor **32** then provides a signal to energize its associated turbidity indicator **40**, the overboard-line valve **34** and indicator **42**, and the by-pass return valve **38**. The energization of the turbidity indicator alarm light **40** warns the operator that the oil-water separator system is not functioning properly.

The exemplary figures given above are based on a flow rate of liquid from the separator system which is 10 gallons per minute, and a value of 10 ppm or greater of oil in the liquid. This requires a volume of about 23 milliliters per hour of oil to trigger the oil-sensing probe and shut off the system. Thus the time period of the electrically timed valve **24** is set for an hour. It should be apparent that the alarm indicators may be bells rather than lights or any other type of indicator which will attract the operator's attention.

A system developed by *U. Cirulis and E.M. Zacharias, Jr.; U.S. Patent 3,973,430; August 10, 1976; assigned to Process and Pollution Controls Co.* is one utilizing a differential sound velocimeter which compares the velocity of sound in the liquid under study with the velocity of sound in the liquid (largely water) from which the oil or other pollutant has been separated by mechanical and/or solvent extraction means. A suitably calibrated frequency-to-analog converter is utilized to indicate the amount of the oil concentration and/or shut down the water transporting system if the concentration exceeds a predetermined value. Figure 48 shows the essential features of such a system.

In the drawing, system **20** is seen to comprise a plurality of input valves **22^1, 22^2. . . 22^n** for selecting the output from any one of the cargo tanks **1, 2 . . . n** for measurement while the tank contents are being discharged. Also shown are a valve **24** connected to a source, of cleansing and flushing solution for cleaning the measuring system, a separator **26** such as a coalescing type filter or a De Laval centrifugal separator and a different transducer assembly **28**.

Upon opening one of the valves **22**, pumps **30** and **32** are turned on. Since a small amount of the sample solution is needed to operate the measuring system, valve **34** is opened, as desired, to deliver the excess, overflow liquid to the source tank or to a storage tank.

A sample solution from one of the tanks, as selected by the appropriate valve **22** is pumped on a substantially continuous flow basis by pumps **30** and **32** to a strainer **36** which stops the large debris but does not preclude the passage of oil therethrough. Under certain conditions (solution without large solids), the strainer **36** may be omitted from the system. The flow of the sample solution next reaches an optional preheater **38** which should be used to raise the solution temperature to about 50° to 65°C. If the solution temperature is already in that range, the preheater is unnecessary. The solution reaches an

FIGURE 48: OIL POLLUTION MONITOR USING ULTRASONIC MEANS FOR OIL DETECTION

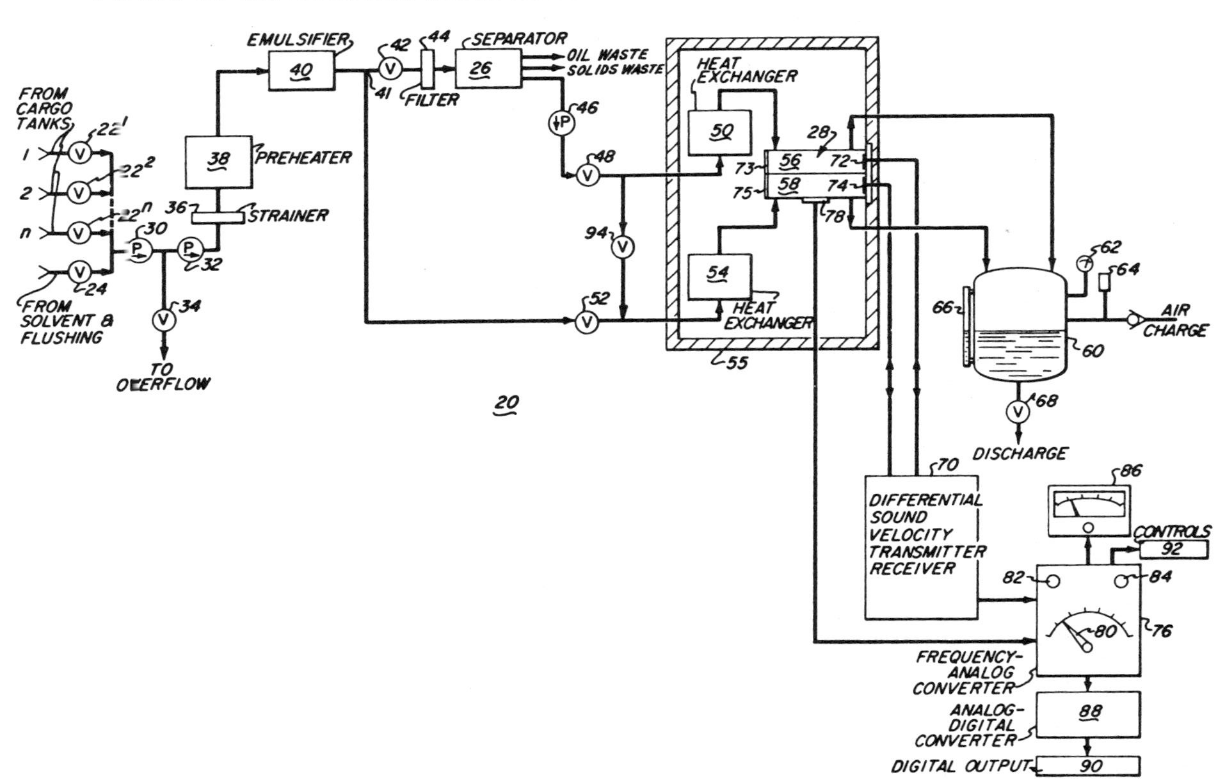

Source: U.S. Patent 3,973,430

optional emulsifier **40** next. The emulsifier **40** is utilized to break the large hydrocarbon droplets into small ones so that the hydrocarbons are more uniformly distributed through the solution. Emulsification is also obtained, in some degree, from the operation of the liquid transporting pumps.

At a junction **41**, some of the solution flows to a valve **42** which, when open, feeds a portion of the solution to a filter **44** and the separator **26**. The filter **44** is preferably of very fine mesh in order to remove the fine solids from the solution. The separator **26**, which is preferably a coalescing type filter or a De Laval centrifugal separator, removes the oil and solids from the water and leaves a standard solution of water to be pumped by a pump **46** through a control valve **48** to a heat exchanger **50**.

The other portion of the solution (unknown) flows through a valve **52** to a heat exchanger **54**. Both heat exchangers and differential transducer assembly **28** are mounted in an insulated tank **55** containing a constant temperature bath for holding the temperature of the standard and unknown at the same temperature, preferably 72°±1°C for reasons which will be evident as this description proceeds. The heat exchangers may also be mounted outside the constant temperature bath. In such cases, the liquid from the constant temperature bath may be circulated through the heat exchangers. In any event, it is important that the temperatures of the two solutions in the differential transducer be equal within ±0.1°C.

The standard solution enters transducer assembly (velocimeter) **56** and the unknown solution enters transducer assembly (velocimeter) **58**. Both solutions are kept under pressure because the outputs of both assemblies **56** and **58** are fed to a tank **60** which is kept under pressure by an air charge and which serves to keep the pressures in assemblies **56** and **58** equal. The usual pressure gauge **62** and pressure relief valve **64** are provided. When the liquid in tank **60** reaches a predetermined level, valve **68** is opened by a liquid level control mechanism (not shown) to discharge some of the liquid from tank **60**. A sight glass **66** is used for visual monitoring of the liquid level in tank **60**.

As the solutions flow through transducer assemblies **56** and **58**, a differential sound velocity transmitter-receiver alternately applies a pulse to transducer **72** and to transducer **74**. The pulse excites the transducer **72** which transmits an acoustic wave to reflector **73**. The wave is returned to transducer **72** and the electrical signal produced is fed to differential sound velocity transmitter-receiver **70**. Similarly, the pulse excites transducer **74** and the acoustic wave produced is reflected back to transducer **74** by reflector **75**. The shift between excitation of transducer **72** and transducer **74** cannot be more rapid than the travel time of the acoustic waves.

The output of differential sound velocity transmitter-receiver **70** is a frequency difference which is applied to a frequency to analog converter **76**. A temperature sensor **78** in combination with a signal from sound velocity transmitter-receiver **70** will cause a ready light **82** to illuminate when the temperature of the bath is 72°±1°C and the differential sound velocity transmitter-receiver **70** is in phase lock condition. Frequency to analog converter **76** may be provided with a control **80** which is positioned depending upon the type of hydrocarbon in the tank solution. It adjusts the sensitivity of the converter **76**. Indicator light **84** illuminates when the oil concentration is above a predetermined value.

The output of converter **76** may be fed to an analog to digital converter **88** and then to digital output **90**. One or more controls **92** to shut down the system and/or actuate an alarm may be connected to converter **76**.

In order to calibrate the system, valve **52** should be closed and a valve **94** is opened. Thus, a standard solution flows through both transducer assemblies making it possible to adjust the system.

A simple method for detecting oil in water has been developed by *P.T. Thyrum; U.S. Patent 4,004,453; January 25, 1977.*

In one variation of the method a sample of water is first filtered at a controlled rate. A dye-impregnated pad is pressed against the upstream surface of the filter element and then removed. The upstream surface of the filter element may now be observed with the eye against colorations of known concentrations for variations in color intensity clearly discernible within the range of 0 to 30 ppm by volume of oil-in-water. The dye-impreganted pad is prepared by submerging a white absorbent material in a saturated solution of an oil-soluble, water-soluble dye, and drying them in a rack under vacuum.

In another variation of the method, where the sample also contains discoloring pigmented contaminants, an intermediate pad is pressed between the filter element and the dye-impregnated pad, and its coloration compared to known standards. In lieu of the intermediate pad method, a thin prefilter is placed on the upstream surface of the filter element during filtering, then discarded, and the filter element is processed as described in the first method above.

A device developed by *R.M. Dille, M.H. Van Stavern, D.L. Shull and D.F. Gripshover; U.S. Patent 4,045,671; August 30, 1977; assigned to Texaco Inc.* provides a sensitive arrangement for determining the presence of oil, at low ppm, in water. It employs infrared absorption measurement, and first mixes the water with an oil solvent that may be separated from the water after taking any oil into solution. Also, the solvent is one which does not have any significant infrared absorbency at a predetermined wavelength, which does have absorbency by hydrocarbons. After separation of the oil solvent, it is continuously passed through an absorption cell in an infrared spectrometer, to monitor the presence of oil in the water. Figure 49 illustrates such a device.

With reference to the drawing, it will be observed that there is a mixing container **11** which has an inlet conduit **12** connected thereto. There is also a branch conduit **15** that joins the conduit **12** prior to its connection into the mixing container **11**.

The mixing container **11** might take various forms, and for example it may be similar to a commercial motorized blender. Thus, there is a stirring rod **16** that has a paddle tip **19** at the lower end thereof for causing a fluid shearing action as the rod is rotated. The speed of rotation of the mixing rod **16** and its tip **19** is empirically adjusted in order to maintain the shear rate less than that which will create a stable emulsion with the solution being mixed.

The mixing container **11** has a circulating conduit **20** which is connected between an overflow outlet **23** in the container **11**, and an inlet **24** near the bottom of the container. Consequently, a body of liquid **27** in the container **11** will tend to circulate through the conduit **20** in a continuous manner.

It will be noted that the conduit **12** is connected to an inlet **28** in the container **11**, which is located vertically in between the outlet **23** and inlet **24** of the circulating conduit **20**. And, there is a settling container **33** that is connected to receive a portion of the mixture from the circulating conduit **20** via an interconnecting conduit **34**.

FIGURE 49: TEXACO DESIGN FOR INFRARED OIL-IN-WATER MONITOR

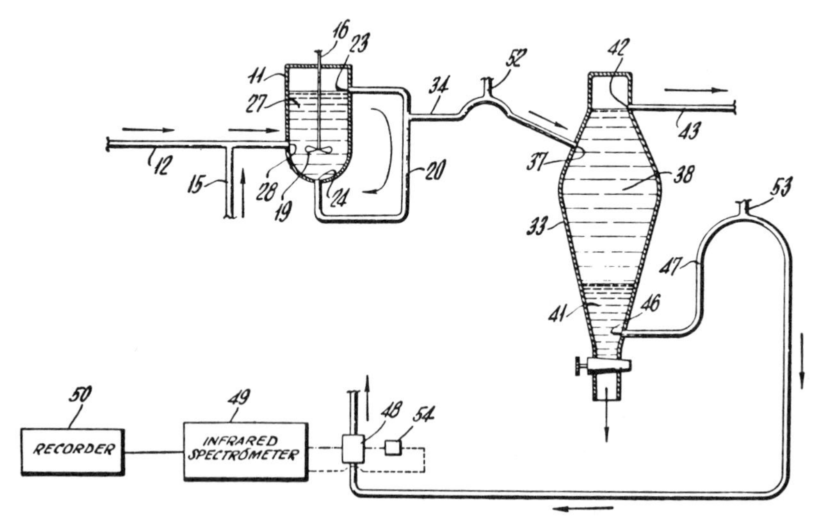

Source: U.S. Patent 4,045,671

The settling container **33** has an inlet **37** to which the conduit **34** is connected. The liquid mixture will settle by gravity into an upper portion of water **38**, and a lower portion **41** which is an oil-solvent solution that has a greater density than the water.

There is an overflow outlet **42** near the top of the settling container **33**. It may have a conduit **43** connected thereto for carrying away overflow water. This would also carry excess oil in the event that the oil quantity becomes excessive for the amount of solvent. There is another outlet **46** for the solution that is located near the bottom of the settling container **33**. A conduit **47** is connected to the outlet **46** for carrying the oil in solvent **41** out from the bottom of the settling container **33** to a cell **48** of an infrared spectrometer **49**. Of course, the spectrometer **49** may include a recorder **50** if desired.

There are siphon breaking standpipes **52** and **53** connected into the high points of the conduits **34** and **47** respectively. There are only schematically indicated since, of course, each must actually extend vertically upward to reach above the highest level of liquid in the connected system.

It will be understood that the spectrometer also includes an infrared detector element **54** that senses the amount of absorption of the infrared energy having a predetermined wavelength which the fluid in the cell **48** absorbs.

A particular example of a method of operation may be described with reference to the drawing. A first step is that of mixing about 10% by volume of the oil solvent (trichlorotrifluoroethane) with the water that is being monitored for oil content. Such a mixture is accomplished by the pumping rates of water being introduced in the conduit **12**, and the solvent flowing through the conduit **15**.

It may be noted that while the particular solvent trichlorotrifluoroethane is preferred, other solvents may be employed as long as they meet the requirements of forming a chemical solution with any oil in the water and at the same time have a density that is greater than water for effecting the separation later on. Also, the solvent needs to be one which does not have significant infrared absorbency at the infrared wavelength which is employed to detect the presence of oil. It may be noted that carbon tetrachloride meets these requirements although it has the disadvantage of being a highly toxic substance.

The mixture of water which is being monitored, and the oil solvent that is being introduced therewith, both flow into the mixing container **11**. There they are thoroughly mixed while employing a shear rate that is sufficient to insure thorough mixing in order to obtain a chemical solution of all the oil contained in the water while not being sufficiently high in shear rate to form a stable emulsion between the solution and the water.

One step which helps to obtain the desired results, is that of employing the circulating conduit **20** so that most of the liquid mixture **27** continues to circulate from the mixing chamber **11** around through the conduit **20**. During the mixing procedure, a portion of the mixture is diverted through the conduit **34** and flows over into the settling container **33**.

In the container **33** the liquid constituents are allowed to separate out under gravity action, so that the water **38** floats on top of the oil-solvent solution **41** that collects at the bottom of the container **33**. Then, solution **41** is carried through the conduit **47** to the cell **48** of the infrared spectrometer **49** where the absorption reading is taken. The reading involves the infrared detector **54** of the spectrometer.

It will be appreciated that the process is continuous so that whenever any oil appears in the water which is being monitored, it will go into solution with the oil solvent and then following the separation, it will pass through the cell **48** of the spectrometer. The spectrometer will measure an absorption change and provide a reading which indicates the presence of the oil in the oil-solvent solution. It may be noted that this is a highly sensitive procedure and is accurate to indicate very small quantities of oil in the water.

A device developed by *P.J. Herzl; U.S. Patent 4,048,854; September 20, 1977; assigned to Fischer & Porter Co.* is a device for metering a fluid stream constituted by a mixture of oil and water to determine the volumetric ratio of oil-to water in the stream. The system includes a vortex meter through which the stream is conducted to produce a meter signal whose frequency depends on the volumetric flow and whose amplitude depends on the mass flow of the stream. Derived from the meter signal is a volumetric signal that is solely a function of frequency and a mass flow signal that is solely a function of amplitude, the volumetric signal being divided by the mass flow signal to produce a density signal. The volumetric and density signals are fed to a computer to which is also applied a temperature signal that depends on the temperature of the metered stream. The computer has stored therein the relationship between water density and temperature and that between oil density and temperature, and it functions in response to the signals applied thereto to determine the respective volumes of oil and water in the mixture.

IDENTIFICATION

The identification of an oil spill in terms of sources can be a valuable aid in fixing responsibility for a spill.

A technique developed by *C.W. Brown, M. Ahmadjian and P.F. Lynch; U.S. Patent 3,896,312; July 22, 1975* is one for comparing the infrared spectrum of an unidentified oil sample to the infrared spectra of a plurality of identified oil samples.

A technique developed by *J.F. Fantasia and H.C. Ingrao; U.S. Patent 3,899,213; August 12, 1975; assigned to the U.S. Secretary of Transportation* is one for the identification, from a remote location, of oil comprising a marine oil spill. The technique includes directing pulses of high energy artificial light onto the spill to cause the oil to fluoresce and the incremental scanning of the frequency spectrum of the thus generated fluorescence energy; there being a distinct fluorescent spectral signature for each type of oil. The technique also includes the rejection of background radiation and the real time presentation of the fluorescence spectrum of the oil comprising a spill. Figure 50 is a block diagram of the apparatus which may be used.

FIGURE 50: AIRBORNE LASER REMOTE SENSING SYSTEM FOR THE DETECTION AND IDENTIFICATION OF OIL SPILLS

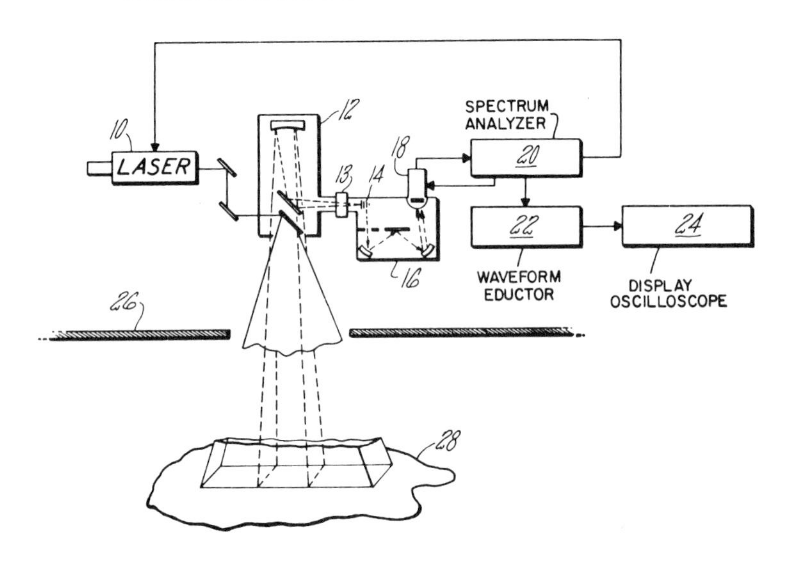

Source: U.S. Patent 3,899,213

The apparatus includes a laser transmitter **10**, a receiver-telescope **12**, an ultraviolet blocking filter **13**, an image slicer **14**, a spectrometer **16**, an image dissector **18**, a spectrum analyzer **20**, a waveform eductor **22** and a display device **24**. All of the above cataloged apparatus is mounted in an aircraft, the frame being indicated schematically at **26**, and is employed in the identification and classification of oil spills; a spill being indicated at **28**. While the apparatus may be employed in both oil spill detection and classification modes, the present discussion is concerned only with classification.

Once the system is airborne and an oil spill detected, the apparatus will be energized in the classification operational mode and the laser **10** caused to emit pulses of ultraviolet energy which are directed down onto an area of the sea surface directly below the aircraft. The incident laser pulses will cause the oil comprising spill **28** to fluoresce in the visible region of the electromagnetic spectrum. A portion of this fluorescence energy is collected by the telescope **12** and focused onto the entrance slit of spectrometer **16** by image slicer **14**.

The fluorescence radiation entering the spectrometer entrance slit is dispersed by a plane grating into a spectrum characteristic of the oil type. That is, the spectrometer separates the fluorescence energy reemitted by the spilled oil, as the result of stimulation by the high energy artifical light source comprising laser **10**, with respect to its frequency; i.e., its color. This oil fluorescence spectrum, which is of course arranged linearly, is imaged onto the photocathode of image dissector **18**.

The image dissector **18** converts the output of the spectrometer **16** into a usable form for analysis. Photoelectrons emitted by the photocathode of dissector **18** produce a deflectable electron image of the oil fluorescence spectrum. A portion of the electron image is dissected by a slit aperture in dissector **18** and amplified in a photomultiplier section. Restated, the spectrometer optical output is divided into segments which are incrementally scanned in a preselected sequence by the image dissector.

Electrical output signals commensurate with each incremental step in the scanning of the spectrometer optical output are applied to the spectrum analyzer **20**. The spectrum analyzer **20** controls the firing of laser **10** and the scanning of the spectrometer output by image dissector **18**. Analyzer **20** also subtracts the background radiation received by the spectrometer from the signals received as a result of stimulation of the oil spill **28**. Thus, the spectrum analyzer **20** processes the signal resulting from the scan of the fluorescent spectral "signature" of the target oil spill and controls the operation of the high energy light source which causes the oil to fluoresce. The output of the analyzer **20**, a signal commensurate with the spectral content of the returned fluorescence in usable form for display and/or computer analysis, is averaged in the waveform eductor **22** and thereafter applied to a display device and/or computer **24**.

CONTAINMENT

Containment and countermeasures are positive actions taken to limit the continued spread and migration of the spill and to effect cessation of flow at the source. These steps are the first corrective actions taken, and they should be initiated as soon as possible after the spill is discovered (9). Typical countermeasures might include:

1) Isolation and evacuation of spill area to protect life or health.
2) "Shutoff" activities at the source of the spill, which might range from simple valve realignment to extensive salvage operations. Tank ruptures, for example, may be sealed with chemicals which foam in place to form reliable seals.
3) Placing booms or other physical or adsorbent barriers so as to prevent contact of the spill with areas of sensitive beneficial uses such as parks, estuaries, tributary streams or water supply intakes.
4) Preplanned construction of trenches or dikes to isolate potential spill areas on land.

Containment is the critical first step of any coordinated spill cleanup activity. The rapidity and effectiveness with which it is applied will limit the adverse impacts of the spill on other beneficial uses of the affected water or land area. Table 19 summarizes some of the containment methods available.

TABLE 19: SUMMARY OF CONTAINMENT METHODS

Type of System	Principle of Operation	Advantages	Disadvantages
Air barriers	Subsurface bubbling to create upswelling of water surface	Do not impede vessel movement	Are costly to install and maintain. Are limited by environmental factors (wind, current).
Piston film or herder chemicals	Surface tension phenomenon	Can be easily applied. Small dose required.	Only provides limited containment for a matter of hours. Government approved products must be used.
Booms	A physical barrier	Can be deployed quickly. Are physical barriers.	Work best in calm waters. May be used in limited currents and waves.
Hose spray	Turbulent barrier to oil	Can be rapidly applied	Is limited to use in confined areas and calm water. Is temporary method.
Sorbent barrier	Both physical barrier and absorbent surface for oil pickup	Can be easily deployed. Can be used for both containment and pickup.	Works best in calm water. Oil is not effectively contained. Slows spreading.

Source: Reference (9)

The Navy preferred containment equipment/procedures focus on the use of piston film chemicals and solid, floating booms which are described as follows.

Piston Film Chemicals: Piston film chemicals have high surface activity and spread rapidly over the water surface. The spreading force of the chemical is sufficient to overcome the spreading forces of the slick. Consequently, these chemicals will push the oil layer back until it reaches a limiting slick thickness, which the piston film cannot exceed. The oil may be moved ahead of the spreading film toward a collecting or containment device as is shown in Figure 51 (left hand view), or the piston film may be quickly spread around the periphery of the spill as is shown in Figure 51 (right hand view). This technique simply slows down the spreading rate.

FIGURE 51: USE OF PISTON FILM CHEMICALS

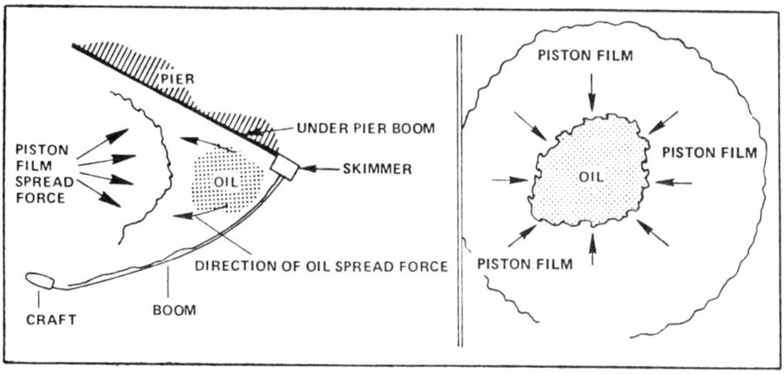

Source: Reference (9)

Containment Booms: Containment booms are solid (floating barrier) booms. They are solid, continuous obstructions to the spread or migration of oil spills. Because they are the most effective containment device, they are preferred for use with Navy related spills (9). This portion is intended to impart a conceptual understanding.

1) Booms are available in various sizes (in fifty-foot lengths) which are joined to form a continuous barrier to the oil. Their freeboard must be sufficiently high to prevent the oil from being washed over the boom, and the skirt length must prevent oil from being swept under it. Consequently, booms are purchased in several height/depth sizes to meet their use requirements in various wind and sea conditions.
2) Booms may be used in either a dynamic (towed) mode or fixed position. Figure 52 demonstrates typical use modes and purposes. It shows the boom being towed in a "vee" configuration in conjunction with a skimming device. The boom directs the oil to the skimming device where it is collected.
3) In the second case the booms are being used to prevent oil from going under the pier and also to direct oil to the skimmer. The slick would

move to the skimmer under the influence of wind and current, or could be pushed toward the skimmer by hose spray, air jet, or piston film, if necessary.

4) In the third instance the booms are anchored in a position that will entrap the oil but leave the channel open for navigation if necessary. The angle at which the boom is set is important in order to avoid loss of collected oil due to entrapment in the current or from being carried under boom skirts.

5) The fourth scene depicts typical use of a boom stretched across a stream. This alignment is feasible in small streams, mild currents, or tidal fluctuations. As depicted, diagonal deployment, in lieu of perpendicular, has been generally found more effective in flowing streams.

FIGURE 52: TYPICAL USES OF FLOATING BOOMS

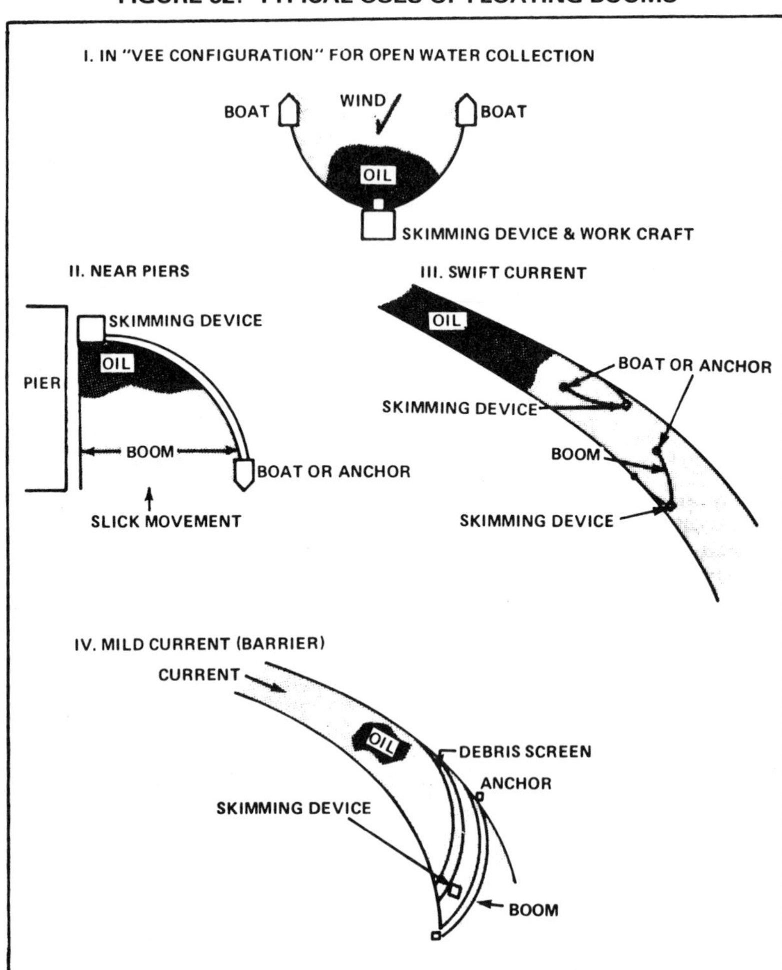

Source: Reference (9)

Spill containment by use of hose spray in confined areas, as listed in Table 19, can be an effective method. This technique is immediately available to ships' forces and can provide the earliest possible containment action when a spill occurs between nested ships against quay walls.

A technique developed by *W.P. Kirk and D.W. Reynolds; U.S. Patent 3,788,079; January 29, 1974;* is one in which a floating liquid such as an oil spill is collected under a sheet with weighted periphery. The sheet is deployed by spreading it horizontally over the liquid surface, after which the periphery submerges, confining and centrally concentrating the liquid. The sheet retains the liquid for subsequent pumping into a vessel.

A similar device is described by *W.P. Kirk and D.W. Reynolds in U.S. Patent 3,963,617; June 15, 1976.*

Chemical Barriers

Collecting agents (specifically one developed by the Naval Research Laboratory and available in one-pint containers through the Federal Supply System as FSN 9G 6810-172-9110) are effective when used around the edges of a slick in the initial response to a spill. Such use retards the spread of the slick and thickens the oil for more efficient pickup by skimmers along the upwind edge of a slick. Collecting agents, as a result, assist in pushing oil to a waiting skimmer; and in the final stages of cleanup, they help to concentrate and direct the "rainbow" sheen into a skimmer. The "piston film" available through the previously mentioned FSN number is a water-soluble organic material and has very low toxicity. For activities which plan to make use of this chemical on a frequent basis for small harbor spills, EPA and local and state agencies should be contacted for blanket prior approval. EPA recommends that this chemical be applied at a rate not to exceed two gallons per linear mile per six-hour period, with no more than three applications in a 24-hour period.

Navy tests and field use have confirmed that dropwise application of the "piston film" results in maximum effectiveness, and in most situations the amount used is well within the EPA recommendations. Figure 53 is a schematic of the effect of piston film on an oil slick and lists the maximum oil lens thickness which the chemical can push relative to various types of oils. Piston film is most effective in situations when the water surface is free of organic (e.g., fish) scum or other floating contaminants, as would be the case in open water areas or harbor pier areas which experience a large amount of tidal flushing action.

In those low current "dead water" areas such as around some piers and nested ships, the chemical is not effective; but in these situations the existing natural film provides the same retarding effect on a slick as the piston film. Whether or not to use piston film in a given situation is easily determined by applying a small amount dropwise and observing the effect on the slick edge. Experience has shown that piston film is normally effective only in calm waters. The film also may not be effective on small quantities of widely dispersed oil which produce a visible "rainbow."

FIGURE 53: TYPICAL SLICK THICKNESS WITH NRL PISTON FILM

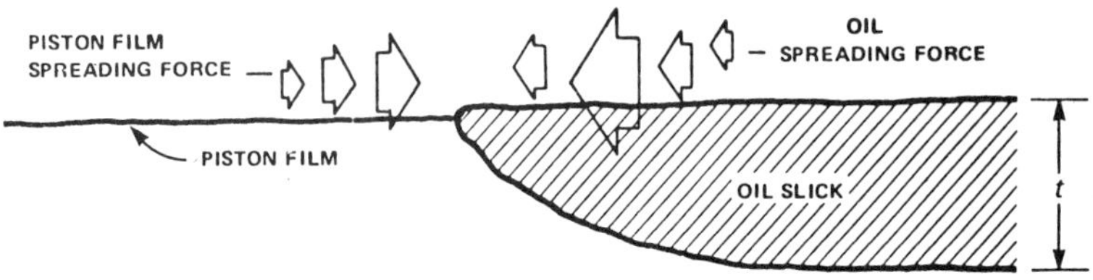

Oil	t (mm)
Marine Diesel (Navy)	5
Navy Special	11
No. 5 Fuel Oil	11.5
Lago Crude (South America)	5

Source: Reference (9)

A process developed by *R.L. Ferm; U.S. Patent 3,810,835; May 14, 1974; assigned to Chevron Research Co.* is one for treating an oil slick to contain it and prevent its uncontrolled spreading comprising applying to open water areas in the vicinity of the oil spill a chemical agent which repulses the oil spill. By judicious application of the chemical agent, the oil slick can be gathered into a limited area which facilitates cleanup. The chemcial agent is selected from the group consisting of (1) N,N-dialkyl amides; (2) n-alkyl and n-alkylene monoethers of (a) ethylene glycol and (b) polyethylene glycol, (3) polyethylene glycol monoesters of n-alkyl acids; and (4) n-alkyl and n-alkylene monoesters of propylene glycol.

A technique developed by *L.W. Jones; U.S. Patent 3,844,941; October 29, 1974; assigned to Amoco Production Co.* is one in which sulfur is spread over spilled oil to hold the oil mass together to permit it to be easily removed from the body of water. In a preferred embodiment, molten sulfur is sprayed as very fine strands over and around the periphery of an oil spill to enmesh the oil in a sulfur web and keep it confined. In another embodiment, sulfur powder is spread over the oil and tends to hold it together.

Sometimes it is desired to distribute the sulfur from an aircraft. An example is shown in Figure 54. A helicopter **70** transports a huge supply of powdered sulfur and spreads it through spraying means **72** to form a sulfur coating on the oil spill. Theoretically, sulfur distribution nozzles for molten sulfur could be used here. However, as a practical matter the weight of the heaters, etc., would ordinarily be too much for the helicopter. Thus, it is contemplated that the helicopter will be used primarily for the distribution of powdered sulfur. Airborne equipment could however be used for quickly applying a limited amount of the plastic sulfur for confining small spills or laying down a barrier for protecting vulnerable beaches from oil encroachment.

FIGURE 54: USE OF SULFUR FOR COMBATING OIL SPILLS

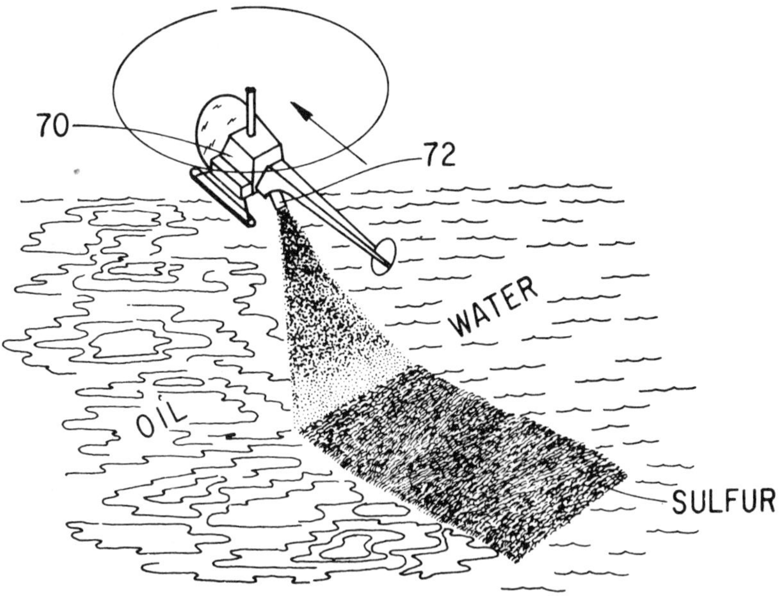

Source: U.S. Patent 3,844,941

A process developed by *W.L. Stanley and A.G. Pittman; U.S. Patent 3,869,385; March 4, 1975; assigned to the U.S. Secretary of Agriculture* involves applying a polyisocyanate and a polyamine to the oil spill. The said reagents react, yielding a polymer which entraps the oil, forming a rubbery gelled mass, and thereby preventing dispersion of the oil. The reagents may be applied to the entire area of the oil spill or, more preferably, to selected parts thereof—for example, to the outer periphery thereof.

In typical practice, the reagents are applied to the outer periphery of the body of oil which constitutes the spill. The application can be accomplished by means of a boat or other vessel which circles the oil spill and concomitantly deposits the reagents on the edge of the oil. Alternatively, the reagents can be applied to the perimeter of the oil spill by low-flying aircraft. To facilitate their application, the reagents may be employed in the form of solutions in kerosene or other inert solvent. Soon after the reagents are applied, a rubber-like gel is formed and this barrier or dam prevents dispersion of the oil. Conventional means such as pumps or skimmers can then be used to harvest the oil which is contained by the so-formed barrier or dam.

A process developed by *G.P. Canevari; U.S. Patent 3,959,134; May 25, 1976; assigned to Exxon Research and Engineering Co.* involves the use of mixtures of C_{10}–C_{20} aliphatic carboxylic acids or the sorbitan monoesters thereof, in combination with nonpolar solvent systems for containing oil slicks floating on a water surface. Oil slicks floating on a water surface may be contracted by applying to the perimeter of the oil slick an effective amount of an oil collection agent as described above, whereby the surface area of said oil slick is reduced. After the size of the oil slick is reduced by means of the oil collection agent, the oil may then be physically removed from the water. The oil collection agents described have the additional advantages of having low toxicities as well as pour points low enough to enable their use during operations where water and/or air temperatures are less than 45°F.

Floating Booms or Fences

Regardless of the particular design, all containment booms are acted upon by the same hydrodynamic forces and consequently lose oil, i.e., fail, in the same general ways. Figure 55, views (a) and (b), illustrate the hydrodynamic instabilities which cause a boom to lose oil in a current. Views (c) and (d) show the physical behavior of booms in situations where both current and waves are present. Such situations can greatly increase the oil loss rate due to the mechanisms of views (a) and (b) and the splashing of oil over the boom freeboard. View (a) shows a mechanism termed Drainage Failure which occurs if the oil slick in front of the boom is deeper than the boom skirt. This failure mode does not happen frequently, unless the boom skirt is pushed up to the surface by the prevailing current as shown in view (c). Loss of oil by a mechanism known as Headwave Failure is illustrated in view (b). This is the most commonly occurring mechanism of oil loss past a boom and takes place when sufficient oil has been stopped on the upstream side of the boom to form a headwave of half-arrowhead shape at the upstream edge of the slick.

Oil loss occurs when water currents relative to the boom reach speeds in excess of 0.75-0.8 knots. At these speeds the action of the water flowing by the stationary oil headwave causes turbulence at the oil-water interface on the downstream side of the headwave. This turbulence causes oil globules to be broken away from the headwave and carried down under the boom skirt by the prevailing current. This oil will resurface again downstream of the boom at a distance depending upon the skirt depth, relative current, oil droplet size and oil density. It should be kept in mind that both the drainage and headwave failure modes result in an oil loss rate (depending upon the oil volume initially upstream of the boom) and that the majority of the oil slick is still being contained on the upstream side.

Consequently, during an oil spill operation, the area in front of a boom that is losing oil should still be worked with skimmers while a secondary containment and skimming effort is deployed downcurrent where the oil resurfaces. The loss rate by headwave failure can be reduced by keeping the arrowhead wedge as small as possible, that is, by rapidly skimming oil from in front of the boom to prevent its damming up and forming a large headwave. Another technique which may be effective is to place sorbent mats at the upstream edge of the slick where the headwave forms. The oil in the headwave will be sorbed into the pores of the mats where the turbulence will be less effective at causing droplet formation and loss under the boom skirt. Figure 55, view (c), shows a boom attitude in a current which can increase the oil loss rate from either the drainage or headwave loss mechanisms. The skirt can be deflected excessively in a current because either there is insufficient ballast weight or the primary tension member is too close to the water line [see view (a) of Figure 55].

FIGURE 55: BOOM OIL LOSS FAILURE MODES

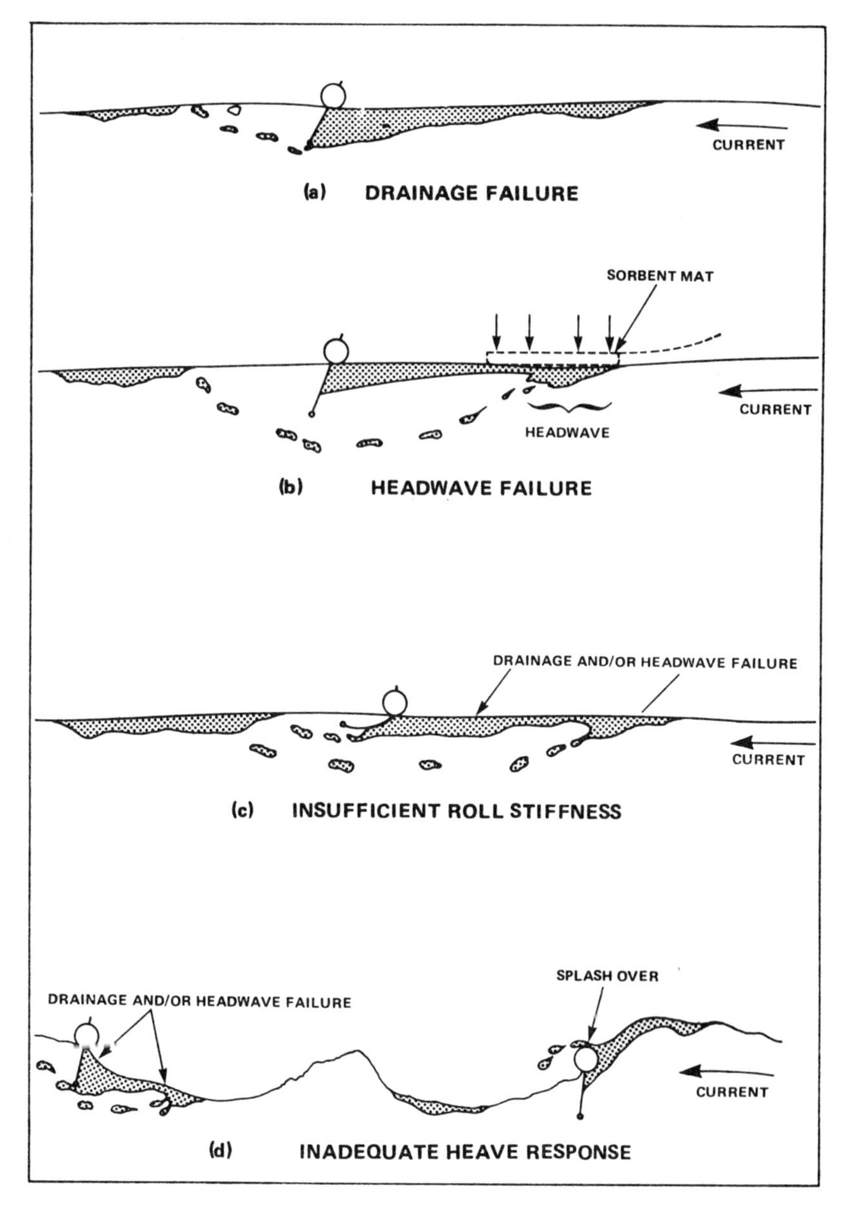

Source: Reference (9)

The sensitivity of a boom to this condition is measured by its roll stiffness, or how many foot-pounds of torque are required to roll the boom to a given heel angle, say, 45 degrees. To prevent the behavior shown in Figure 55, view (c), a boom should have a high roll stiffness. Good roll stiffness is also important to prevent loss of freeboard due to wind loads. Figure 55, view (d), shows oil loss over or under a boom when it has inadequate heave response for the prevailing waves and relative water current. Heave response depends upon the buoyancy available above the still waterline, the boom mass and the float waterplane area. A boom with good heave response is one that can closely follow the water surface as a wave train passes by the boom. Sufficient freeboard is also required to prevent splashover of oil due to short wavelength waves, or wave chop, impacting the boom.

The physical properties of any boom which influence its response in current and waves are illustrated in Figure 56. Sufficient reserve buoyancy (that above the still water mark), freeboard, skirt depth, float waterplane area and low total boom weight are important to good heave response in waves to minimize the loss mechanisms of Figure 55 (d). Sufficient roll stiffness, or the torque required to roll the boom to a given angle, is needed to prevent excessive skirt deflection with subsequent oil loss as in Figure 55 (c). A high roll stiffness can be obtained by sufficient ballast weight and/or location of the flotation members outboard of the boom centerline. The location of the primary tension members (chains or cables) and the resultant force vector holding a boom in a current also have an impact upon the roll stiffness of boom during field use.

FIGURE 56: PHYSICAL PROPERTIES OF A BOOM

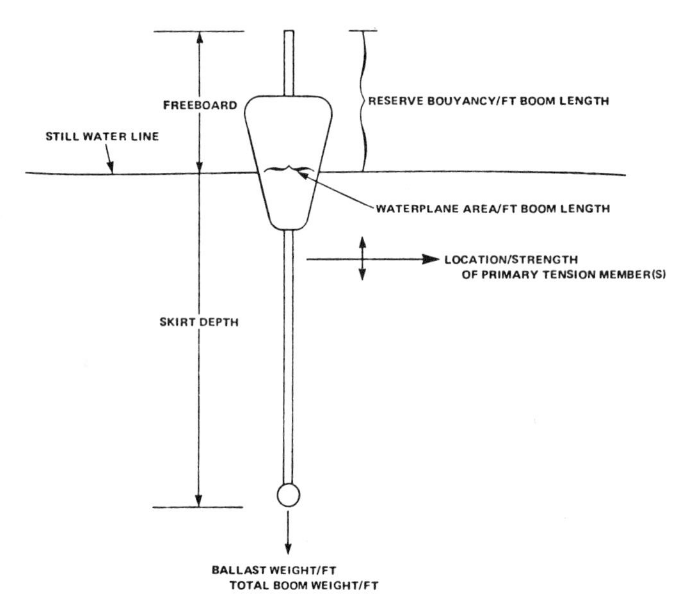

Source: Reference (9)

Figures 57 and 58 illustrate the effect of these properties when a boom is subjected to currents or waves. As shown in Figure 57, hydrodynamic drag on the submerged portion of a boom can cause the loss of skirt depth, Figure 57 (a), or freeboard height, Figure 57 (b). If the line of action of the boom towline force (due to tow or anchoring lines) is above the line of action (center of pressure) of the drag force, as in (a) of Figure 57, the boom must have sufficient ballast weight or float-righting moment, or the skirt will be pushed up toward the surface. Similarly, if the line of action of the boom towline force is below the center of pressure due to drag, as in (b) of Figure 57, the boom must again have sufficient float-righting moment and ballast weight (i.e., roll stiffness), or the freeboard will be lost.

FIGURE 57: BOOM ROLL IN CURRENTS

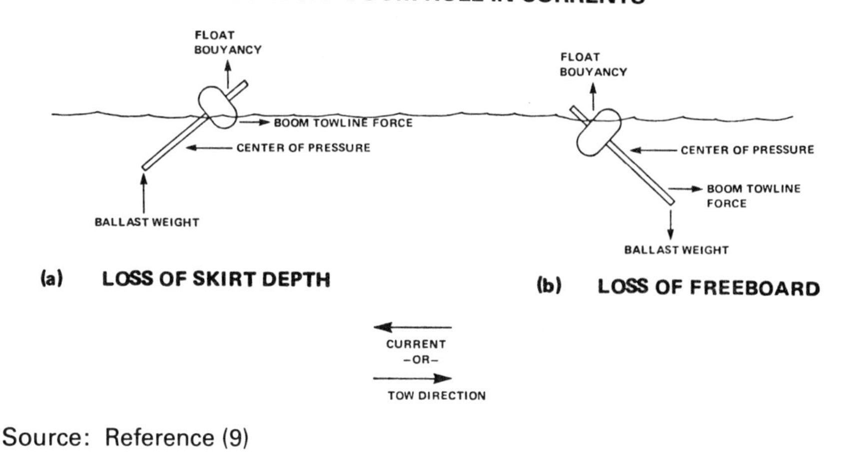

Source: Reference (9)

FIGURE 58: BOOM HEAVE IN WAVES

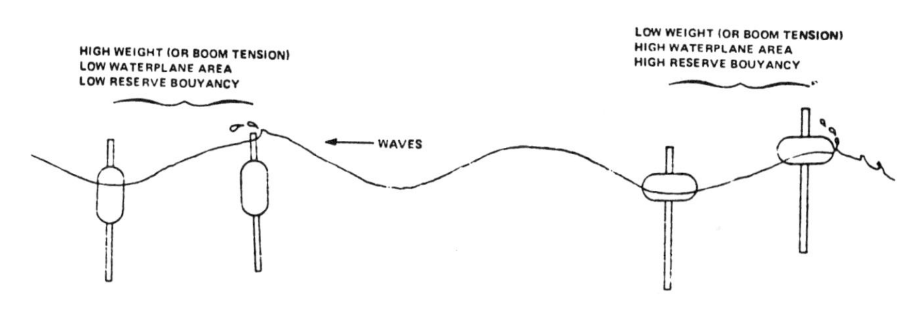

(a) POOR HEAVE RESPONSE (b) GOOD HEAVE RESPONSE

Source: Reference (9)

Since the roll stiffness of a given boom cannot be varied during field use, the skill of operating personnel is required to minimize hydrodynamic drag.

This can be achieved by using the angular deployment techniques discussed in Chapter 3 and/or moving the location of the boom thereby restoring force toward the center of pressure by adjusting the depth of the tow point at the boom ends. Figure 58 is a qualitative schematic of boom heave response. A boom with small waterplane area, high weight or little reserve buoyancy, as in Figure 58 (a), will suffer sluggish response to the change in water surface due to a passing wave. A high tension force in the boom (due to high currents or a too-fast tow speed) has the effect of increasing the effective weight of the boom for heave response.

A boom with a higher waterplane area, lower weight and higher reserve buoyancy, as in Figure 58 (b), will have a better heave response and will minimize loss of oil as shown in Figure 55 (d). As in the case of roll stability, the physical properties of a given boom cannot be changed during field use so operator skill is required to either angle the boom to the incoming waves and/or reduce the tensile load on the boom by angling the boom relative to the current or reducing the boom tow speed (e.g., when boom is towed in a "U" shaped catenary).

The effectiveness of a boom can best be measured by defining requirements in terms of the following criteria:

1) Oil Containment Criteria
    a. Roll stiffness in currents
    b. Heave response in waves
    c. Freeboard height, skirt depth
    d. Total drag under catenary and straight line tows
2) Reliability Criteria
    a. Fabric strength: tensile, tear, puncture, abrasion, weathering, low-temperature flexibility
    b. Overall boom tensile strength
    c. Distribution of tensile loads over boom height
    d. Ease of field repair
    e. Probability of mechanical damage
3) Storage and Deployment Criteria
    a. Storage volume
    b. Weight per foot
    c. Deployment time
    d. Equipment and manpower required for deployment

All booms can lose oil by any one of the four causes shown in Figure 55. However, one boom design can be more effective than another at minimizing the oil lost by these mechanisms. The oil containment criteria are measures of effectiveness for containing oil in view of the four failure modes of Figure 55. Booms with a high roll stiffness, i.e., a large force required to roll them about their still water mark, are better than those with low roll stiffness. A high roll stiffness is achieved either by placing heavy ballast weights on the bottom of the skirt or by moving the flotation off the boom centerline so that the float can exert a large torque to resist the forces trying to roll the boom.

A boom with good heave response in waves is one that will stay on the water surface as waves pass by and not be momentarily submerged by the wave crest. Good boom design takes the properties of roll stiffness into consideration in choosing an adequate freeboard height to prevent splashover of wind driven wave "chop." The freeboard height should be adequate yet low enough to eliminate tipping in a high wind. The selection of skirt depth is likewise a compromise depending upon the roll stiffness desired and the limitations on ballast weights for easy handling. A boom with a high drag force in catenary or straight line tow is not necessarily bad if boom strength is adequate. The total drag force is due to the relative speed between the boom and surrounding water. Where low currents exist, the speed of tow boats can simply be reduced to lower the overall drag force.

In a high current situation, however, tow boats may be of insufficient horsepower to counteract the boom drag, and the boom must be angled to the current or anchored. Also, when using booms with high drag factors, tow boats may lose steerageway at low speeds if the tow boat is not powerful enough. A boom with a low drag force is desired, consistent with the other oil containment criteria of high roll stiffness, good heave response and adequate skirt and freeboard. The primary criteria for boom reliability are the mechanical properties of the boom fabric. A boom fabric must be sufficiently resistant to tearing, puncture and abrasion inasmuch as the boom is dragged over concrete aprons and splintered dock edges during deployment and recovery. High values of ultimate tensile breaking strength and resistance to weathering and brittle cracking at low temperatures are also desirable properties of the boom fabric.

A reliable boom should have adequate overall tensile strength but, equally important, should have this load distributed properly over the boom height. Past field experience has shown that when a boom is designed to have the majority of the tensile load taken by cables or chains, care must be exercised in rigging tow or anchor points lest this load be accidentally transferred to the fabric, causing tearing, or lest the load be applied off the line of action of center of pressure to alter the vertical attitude (Figure 57). It is better to have the entire tensile load carried by the fabric without the need for cables or chains. Ease of field repair and the probability of mechanical damage are also important considerations. Storage and deployment are important aspects of efficient use of boom by naval activities.

The bulkiness of a boom, as measured by its storage volume and weight per foot, has a great impact on transportability of boom within a naval activity and on the time, equipment and manpower required for boom deployment. The Civil Engineering Laboratory, Naval Construction Battalion Center, Port Hueneme, California, has tested different boom stacking configurations and power assist equipment and all-weather containers to store 500 feet of boom. The results of this effort will provide technical requirements for procurement of boom support equipment to give naval activities a greater flexibility in positioning, transporting and deploying boom during oil spill emergencies. All of the effectiveness criteria listed above are used by NAVFAC headquarters in a continuing effort to procure on a competitive basis, for use by naval activities, oil containment booms reflecting the most current level of technology.

While there are a large number of booms commercially available, they can all be grouped into one of the four general designs shown in Figure 59. Three

FIGURE 59: BOOM TYPES

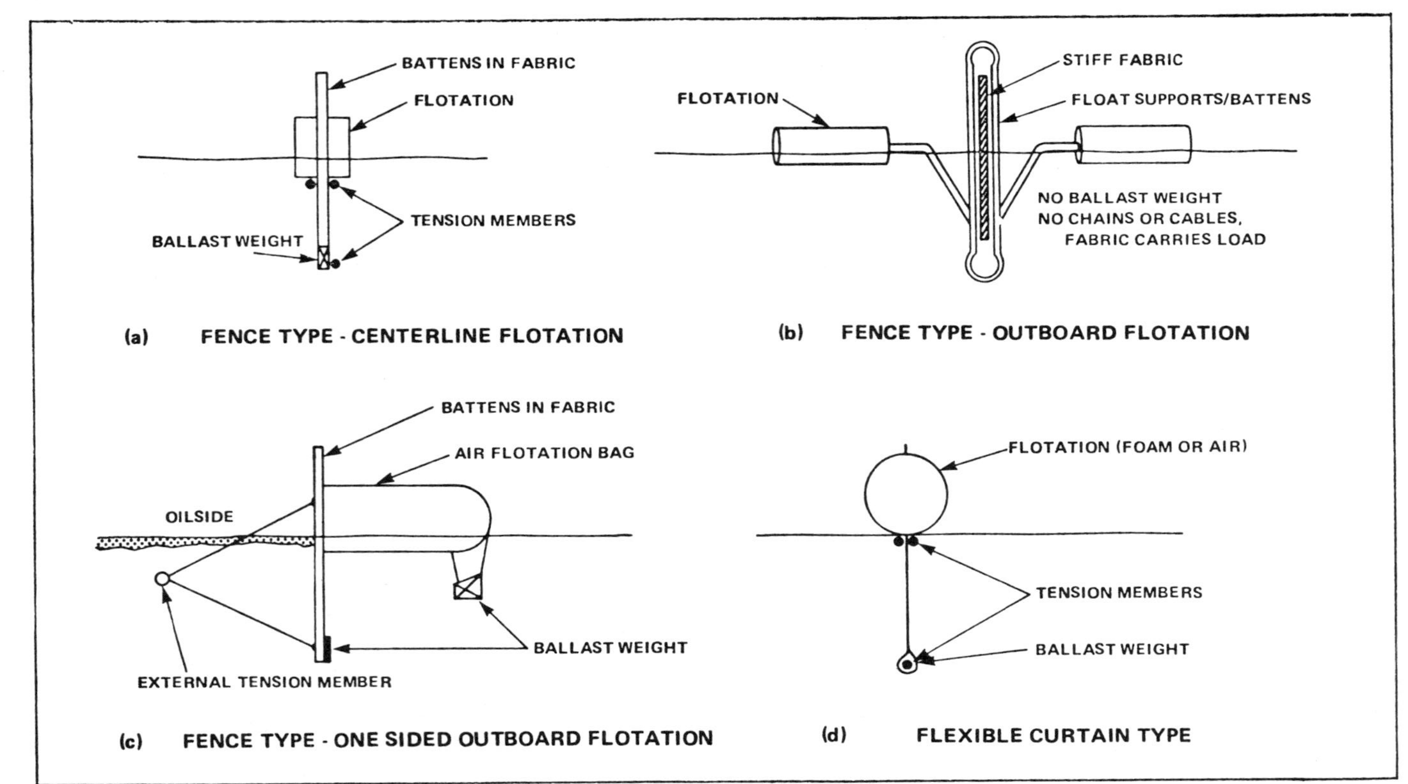

Source: Reference (9)

of the four designs, views (a), (b) and (c), are termed fence-type booms since they use rigid battens to mechanically couple the skirt and freeboard together, causing them to roll as a single unit. In the flexible curtain type boom shown in Figure 59 (d), no battens are used; and the skirt is free to move independently of the flotation and freeboard. The flotation on fence-type booms is attached near the centerline, as in Figure 59 (a), or away from the centerline, either on both sides, view (b) of Figure 59), or only on the nonoil side, view (c) of Figure 59.

Lead, or chain ballast, is required on the design of view (a) to locate the center of gravity low enough for good roll stiffness. With the flotation off the centerline as in view (b), roll stiffness is achieved by the submerging floats through their lever arms and rigid battens directly to the boom skirt and freeboard. No ballast weights are required. The design of view (c) is that developed by the Coast Guard for the specialized operation of a catenary sweeping tow mode (with skimmer operating independently inside the catenary) in a high seas environment. This special design has limited application for Navy use since it is intended for oil to be contained on the nonfloat side only. This cannot be guaranteed in tidal situations.

Also, although location of the primary tension member away from the boom does provide for good boom response in waves, the proper rigging of these loose exterior lines is a major undertaking during deployment and recovery. During a test and evaluation project at the Civil Engineering Laboratory, Port Hueneme, in which the design of view (c) was not tested, the design of view (b) was found to have better roll stiffness and heave response, as evidenced by its ability to maintain a vertical attitude and to contain oil in higher waves and currents, than either of the boom types of views (a) and (d). An additional advantage of the boom type of view (b) is the lack of chains or cables as tension members; the tensile load is carried by the heavy fabric forming the skirt and freeboard.

The flexible curtain type design of view (d) has centerline flotation of either air or solid foam elements. Tension members are placed at the bottom of the skirt and/or just under the flotation near the waterline. Lead or chain ballast at the bottom of the skirt is used to provide roll stiffness, which is much less than either of the designs of views (b) and (c). Proper rigging to place a portion of the tensile load on the chain at the bottom of the skirt is also used to provide roll stiffness.

NAVFAC has defined three sizes of booms for central procurement on a competitive basis using military specification MIL-B-28617B. These booms, their sizes, and the environment in which they are designed to be effective, are listed in Table 20. Class I boom is intended for quiescent water situations where current and waves are low; Class II boom is designed for open water harbor environments with moderate waves and current; Class III boom, with its 12 inches of freeborad, is intended for use in water situations where higher waves are present. Figure 60 is an illustration of the boom being deployed from a dockside container around a spill.

For ruggedness and durability the MILSPEC requires that the boom flotation be closed cell foam or other punctureproof material. For the same reasons, the boom fabric must pass a series of strength tests including overall tensile strength, tear, abrasion and weathering tests. Boom procured under this MILSPEC is subjected to a first article field test to verify its strength and upright stability during straight line and "U" shaped catenary tow modes. An important part of the boom procurement is the associated boom hardware of male-female connectors, floating single point tow assembly and sliding bulkhead attachment. The rigid boom connectors allow sections of boom to be coupled or decoupled while in the water. In this way damaged boom sections can be replaced or sections of different size or length can be brought into service during a spill with minimal manpower and loss of oil.

The tow assembly has a single above-water tow point to facilitate towing, boom repositioning or mooring. The sliding bulkhead attachment allows for an effective leakproof oil seal against a piling, quay wall or other structure where tidal variations exist. It is well known that all booms begin to lose oil at perpendicular currents of approximately 0.7 knots. Therefore, in field use, booms must be angled to the prevailing current to reduce the perpendicular current and minimize this oil loss rate. Because the commercially available boom produced to date has not totally met Navy-wide needs, NAVFAC is currently developing a Navy designed boom for ultimate use. It is expected to be much more durable and effective.

Rapid deployment of containment booms in an oil spill situation often means the difference between a controllable spill and a major pollution incident. To assist in boom deployment, NAVFAC is outfitting specific shore activities, which have been designated as having oil spill cleanup responsibility, with oil spill control utility boats. These 20-foot utility boats are trailer-mounted and have a multiple "vee" fiberglass hull for added stability. They are also equipped with a Sampson Towing Post rated at 6,000 pounds static pull. Each post is fitted with an expanded metal safety screen to prevent injury to personnel should a tow line part. The boat is powered by an 85-horsepower outboard motor with remote controls located on an operator's console; it comes equipped with standard USCG-approved safety equipment including paddles, boat hook, dry chemical fire extinguisher, first aid kit, emergency flares, anchor, life preservers, and foghorn. Energy absorbing fenders have also been provided to lessen any damage sustained in the severe environment of oil spill cleanup operations.

TABLE 20: NAVFAC BOOM TYPES

Class	Skirt Depth (inches)	Freeboard (inches)	Current Velocity Perpendicular to Boom (knots)	Wind Velocity Perpendicular to Boom (mph)	Wave Height to Length Ratio
I	8	4	1.0	15	0.08
II	16	8	1.5	20	0.08
III	24	12	2.0	25	0.08

Source: Reference (9)

At this point, a number of proprietary boom designs will be described and illustrated.

FIGURE 60: OIL SPILL BARRIER SYSTEM COMPONENTS

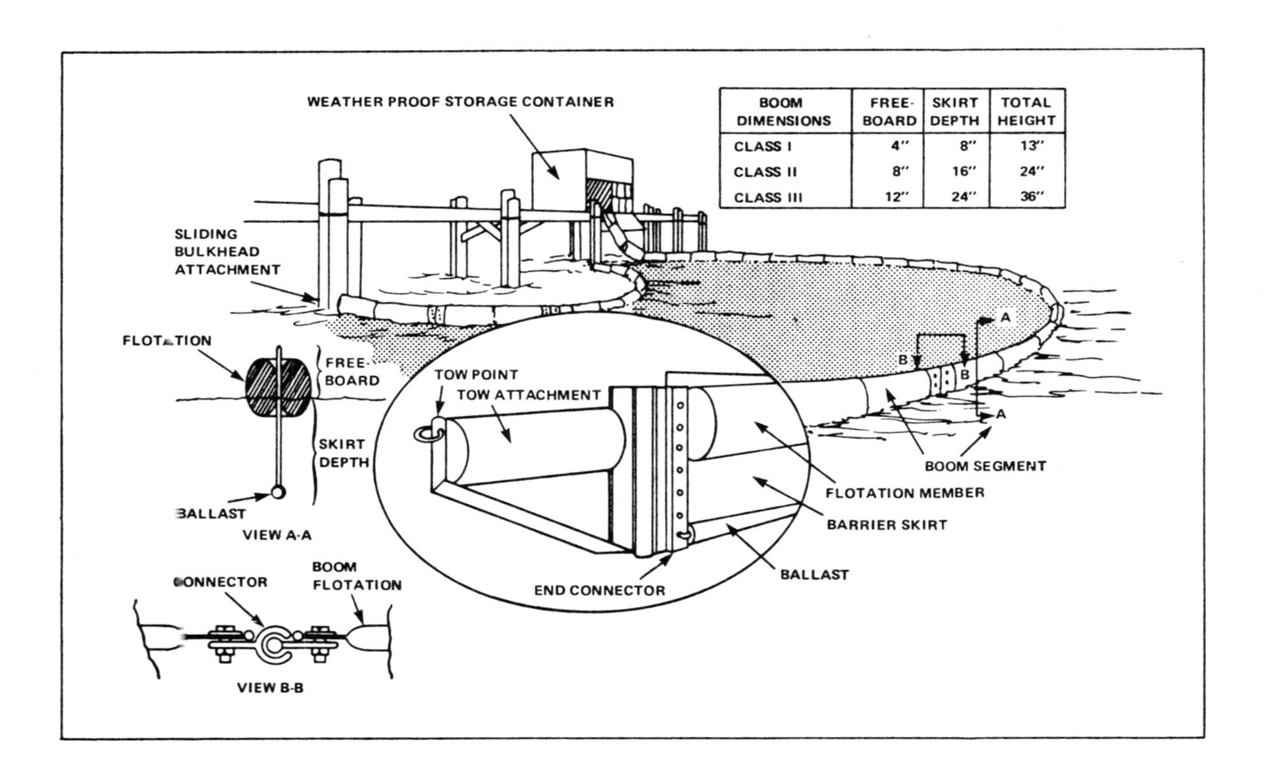

BOOM DIMENSIONS	FREE-BOARD	SKIRT DEPTH	TOTAL HEIGHT
CLASS I	4"	8"	13"
CLASS II	8"	16"	24"
CLASS III	12"	24"	36"

Source: Reference (9)

A design developed by *P. Preus and J.J. Gallagher; U.S. Patent 3,783,621; January 8, 1974* provides a barrier for substances floating on water having a flotation member and a liquid pervious and a liquid impervious skirt depending therefrom. The liquid impervious skirt is deflected at currents greater than about one knot and the oily substances may be treated with a particulate oleophilic-hydrophobic substance less dense than water for retention by the liquid pervious skirt.

As shown in Figure 61, the barrier, shown generally at **10**, comprises a flotation collar **12** with a liquid impervious skirt **14** and a liquid pervious skirt **16** depending therefrom in parallel-planar, coextensive relationship. The flotation collar may be an inflatable tube as illustrated or may be of a flotable material such as polystyrene or polyethylene foam. The skirt **14** may be made of any liquid impermeable material having suitable resistance to the environment in which the barrier is likely to be used and suitable flexibility to follow wave and current action and to allow handling and storage thereof.

FIGURE 61: PREUS-GALLAGHER OIL SPILL BARRIER DESIGN

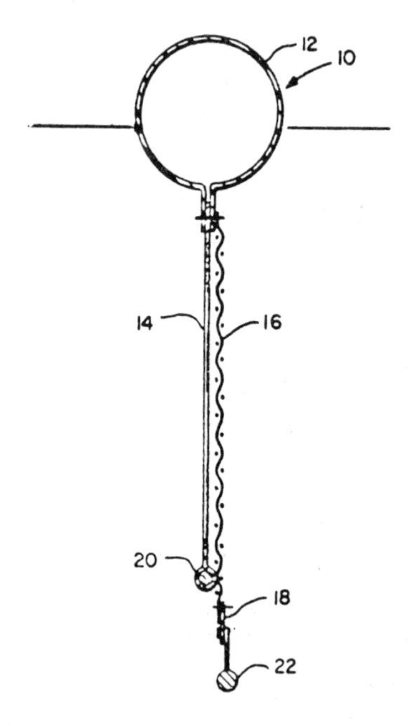

Source: U.S. Patent 3,783,621

The skirt **16** is fabricated from a liquid pervious or permeable material such, for example, as a net having a mesh of between one-eighth inch and one-fourth inch. Such a skirt may be reinforced with horizontal and vertical structural members (not shown) of some strong, flexible material such as nylon webbing or the like. Any material permeable to water, whether perforated sheet material or woven netting, which has suitable flexibility and is compatible with the environment in which the device is to be used is suitable for use in the skirt **16**. A horizontal strip of webbing **18** provided with grommets is attached to the lower edge of the skirt **16**, and a series of ballast elements **22**, suitable to maintain the skirt **16** in a horizontal configuration against all current action and water force are attached to the lower edge of the skirt **16**.

The skirt **14** has, formed along the lower edge thereof, a ballast member **20**. This ballast member may take the form of sand, lead shot or segments of heavy material either incorporated within the skirt or added onto the lower edge, as desired. The size and disposition of the member **20** is important in that water tends to vortex beneath liquid impermeable skirts at current velocities much in excess of one knot normal to the skirt. The ballast member should therefore be designed to maintain the verticality of the impermeable skirt **14** at current velocities normal to the skirt below about one knot. Such design, dependent upon factors such as density and quantity of the ballast material, height and inherent weight of the skirt, and the total presented area of the skirt, is well within the capabilities of those skilled in the art.

The ballast elements **22** should be of sufficient weight to maintain the verticality of the permeable skirt **16** under all current and sea conditions. Due to the permeability of the skirt **16**, and the reduced water forces acting thereon, it is conceivable that, with certain design parameters, the ballast elements **22** and the ballast members **20** could be of about the same unit weight and still accomplish their respective objectives.

A design developed by *C.L. Gambel; U.S. Patent 3,783,622; January 8, 1974* is a rigid barrier unit for use in assembling an enclosure around a surface area of a body of water which is provided with adjustable buoyancy and ballasting chambers so that a nearly neutral buoyancy condition can be established with substantially all of the mass of the unit below the turbulence level of a body of water in which the barrier unit is placed. A method of deployment of such units involves a floatation of the units in horizontal attitudes to the area to be enclosed, followed by a ballasting of the units into vertical attitudes so as to extend around an oil or chemical spill area. The elements of such a design are illusted in Figure 62.

The individual barrier units **10** are linked together by linking means **12** carried at the ends of each barrier unit **10** so that a series of such barrier units can be coupled or linked together in end-to-end relationships. The relative positions of the barrier units show the high degree of flexibility of the linking means **12** about vertical axes at points of coupling of adjacent barrier units. Typically, a long chain of barrier units would be linked together and drawn around a water area to be enclosed to form an encirclement or enclosure. The dashed line illustrates a mean waterline which for purposes of this discussion may be considered as the surface level of a calm body of water. The barrier system is intended to float in a generally vertical orientation in a body of water with a substantial portion of the mass of the system below the surface level of the body of water. The purpose in this arrangement is one of stabilizing individual barrier units of a series against turbulence encountered just above and below the mean waterline so that the preferred orientation of the barrier system is maintained over a wide range of sea conditions.

FIGURE 62: GAMBEL RIGID BARRIER DESIGN AND MEANS OF DEPLOYMENT

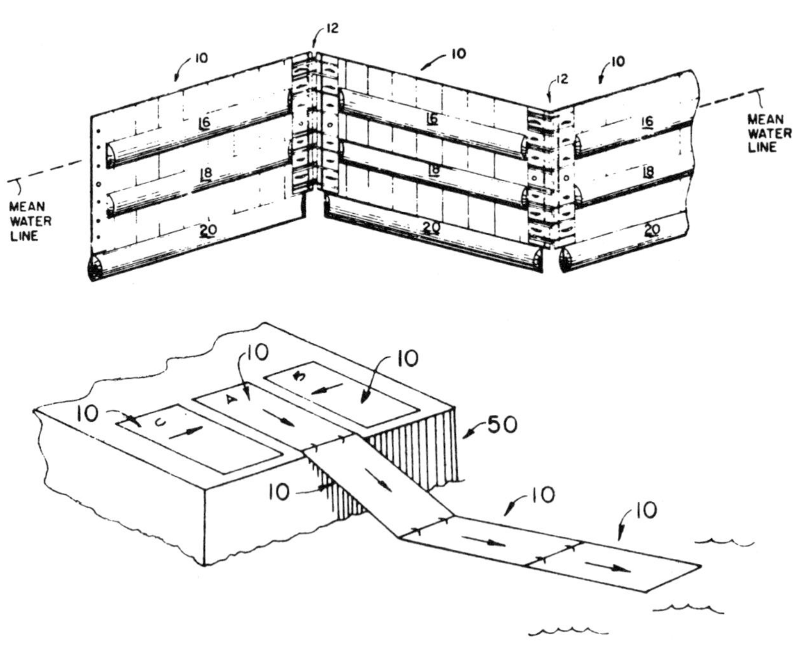

Source: U.S. Patent 3,783,622

Generally, each barrier unit **10** comprises a relatively rigid member having a rectangular shape and which is of a sufficient size and strength to maintain its shape and integrity against wave action and other sea conditions where it is to be used. The main body of construction, and the main body section is fabricated from plate steel, aluminum, or other metal by known techniques. The bulkhead may be constructed as a single layer of plate metal to which reinforcing ribs **14** are welded or secured, or it may be formed as a double walled structure for its below-water portions with a single plate extending upwardly above the water as a wave and sea shield.

Each barrier unit **10** is provided with adjustable buoyancy means for controlling floatation and ballasting of the barrier unit. The function of the adjustable buoyancy means is to permit establishment of a nearly neutral buoyancy for the entire barrier unit **10** with substantially all ballasting of the unit carried at a level which is below the turbulence level of a body of water in which the barrier system is to be utilized. The buoyancy means comprises a series of airtight buoyancy chambers or tanks **16, 18** and **20**. It has been found that a series of three such tanks, as illustrated, provides for desired controls of buoyancy, attitude, and ballasting for the barrier units **10** if the buoyancy tanks are arranged at different levels along the length of the main body of the barrier unit, as shown. The preferred arrangement provides for an uppermost buoyancy tank or chamber **16** positioned close to the expected waterline (which for purposes of calculation would be the surface level of a calm sea) so that its major mass is just below that waterline.

A second buoyancy tank or chamber is located at an intermediate level between the uppermost tank **16** and the position of a third buoyancy tank **20** carried at the lowermost level of the barrier unit **10**. Each of the illustrated buoyancy tanks is fabricated from metal or other material to provide an airtight cylinder. Valving and control means are provided for flooding and blowing the interior space of each cylinder in accordance with known techniques. Flooding may be accomplished by water inlets (not shown) which include known valve means for being opened and closed by remote control, and blowing is provided by air inlets in the tanks connected to a source of pressurized air by way of hoses or conduits which can be interconnected and controlled between individual barrier units of a system of such barrier units deployed in a body of water. Of course, means are provided for venting the individual buoyancy tanks during flooding, and water is exhausted from the tanks when they are being filled with air.

The lowermost buoyancy tank **20** may be considered a ballasting means for the entire barrier unit **10**. The lower tank **20** is typically completely flooded when the barrier unit is deployed and oriented to the vertical orientation, and this provides for a very low level of ballasting of the entire barrier unit at a level substantially below the active interface between the body of water and the atmosphere. The upper buoyancy tanks **16** and **18** function to control buoyancy and floatation of the barrier unit so that a nearly neutral buoyancy can be established for the entire unit. With these features of low level ballasting and neutral buoyancy control, it is possible to construct a rigid barrier unit **10** of substantial size and strength which can be stabilized in very active sea conditions to maintain a preferred vertical orientation of a series of such barrier units linked together. The lower part of the figure illustrates means for deployment of such a barrier.

A barge or other surface vessel **50** can carry stacks of individual barrier units **10** for deployment onto the surface of a body of water. The stacks of barrier units can be carried on a deck of the vessel **50**, and one arrangement would provide for a lower level deck storage and a launching ways astern of the vessel. Typically, racks of barrier units could be arranged abreast of each other as shown so that individual barrier units can be moved to a centerline position for connection to the last of a string of barrier units being deployed. Once a series of such barrier units are deployed, they can be towed to the enclosure area and ballasted to a vertical orientation for final movement around the oil spill or whatever area is to be protected. Preferably deployment will be carried out in a manner to avoid broadside buffeting of any given barrier unit during placement and reorientation.

Alternatively, the individual barrier units may be linked together in calm waters and towed to an offshore site in their horizontal attitudes until the site of a spill is reached. Then, the lead barrier unit can be anchored, and the ballasting tanks **20** of all units flooded so as to change the orientation of the series of articulated units. Pneumatic and other control connections are made when the individual units are linked together and prior to positioning at the enclosure site. When the lead barrier unit has been anchored, the last unit of the series can be towed around the enclosure area for a linkup with the lead barrier unit. This linkup should be made at an upwind area of the site, and it is feasible to keep the enclosure open until oil accumulates in the enclosure area and so that service equipment can move in and out for salvaging and recovery operations. Typically, the barrier units will be anchored once the enclosure is completed.

When the containment operation is completed, the units can be unlinked and their ballasts blown by displacing water from the ballasting tanks with air under pressure. Air can be supplied from a source common to a number of units, or each may carry its own supply of pressurized air for purposes of blowing ballast.

It is further contemplated that in the event of extreme weather and sea conditions, all of the buoyancy tanks of the barrier units can be flooded for sinking the entire barrier system to avoid loss or damage to the barrier units making up the system. Pneumatic lines can be buoyed to the surface for this purpose so that the buoyancy tanks can be blown after a sinking operation to raise the barrier system.

A design developed by *T. Muramatsu, K. Aramaki and Y. Kondo; U.S. Patent 3,786,637; January 22, 1974; assigned to Bridgestone Tire Co., Ltd., Japan* is an oil fence comprising an elongated resilient belt member and a float secured to it for keeping at least a part of the belt member floating above water level. The floating portion of the belt member is kept substantially upright by a weight means and reinforcing means. The oil fence so flexes as to conform with the profile of water surface.

A standard oil fence consists of a flexible tubular float, a wall member depending from the lower edge of the float, and a weight member secured to the lower end of the depending wall member for spreading the wall member in the water. With such conventional oil fence, the tubular float member extends above the water level for preventing the floating drifts from overflowing across the oil fence, and the depending wall member acts to prevent the floating drifts from passing underneath the oil fence together with wave and turbulence.

The conventional oil fence has a shortcoming in that in order to ensure effective prevention of overflow of floating drifts across the fence, it is necessary to provide a high blocking wall above water level by using a tubular float of large diameter. The increased diameter of the tubular float makes the oil fence bulky and difficult to handle. Furthermore, with the increased diameter, the time necessary for inflation and deflation of the tubular float increases. To provide an oil fence covering a large area, a number of oil fence sections must be joined, and if tubular floats of large diameter are used in the oil fence sections, joint means for connecting the adjacent sections of tubular float becomes bulky and heavy. Such heavy joining means acts to further reduce the buoyancy of the float, so that the float diameter must be increased again to compensate for the weight of the heavy joining means. This is a kind of vicious cycle, which results in a very large float diameter.

Another shortcoming of the conventional oil fence with a tubular float of large diameter is that, as the float diameter increases, the oil fence becomes less flexible, and the oil fence may not flex along the profile of the water surface, so that the top of the oil fence may be partially washed by wave and turbulence. As a result, floating drifts may overflow across the oil fence at such washed portions. Figure 63 shows a rough schematic of the construction and function of the improved Bridgestone design.

FIGURE 63: SCHEMATIC OF CONSTRUCTION AND OPERATION OF BRIDGESTONE TIRE CO. BARRIER DESIGN

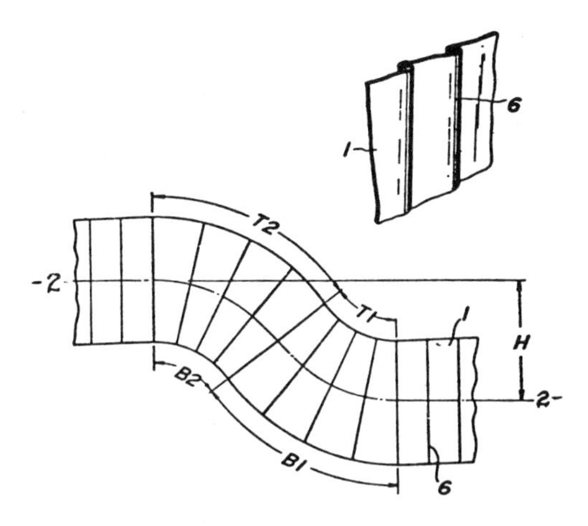

Source: U.S. Patent 3,786,637

The view at the upper right of the figure shows a part of the belt member **1**, in which a plurality of folds or wrinkles **6** are formed in the belt member **1**. The folds or wrinkles **6** provide allowance to the belt member **1** to flex in response to change of the water surface profile caused by waves and turbulence. More particularly, let it be assumed that a part of the belt member **1** is kept substantially vertically in water with the top edge held above water level with the bottom edge dipped in water and that the left-hand portion of the belt member **1**, is raised by a height **H** relative to the right-hand portion of the belt member, for instance, due to wave in a sea, as shown in the lower view in the figure. The dash-dot line in the lower view in the figure represents the position of tubular float member. In the intermediate portion between its comparatively high left-hand portion and its comparatively low right-hand portion, the belt member **1** flexes, and such intermediate portion has two parts: namely, a right-hand part includes a compressed top edge **T1** and a stretched bottom edge **B1**, and a left-hand part including a stretched top edge **T2** and a compressed bottom edge **B2**.

Such compression and stretch of the top edge and bottom edge of the belt member **1** are exaggerated in the lower view in the figure. It is an important feature that, at any part of the belt member **1**, the top edge and the bottom edge can flex in an independent fashion; for instance, the top edge portion **T1** shrinks while its corresponding bottom edge portion **B1** stretches, and another top edge portion **T2** stretches while its corresponding bottom edge portion **B2** shrinks, as shown. Such independent flexibility of the top edge and the bottom edge of the belt member **1** is indispensible in order to use the belt member **1** as an oil fence component, because it provides for the flexing of the belt member **1** along the water surface profile. For brevity, the aforesaid flexing of the belt member **1** along the water surface profile will be referred to as the "wave-profiling flexibility."

The upper portion of the belt member **1** extends above the water level, which upper portion is continuously exposed to winds, waves, direct sun beams, and collison with floating drifts. Thus, the elongated belt member **1** should be made of a resilient material having a high weather-resistivity, a high abrasion resistivity, and a high corrosion resistivity to floating drifts which are expected in the service of the oil fence. A high resistivity to aging is, of course, one of the preferable properties of the elongated belt member **1** of the oil fence.

On the other hand, the lower portion of the elongated belt member **1** is dipped in water, e.g., sea water, so that the belt member **1** should resist against corrosion by sea water and various chemicals dissolved in the water. In addition, the belt member **1** is required to resist physical and chemical attack by fish and the like in water. The belt member **1** is preferably repellent to shellfish and the like, because any deposits of comparatively heavy matter, such as shellfish, on the surface of the belt member **1** tends to reduce the buoyancy.

Thus, typical examples of the material for the elongated belt member **1** are vinyl chloride and neoprene. It is also possible to produce an elongated belt member with desirable properties by applying a suitable coating, such as shellfish-repellent paint.

The reinforcement of the belt member **1** in the direction perpendicular to its longitudinal direction can be accomplished by embedding cords or by forming folds or bent portions including corrugation. Such reinforcement of the belt member **1** can also be accomplished by forming suitable ribs in it; for instance, by forming solid thick ridges or linear gas-filled cylindrical portions extending at right angles to the longitudinal direction of the belt member.

As regards the material for the tubular float member **2**, one or two flexible rubber tubes reinforced by rubberized cords are used in the foregoing embodiment, but the process is not restricted to such materials for the tubular float member. For instance, soft vinyl chloride tubes with suitable reinforcement or light synthetic resin blocks may be used for purposes of providing the buoyancy necessary for floating the elongated belt member **1**. The reinforcement of the tubular float member **2** may be made by using metallic wires.

In order to prevent the float member **2** from being punctured due to collision with foreign matter, e.g., sea drifts or floating buoys, it is desirable to provide suitable protective covering to the float member, at least at those portions which are most frequently exposed to such collisions.

A barrier design developed by *J.A. Sayles; U.S. Patent 3,792,589; February 19, 1974; assigned to Chevron Research Co.* is particularly designed and constructed to be of relatively light weight and to have a compact form when not in use so that it may be stored in a relatively small space. It is designed and constructed with a smooth, tough exterior surface to be handled conveniently by manpower to be set out in operating condition in the water and to be retrieved therefrom without unduly exposing the men to injury by sharp or projecting parts or hazarding damage to the boom structure by snagging such parts. Particular attention is given in the design and construction of the boom to cause the stresses to which it is exposed during use to be imposed on stress members incorporated in the boom for this purpose while preventing undue stress from being concentrated, accidentally or otherwise, on portions of the boom which could be damaged or destroyed by such stresses.

Provision is made for joining boom sections together by a means which will prevent the leakage of contaminants between successive sections to form thereby a barrier of the desired length necessary to collect the contaminants in a spill which may be spread extensively over the surface of the water.

The boom employs inflated air chambers as the buoyant medium and provision is made for maintaining the effectiveness of the boom as a barrier even though some of the air chambers may accidentally be deflated.

To achieve these goals, the boom is constructed of a single sheet of reinforced rubber-like material which is folded along a longitudinal median to form the two sides of it. A tensile stress cable, stress plates, ballast weights, and, in one modification, stiffening members and end plates, are placed between the two sides of the boom in appropriate operating relationship and the sides are then joined together and to at least some of the elements placed between them throughout the area of the sides except for the portions along the top edge of the boom which will form the floatation chambers. However, in the preferred embodiment the sides are vulcanized together and to at least some of the elements enclosed between the sides. Therefore, in the following description vulcanize will be used by way of illustration but not as limitation on the means for joining the parts in coherent relationship.

Preferably a series of relatively short inflatable chambers are formed along the top edge of the boom and each chamber has a valve at one end which enables it to be inflated and deflated. The opposite sides of the boom are vulcanized together between flotation chambers to form the chambers as separate air pockets and this construction gives the boom a degree of stiffness in a lateral direction from the top edge to the bottom edge which will function to hold the top edge of the boom above the surface of the water in a region where an air chamber may inadvertently become deflated by being punctured or otherwise damaged.

For some conditions of use, particularly in rougher water, it will be desirable to incorporate stiffening elements in this lateral region between air chambers to insure that a damaged section will remain above the surface of the water and be effective as a barrier.

When the boom is to be retrieved aboard a vessel, or pulled upon a wharf, or otherwise recovered, the successive air chambers are deflated as they approach the retrieving station so that the boom collapses into a flat sheet of material which, as mentioned above, has a smooth, tough surface free of protruding elements, such as bolts, weights, cables, and so forth, throughout the major portion of its length. Hence, the boom can be drawn over hard edges of wharfs or ships without danger of such protruding elements being caught on them and tearing or otherwise damaging the boom structure. Figure 64 illustrates the construction of this type of barrier.

FIGURE 64: CHEVRON RESEARCH CO. FLOATING BARRIER DESIGN

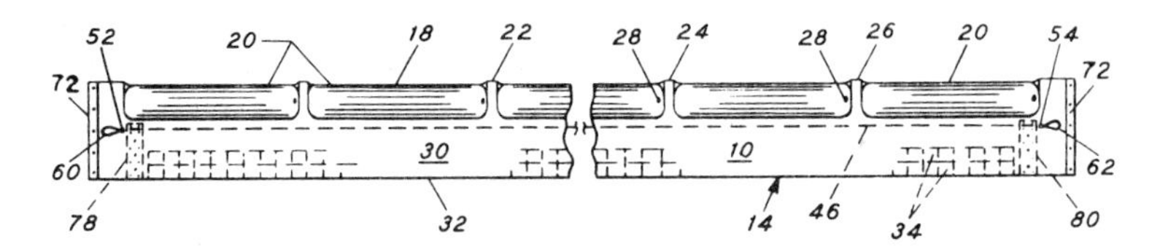

Source: U.S. Patent 3,792,589

As illustrated, the sides **10** of the boom **14** are vulcanized together at a location spaced laterally apart from the top edge **18** of the boom and extending for a distance along it parallel to the edge **18** to form an air chamber **20**. The sides also are vulcanized together at spaced locations, as **22, 24** and **26**, along the length of the boom to form separate air chambers **20**, each of several feet length. Each air chamber has its complementary valve means **28** through which the chamber may be inflated with air and through which the air may be released from the chamber to deflate the boom.

The air chambers form the buoyant means which causes the boom to float on the surface of a body of water and to extend sufficiently above the surface to form a barrier for contaminants floating on the surface. The sides of the boom which extend below the flotation chambers **20** are vulcanized together and to elements incorporated between them which are described hereinafter to form the skirt **30** of the boom which remains submerged in the water during use.

The bottom edge portion **32** of the skirt **30** has incorporated in it ballast weights **34**. These weights preferably are formed from segments of thin lead sheeting which are distributed along the bottom edge portion of the skirt as illustrated.

Although the impregnated fabric of which the boom is made is tough and durable for ordinary handling and shipboard use, it is not designed to withstand directly the high tensile stresses which will be imposed on the boom when it is deployed in the water and exposed to wave action and strong wind and water currents. Therefore, means are incorporated in the boom structure for accepting and distributing the stresses imposed on it in a manner to prevent damage to the fabric of the boom To this end a tensile cable **46** is placed in the skirt **30** of the boom between and enclosed by the sides **10** thereof which are vulcanized to it.

The cable is located below the air chambers **30** and placed approximately along the longitudinal median of the boom and extends within the skirt longitudinally of the boom to adjacent each end, respectively, thereof. At a location approaching a respective end of the boom the cable is extended outwardly from between the wall of the skirt **30** through a respective opening **52** and **54** in one side **10** of the boom as illustrated to provide ends of the cable which are accessible from the exterior of the boom. Each terminal end of the cable is formed as an eye or loop **60** and **62**, respectively, to which a similar eye of a corresponding tensile cable of another section of boom can be connected as by a bolt and nut.

Strips of stiffening material such as strips of synthetic resin which is resistant to the effects of water is secured to each side of each end flap of the section of boom as illustrated at **72**, and the end of the boom in these regions are pierced by holes which provide means for fastening two sections of boom together.

It has been found that the tensile cable which, as described above, is embedded in the skirt **30** can, under adverse conditions, be brought to bear against the side of the boom with sufficient force to cause the cable to tear out of the skirt of the boom in a direction normal to the axis of the cable, thus damaging the boom. The boom is particularly exposed to such damage if it snags while it is being towed or is being lifted by a tensile cable out of the water. It is an important part of this process to provide a construction which will prevent such damage from occurring. To this end a respective stress plate **78** is placed in the skirt of the boom adjacent each corresponding end portion thereof and the end portions of tensile cable **46** are securely fastened to a respective stress plate so that the forces which tend to tear the stress cable from the skirt of the boom will be transmitted to the stress plate and distributed over an area of the skirt sufficient to prevent such forces from being concentrated destructively where the tensile cable bears against the side of the boom.

A technique developed by *P. Preus; U.S. Patent 3,795,315; March 5, 1974* is one in which an oleophilic-hydrophobic fibrous substance is introduced into the slick to absorb the oil and render it impenetrable of the skirt.

It is used in conjunction with a barrier design such as that described by Preus et al in U.S. Patent 3,783,621, described earlier in this section.

The substance is one which is oleophilic or absorbs oil and which in itself cannot pass through the openings in the barrier segments. Such a substance must also be water repellent (hydrophobic) or at least non-water-absorbent such that it will continue to float for an indefinite period time. A material which suitably meets these requirements is a fibrous compound of expanded perlite with clays and fibrous material (Sorbent Type C, Clean Water, Inc.). This material is spread on the water with the barrier and acts to absorb and substantially coagulate the oil such that it cannot pass through the openings in the barrier segments thereby providing for a reduction of hydrodynamic forces on the barrier through the water permeability thereof while providing for retention and "filtration" of oil within the confines of the barrier.

A boom structure developed by *P.O. Oberg; U.S. Patent 3,798,911; March 26, 1974; assigned to Sanera Projecting AB, Sweden* includes a plurality of elongated sections which can be interconnected to permit formation of a boom structure of selected length. Each boom section includes an elongated bouyant member provided with internal partitions for dividing same into a plurality of isolated bouyant compartments. The bouyant member, and the compartments contained therein, can be filled with air to provide the boom with the desired bouyancy. A weighted curtain is secured to the bouyant member for suspension into the water. The bouyant member can be collapsed to permit compact storage of the boom sections. Resilient expander devices are disposed within the bouyant member for permitting expansion or collapsing of same.

Figure 65 illustrates in perspective and partially in section a portion of an

FIGURE 65: SWEDISH FLOATING BOOM DESIGN SHOWING BOOM IN NORMAL POSITON (ABOVE) AND SUBJECTED TO WIND AND WAVE PRESSURE (BELOW)

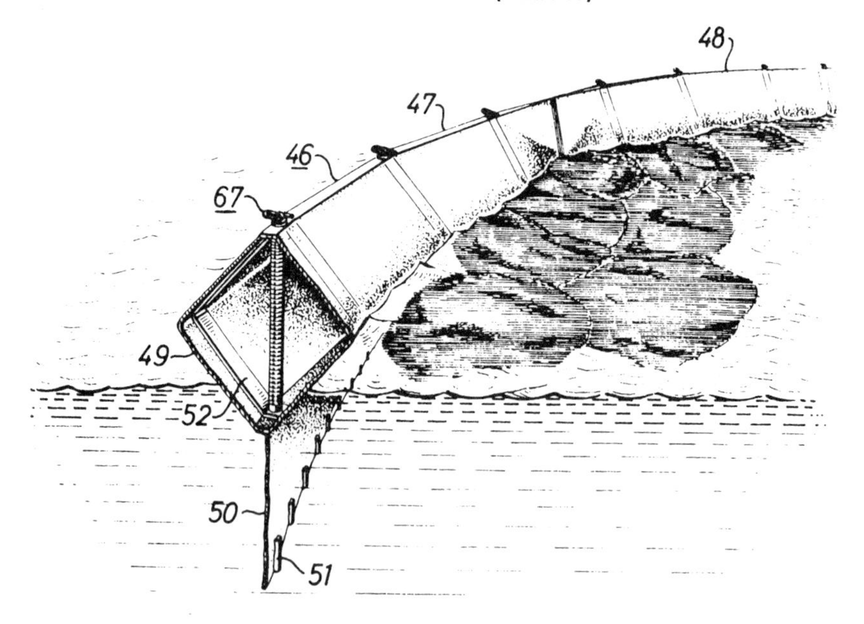

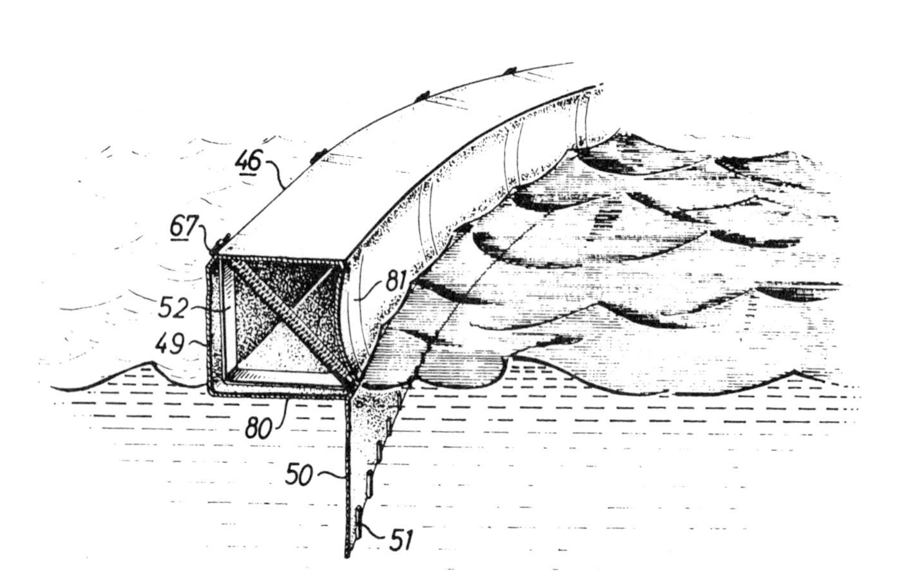

Source: U.S. Patent 3,798,911

expanded boom placed on the surface of the water. The figure shows the normal position of the boom, on the surface of the water when not subjected to lateral forces as a result of wind or wave pressure as well as the position of the boom when the boom is subjected to lateral forces resulting from wind or wave pressure.

The reference numeral **46** indicates generally a boom consisting of a number of sections joined end to end, two sections **47** and **48** being shown. As previously described, each section **47, 48** includes a buoyant body **49** and a curtain **50** having attached at the bottom thereof weights **51**.

Arranged within each section **47, 48** in the manner aforedescribed are expander means **52**. The upper view illustrates how the boom behaves in quiet weather, while the lower view illustrates how in rough weather the boom is liable to tilt and lie on one side thereof, while at the same time another side of the boom is pushed inwards by wind and wave pressure. The thus occuring forces attempt to compress the hose, and the expander means **52** must be so arranged that compression of the hose is prevented. At the same time, however, it must be possible to eliminate the effect which counteracts the compression of the boom in a simple manner, so that the boom can be easily rolled up.

The expander means **52** is provided with a latching means, generally indicated at **67,** which prevents the expander means, and thereby also the boom **46**, from being pressed together as a result of lateral wind and wave pressure forces.

The quadratic or rhomboidal shape of the boom **46** in the expanded position is advantageous with respect to expansion of the boom but also affords a significant extra effect with regard to the ability of the boom to hold oil or other impurities of the water surface enclosed. This extra effect is illustrated in the lower view in the figure, which shows how a boom has been capsized by wind and wave pressure and/or while being towed and lies on a substantially flat side **80**. An adjacent side **81** is subjected in this position to the pressure from wind and waves and becomes somewhat arched as illustrated in the figure and, together with the curtain **50** forms a screening wall which prevents the impurities and the water driven against the wall **81** from breaking over the same.

A device developed by *B.J. Dubois; U.S. Patent 3,798,913; March 26, 1974; assigned to Gamlen Maintre SA, France* is characterized by providing, in combination with a floating body of the type used to control oil slicks and the like hollow outboard stabilizing means open at the bottom thereof and being movable from a normally outboard, stabilizing position during use to a closed position for storage.

A barrier design developed by *D.P. Hoult and J.H. Milgram; U.S. Patent 3,802,201; April 9, 1974* is particularly designed to contain oil floating on the surface of the sea at a location in which relatively rough seas can be anticipated.

When a barrier, comprising a vertical sheet of fabric of length L and draft D is placed in current, the drag force of the current causes the barrier to assume a roughly parabolic shape, and the barrier carries a certain tension. The tension is transferred to the moorings of the barrier in such a manner as to balance the net drag force acting on a barrier. This tension can be substantial; for a barrier 1,000 feet long and 3 feet deep in a 2 knot current with the ends of the barrier on a line perpendicular to the current and about 400 feet apart, the tension is about 10,000 lb.

If the tension is carried in the vertical sheet of the barrier, then, as the barrier rolls in a wave field, first the lower edge, and then the upper edge, of the barrier becomes slack. Thus, the location of the tension carrying force shifts from the lower to the upper edge of the barrier. This constantly shifting tension force destabilizes the barrier, causing it to roll over in the waves.

In view of the foregoing, it is desirable to provide a barrier of the type described which is hydrodynamically stable in strong currents and high waves. (A typical barrier of the design described below, having a 2½ foot draft, is stable in waves 5 or 6 feet high, and currents in excess of 2 knots.) Figure 66 shows one form of such an improved barrier design.

The cylindrical flotation elements **16′** are air inflated by compressed air bottles **118** and are attached to the barrier sheet **22′** adjacent support members as by a flange **120** on flotation element **16′**.

Barrier curtain **22′** is provided with support members which surround a flotation element **16′** in the form of a rigid ring **150** having a lower rigid extension arm **152** and an upper rigid extension arm **154**, extending from the top to the bottom of the barrier curtain and attached thereto. Ring **150** is attached to flotation element flange **120**, providing a sufficiently rigid connection so that tie elements are not necessary.

A slack controlling cable **132** is strung between the series of pairs of support members or on the lower extensions **152** of the ring supporting members. Cable **132** is attached to each such member in the lower portion thereof. Cable **132** is located about 0.45 D (D, as defined above, being the draft of the barrier) above the lower edge of the barrier, the location of the steady state hydrodynamic center of the barrier design.

Added elasticity in the tension carrying cable **14** keeps the bridles taut more of the time. By constructing the tension carrying cable of relatively elastic material such as nylon, the jerky motion of the barrier causing oil to splash over it or under it is prevented. Increasing the elasticity of the tension carrying cable also improves the roll stability of the barrier.

A further advantage of the present design is that when the design current of four knots is exceeded, the barrier pulls out of the water, and rides on its lower edge, so as to reduce the forces acting on it. This action, attained by properly adjusting the lengths of the lines **156** and **158**, prevents destruction of the barrier due to excessive forces.

If desired, top and bottom grab lines **142, 144** respectively, may be connected between the tops and bottoms of the pairs of support members, or extensions **154, 152**, leaving sufficient slack therein so that they do not interfere with the operation of the barrier. Lower grab line **144** may be of chain for weighting. As shown the oil **146** floating on water surface **148**, with a left to right current, is kept to the left of the barrier by the sheet or curtain **22′**. The draft D of the sheet **22′** must be sufficient to extend below the floating oil and into the water itself.

It is a particular feature of this collapsible barrier that it may be deflated

FIGURE 66: TRANSVERSE AND LONGITUDINAL ELEVATIONS AND TOP VIEW OF HOULT-MILGRAM BARRIER DESIGN

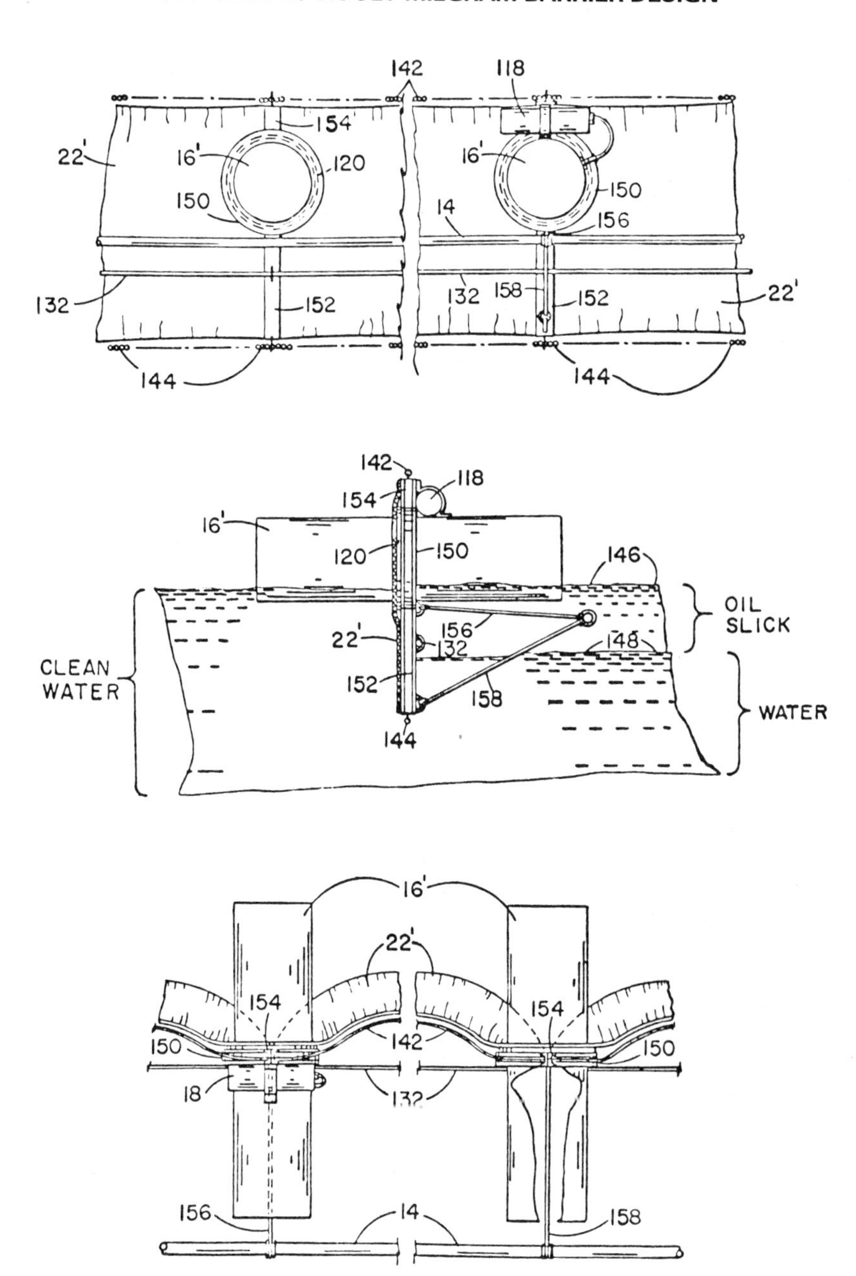

Source: U.S. Patent 3,802,201

and collapsed into a relatively flat foldable assembly, which may be packaged into a reasonable volume. As deflated, a barrier of the type described will have a maximum thickness of about 3 inches at the flotation and supporting portions, and a 1,000 foot length thereof containing about 200 flotation and supporting modules can be flake folded into a package about 25 feet long and 5 feet high, which package will readily fit into a C-130 aircraft for air deployment.

An oil confining boom design developed by *G. Van't Hof; U.S. Patent 3,803,848; April 16, 1974* is a vertically floating wall having a generally arcuate plan for confining oil spills or the like in open water. Several such arcs may be joined to form a sinuous wall. A pair of walls may be connected to form a V to be towed across an oil spill. A plurality of such walls may be joined end to end to form a polygonal enclosure for oil. The wall is preferably made by a plurality of vertically extending bar-like members having a trapezoidal cross section arranged side by side with the nonparallel sides of the trapezoid in abutment to form a substantially liquid-tight wall. A series of the bars have their narrower sides facing in one direction to form an arc concave in that direction. An adjacent series of bars are similarly abutted to form an arc concave in the opposite direction and such series are successively alternated to form the sinusoidal plan. A prestressing cable extends through in the trapezoidal members along the length of the sinusoidal assembly for applying a force towards the concave side of each of the arcs to thereby form an elastically stable structure.

A floating boom design developed by *P.O. Oberg; U.S. Patent 3,807,177; April 30, 1974; assigned to Sanera Projecting AB, Sweden* has a curtain of cloth-like material depending substantially vertically in the water and supported by buoyant bodies and maintained in a substantially vertical position by weights attached preferably to the lower edge of the curtain. The buoyant bodies comprise hermetically sealed buoyant bags attached to one or both sides of the curtain and constructed of soft cloth material which is impermeable to liquid. The bags when not influenced by water pressure are flat and of substantially uniform thickness. The buoyant bags enclose a constant quantity of gas, preferably air, and are attached to the curtain such that the quantity of gas enclosed in each buoyant bag when subjected to water pressure as the boom is placed in the water can be freely pressed up towards an upper portion of the buoyant bag to create in said portion a balloon-like swelling.

Such a boom design is shown in Figure 67 which illustrates a portion of a boom structure, generally illustrated by the reference numeral **1**, laid out in the water, a section surface being shown through the boom and the water by the level line **2** principally marked with the height of the boom structure in relation to the surface of the water **3**. The boom structure includes a curtain **4** which is located substantially vertically in the water and on the sides of which buoyant bags **5** are arranged in substantially uniform spaced relationship. The buoyant bags **5** are hermatically sealed and are made of an air-tight and water-tight, substantially thin and soft cloth material, for example a fabric material or the like impregnated with a rubber or plastics composition and attached substantially perpendicular to the longitudinal direction of the boom **1**.

The boom curtain **4** also comprises a soft cloth material impermeable to water and oil, for example suitably a cloth, fabric or the like impregnated with a plastic or rubber composition and may be reinforced at the top by means of a folded seam or a glued or welded edge (not shown). The curtain is provided at the bottom thereof with a hem **6** in which the weights are placed, the weights preferably being in the form of a chain **7** intended to maintain the boom **1** in a substantially vertical attitude.

FIGURE 67: SWEDISH FLOATING BOOM DESIGN

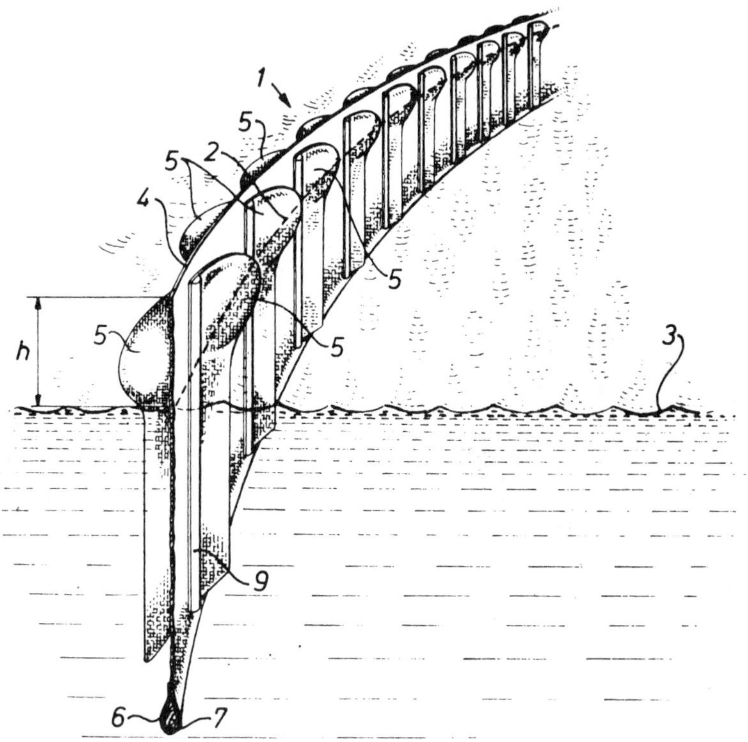

Source: U.S. Patent 3,807,177

The total supporting power afforded by all the bags located on a portion of a boom enables the boom to protrude above the surface of the water **3** to a height **h** sufficient to effectively prevent oil trapped within the boom structure from washing from one side of the boom to the other. Practical experience has shown that a suitable volume of air enclosed in each buoyant body at atmospheric pressure is approximately 1.5 dm^3.

In a free state, i.e., when not subjected to the pressure of the water, each buoyant bag **5** forms a rectangular air cushion of uniform thickness. The buoyant bags are suitably attached in pairs to the boom curtain **4** with one bag in each pair located on either side of the curtain. The distance between each pair of buoyant bags may be approximately equal to the height of the curtain **4**, i.e., in practice approximately 1 meter. To provide for a certain degree of rigidity of the curtain **4** in a vertical position, substantially vertically extending pairs of stiffening ribs **9** made of aluminum for example and having a semi-cylindrical shape are arranged in the longitudinal direction of the curtain with the same spacing as the pairs of buoyant bags, the stiffening ribs **9** being arranged opposite each other on either side of the curtain **4** and attached to the curtain and held together preferably by means of rivets.

A floating boom design developed by *N.D. Tanksley; U.S. Patent 3,807,178; April 30, 1974; assigned to Pacific Pollution Control* consists of a sheet-like partition member or barrier to which a plurality of floats are demountably attached. The floats include the frame on which float members are mounted and a movable portion of the frame which frictionally engages or clamps to the partition member. The partition member preferably is a flexible sheet deployed from a roll for convenience in storage, with the float members being periodically attached to the partition as it is unrolled and draped into the water. The float means which are clamped onto the partition are preferably constructed with float elements cantilevered off of each side of a central frame which is clamped to the partition. Such mounting of the floating elements provide an outrigger effect which insures great stability of the boom in water and allows passage of oil and water between the partition and the floats. The floats are constructed to allow convenient splicing of one partition member to another.

In actual use, it has been found that 300 foot lengths of such a floating boom can be deployed in 12 minutes or less. Retrieval of a 300 foot length of floating boom requires about 8 minutes. The partition which has been found to be particularly advantageously employed in the floating boom has a height of about 24 inches, about 9 to 11 inches of which protrude above the water. The float elements have a diameter of about 6 inches and a length of about 10 inches and are secured to the upright members at about 15 inches from the lower end of the frame. The frame including the arm members may be formed of 304 stainless steel, with the upright members preferably having a U-shaped channel cross section for additional rigidity to ensure good frictional engagement of the partition.

In addition to hollow nylon float elements, closed cell polystyrene foam having a density of about 2 pounds per cubic foot and formed in a 6 by 10 inch cylinder will provide satisfactory float elements for support of the partition member. These float elements are typically attached at about every 4 to 5 feet along the length of a ¼ inch thick nylon reinforced polyurethane covered partition sheet.

A deployment apparatus developed by *N.D. Tanksley; U.S. Patent 3,807,617; April 30, 1974; assigned to Pacific Pollution Control* is designed to be used in conjunction with the boom design described above in U.S. Patent 3,807,178.

A barrier design developed by *L. Ballu; U.S. Patent 3,811,285; May 21, 1974; assigned to Pneumatiques Caoutchouc Manufacture et Plastiques Kleber-Colombes, France* consists of a skirt-like element immersed in the water and supported by floating elements which are constituted by air-tight pockets whose openings face in the downward direction when installed in the water; the pockets are thereby connected with one another by bands of flexible fabric.

A barrier design developed by *R.A. Benson; U.S. Patent 3,818,708; June 25, 1974; assigned to Submarine Engineering Associates, Inc.* is a durable, stable oil containment boom which will remain functional at all times for extended periods of time. It has sufficient buoyancy and stability to permit the attachment of secondary systems such as hoses, small boats, ancillary clean-up devices, and the like, while still being sufficiently flexible to withstand shocks and strain without damage. Sections can be quickly connected or released without tools for emergency purposes, and a positive seal is maintained between sections. Convenient lifting points are provided at frequent intervals along the boom for handling of the boom and for use by clean-up personnel in tying-off clean-up equipment and boats as well as for a steadying point for personnel working near the boom.

End connections seal the barrier to the pier or other permanent structure while allowing for vertical tide and wave action as well as providing a connecting/disconnecting terminal for normal opening and closing of the boom. The boom will not interfere with or damage a ship or boat operating in its vicinity; e.g., there are no cables, lines or chains to foul ship propellers and shafts.

The dam is constructed of a material having Shore A scale durometer hardness between 60 and 90, the barrier section having its center of gravity below its center of buoyancy, and having at least one end constructed and arranged for connection to an adjacent section. In preferred embodiments the dam is of polyurethane material and has an increased average weight density in its lower 20%; the floats are D-shaped foam filled tubes which extend continuously substantially (at least 99%) the length of the barrier section to provide maximum strength and continuity; vertical stiffening ribs of less flexibility than the dam provide ridigity; adjacent sections are hinged to each other with plastic pins; at least one-fourth of the height of the dam extends above the floats; and the buoyancy and weight of the barrier are related so that in water at least 50% of the height of the floats will be submerged.

As shown in Figure 68, such a typical boom section **1** consists of a vertical dam **6** sandwiched between two D-shaped flotation tubes **8**. The overall width **W** of tubes **8** is one-fourth to two-thirds (preferably one-third) the overall height **H** of the dam, and is greater than the height **h** of the tubes (preferably **W** = 3**h**). As a result, the center of buoyancy moves significantly towards the outboard side of the flotation tubes **8** when the boom tilts about an axis approximately through the geometric center of the tubes. The weight distribution yields a center of gravity located below the bottom of tubes **8** and below the center of buoyancy. One-fourth to one-half the height of dam **6** extends above tubes **8** to provide an above water barrier.

As the boom tilts or heels about its rotational axis the righting moment and metacentric height increase substantially since the center of buoyancy moves significantly towards the tilted down side of the tubes.

Since the formulation of the basic dam **6** and tubes **8** yields a specific gravity of at least 1.2 and is generally one-fourth inch or greater in thickness thereby resulting in a weight of greater than 10 pounds per linear foot for a three foot dam, not only is the center of gravity at a point below the center of buoyancy but the magnitude of the righting moment is significant.

FIGURE 68: SUBMARINE ENGINEERING ASSOCIATES BARRIER DESIGN

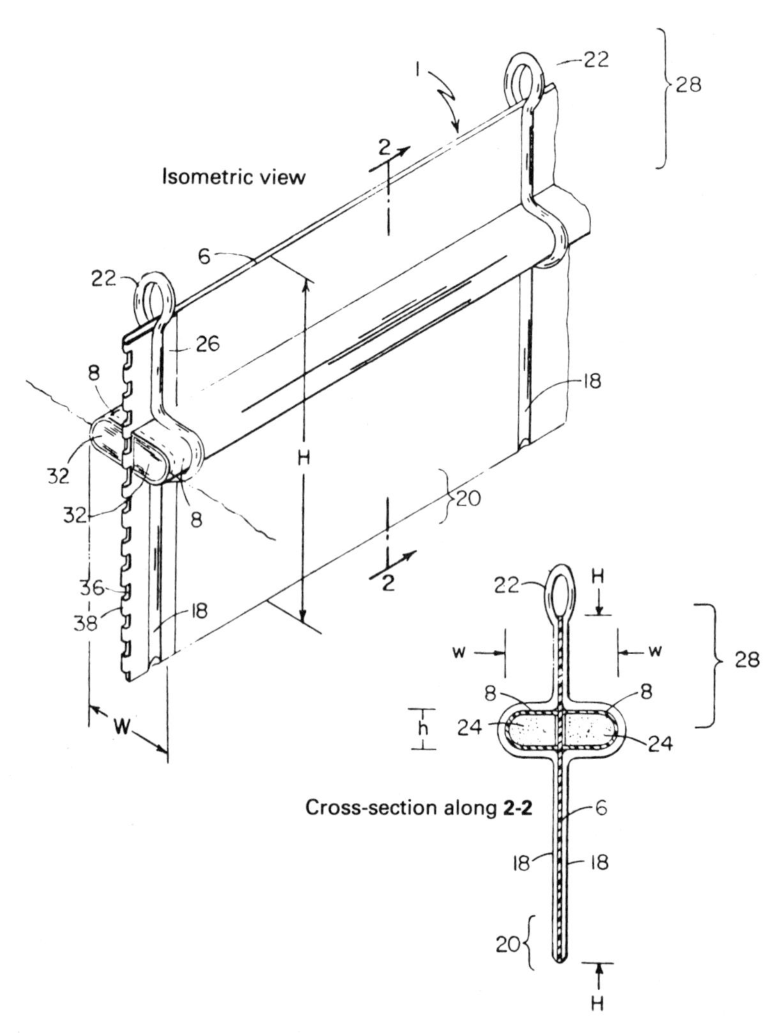

Source: U.S. Patent 3,818,708

The flotation tubes **8** are substantially continuous along the total length of the boom, except for less than one-fourth inch of space (less than 1% of the length of section **1**) between the tubes of contiguous sections, thereby providing a uniform configuration and stability along the entire length of the total boom regardless of the total length of the boom which will vary with the use.

In order to further increase stability, vertical plastic ribs **18** of between 90 Shore A scale and 40 Shore D scale durometer hardness are located at 6 foot intervals along the boom. The stiffness of the ribs closely maintains the relative position of the dam **6** and tubes **8**, thereby also closely maintaining the relative positon of the center of gravity and center of buoyancy. If the above or below water segments of the dam **6** were to bend under the force of water, wind or other pressures, both the center of gravity and the center of buoyancy would shift to points resulting in lower stability.

The ribs **18** extend from a point at the bottom edge of one side of the dam **6** vertically upward, against the dam **6**, around tube **8**, upward against the dam **6**, over the top of the dam **6** in a hand-sized loop **22**, downward on the opposite side of the dam **6**, around tube **8** and on down to the bottom edge of the dam **6** opposite the starting point. The rib **18** is bonded, sealed, or otherwise tightly fastened against the dam in order to provide a semi-rigid stiffener and to serve other purposes which are described subsequently.

The cross-section at any point of the rib **18** is generally of a half-round shape with a base no less than 1 inch and a radius of no less than ½ inch or of a rectangular shape with a base of no less than 1 inch and a width of no less than ¼ inch, or of similar cross-sectional shape. In the case of rectangular ribs, the outside edges are beveled to provide fairing.

In order to further increase stability additional weight is generally added internally within the lower portion **20** of the dam **6** by adding lead or similar weighting materials near the bottom edge of the dam. The result of adding weight is to lower the center of gravity, thereby yielding an even larger righting moment. The additon of weight to the lower part **20** of the dam **6** is possible because of the significant amount of reserve buoyancy which the configuration has, generally in the 6 pounds per linear foot range for a 3 foot dam. The relationship of weight and buoyancy is such that in use, at least 50% of the height of the tubes **8** will be submerged, providing excellent stability.

The center of gravity can be lowered further by reducing the specific gravity of the upper one-third **28** of the dam **6**.

The tubes **8** are bonded to the dam **6**. They may also be cast or molded as an integral part of the configuration. Likewise, the ribs **18** and boom section joiners **26** are bonded to the dam **6** and tubes **8**. The tubes **8** as well as the ribs **18** and section joiners **26** may also be cast or molded as an integral part of the cross-configuration. Additionally, the ribs **18** may be located internally as a part of the basic dam and tube configuration, rather than being positioned on the exterior.

The unique fairing or smoothness of the entire boom is also an important aspect of the device's suitability for its water environment. Specifically, the tubes **8** are essentially continuous throughout the entire length of the assembled boom with epoxy, cast or molded fillets, fairing them to the dam **6** and ribs **18**. Rubberlike plugs **32** or solid ends **32** of the tubes **8** at the ends of each boom section are set so that a close butting occurs when sections are joined with the section joiner **20**.

The joiner **26** utilizes a hinge-like plastic fitting which is attached to the ends of each dam section. The hinge plate **26** is folded around the end of the dam **6** and bonded to the dam. Close tolerance openings **36** are cut, cast or molded into the plate **26** with a vertical separation between hinge projections **38** of generally greater than one inch and a hole through the projecting piece **38** preferably one-fourth inch or greater in diameter. There are right and left hand joiners **26** which when joined bring the sections or components closely together in the same plane vertically and horizontally. The snugging of joiners provides a close fit which prevents oil or other contained materials from leaking or passing through the joined sections of boom or its components.

A boom design developed by *L.G. Green; U.S. Patent 3,839,869; October 8, 1974; assigned to Metropolitan Petroleum Petrochemicals Co., Inc.* comprises a number of nonpneumatic floats arranged in spaced alignment and a flexible web wrapped completely around the floats and bridging the spaces therebetween. The side sections of the web beyond the floats are secured together face to face and weights are attached to the web near the marginal edge of these side sections to form a downwardly extending weighted ballasting fin. The fin forms a hinge connection near the floats to permit the fin to swing about the hinge axis without transmitting its movements to the floats.

A boom design developed by *M.F. Smith and A.V. Anusauckas; U.S. Patent 3,848,417; November 19, 1974* is a unique self-righting and quickly deployable floating boom capable of enduring strong winds and waves which comprises a series of polymer floats each incorporating a horizontally extending shelf securely attached at spaced intervals to an integral composite fin of vinyl sheet reinforced by woven polyester fibers incorporating an interwoven core of two characteristically different fibers that provide the vinyl sheet with different vertical and horizontal flexing capabilities. An extension shelf formed on each of the polymer floats provides the boom with additional buoyancy while also serving as a barrier effectively containing oil and other floating materials despite wind, choppy water and strong waves.

Furthermore, the extension shelves incorporate fore and aft lifting surfaces which tend to induce "planing" and counteract the forces which tend to draw the floating boom beneath the water surface during fast end-wise deployment. The dual fiber core of the vinyl sheet is manufactured with relatively thin horizontal fibers interwoven with relatively stiff, thick vertical fibers to allow the vinyl sheet to flex easily about vertical flexing axes while strongly resisting horizontal flexing about horizontal flexing axes.

Such a boom is shown in Figure 69 which depicts the overall boom, a section of the skirt joining the floats and the posture of the float under rough water conditions.

Floating boom **20** incorporates a plurality of floats **21** securely mounted to fin **22**. Each float **21** incorporates attachment strips **23** molded in position during the formation of float **21**, ready for rapid and simple mounting to fin **22**.

The fin **22** incorporates a woven, dual-fiber web core **26** which is surroundingly enclosed by vinyl **27**. Web core **26** comprises relatively thin stranded horizontal fibers **28** interwoven with substantially thicker vertical fibers **29**. Preferably, fibers **28** comprise fine denier spun polyester yarn, such as "dacron" or "terylene," and fibers **29** comprise higher denier monofilament polyester fibers. The thin fibers **28** will horizontally traverse fin **22**, substantially parallel along the length of the fin, while comparatively thick fibers **29** will traverse fin **22** vertically and substantially parallel over the height of the fin.

Preferably vinyl **27** comprises buna-polyvinylchloride and is approved "international orange" in color to improve the visibility of boom **20**. In manufacturing fin **22**, dual fiber core **26** is "oriented" by being simultaneously heated and stretched out on a frame. Then it is laminated or coated with vinyl

27. This fabrication technique improves the bending characteristics of fin **22** by substantially reducing or eliminating any tendency of fiber core **26** to stretch or elongate.

FIGURE 69: SMITH-ANUSAUCKAS DESIGN FOR SELF-RIGHTING FLOATING BOOM

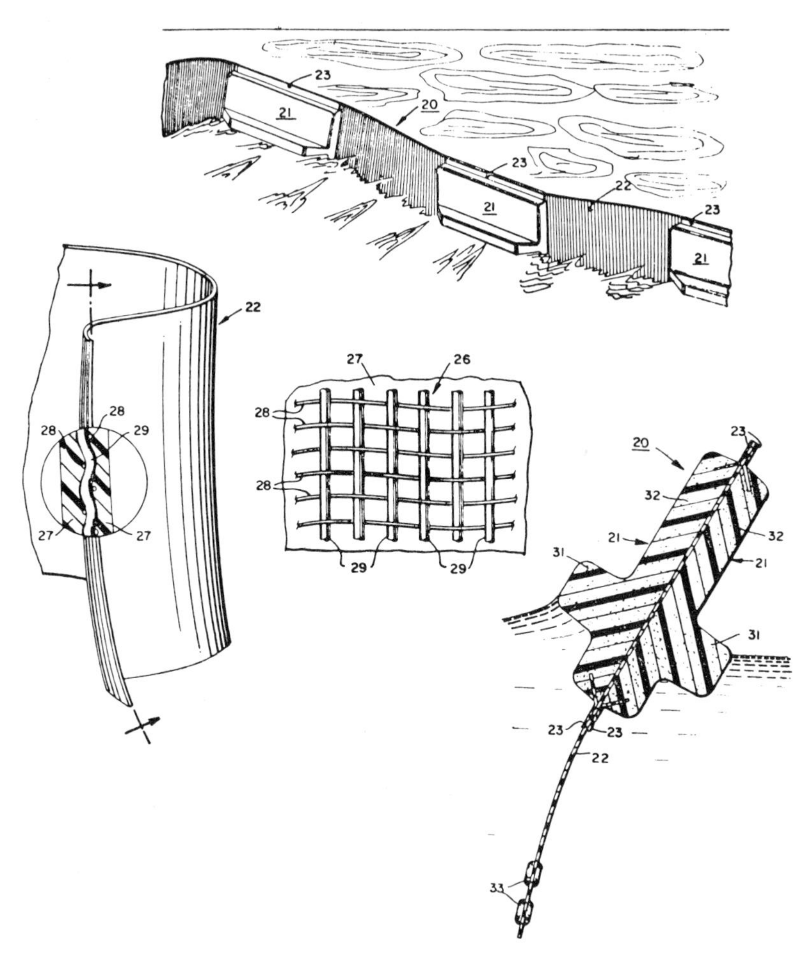

Source: U.S. Patent 3,848,417

The incorporation of dual fiber web core **26** in fin **22** arranged in the manner discussed above provides fin **22** with unique, selective bending characteristics. Since the only resistance to the longitudinal flexing of fin **22** about vertical axes is provided by vinyl **27** and the relatively thin horizontal fibers **28**, fin **22** can be easily flexed about vertically oriented lines.

Fin **22**, however, has a vastly different flexing characteristic about horizontally extending lines. When fin **22** is flexed or "bent over" about a horizontal line, thick fibers **29** must be bent along with vinyl **27**. Since the mono-filament fibers **29** are substantially heavier than fibers **28**, fibers **29** strongly resist forces tending to bend them and, consequently, they rigidify fin **22** against bending about horizontal lines.

This unique dual-flex characteristic of fin **22** is extremely important in the stability and overall performance of floating boom **20**. Since fin **22** can be easily flexed about vertical lines, the fin will respond to normal water currents and waves as presently existing booms do now perform, and, also, the boom can be easily folded accordion-fashion for storage. However, when strong winds and driving waves impact against fin **22**, attempting to bend the fin about horizontal lines, semi-rigidified fin **22** resists these forces with great effectiveness.

As indicated, the high buoyancy of floats **21** coupled with the relative stiffness of fin **22** against bending about horizontal lines maintains the float portion of the boom well above the sea surface, refusing to be horizontally flexed and driven into the water or to allow the contained oil to escape. Furthermore, the controlled rigidity of fin **22** allows the floats **21** to be spaced 28 to 24 inches apart without the necessity of having the floats linked together in a continuous "chain." Also, dual-fiber core **26** of fin **22** substantially increases the tensile strength of fin **22**, thereby essentially eliminating the need for incorporating tension carrying means on boom **20**, if desired.

The relative stiffness of the fin **22** against bending about horizontal axes assures the depending downward deployment of the fin **22** projecting well beneath the surface to guarantee effective containment of oil or other floating material. The ballast weights **33** help to maintain this substantially vertical downward deployment of fin **22**, even against lateral current, wind or wave loads tending to bend over slightly the upper portion of fin **22**.

This semirigid downwardly depending barrier fin, coupled with the skimming "planing" floats, together assure that these booms will "track" readily behind a towboat, and their extremely low drag avoids sweeping or "corner-cutting" as the towboat changes course; thus, during deployment the boom readily moves endwise "in trace" behind the towboat along each leg of its course, while depending fin **22** remains downwardly deployed at all times to serve as a guiding keel. This assures that boom **20** will follow or "track" behind the towboat, substantially following the identical deployment path of the towboat.

The position of extension shelves **31** along the lower half of the front surface **32** of float **21** provides assurance that floating boom **20** will possess the required amount of buoyancy while also providing a substantially greater oil retaining surface area above the water surface. In the preferred embodiment, fin **22** is about 36 inches wide and about 12 inches of fin **22** is maintained above the water surface. Maintenance of the substantially vertical orientation of fin **22** in the water is assured by ballast weights **33** which are mounted along the lower edge of fin **22**, imparting added weight thereto.

By analogy, the booms exhibit the high lateral stability against "initial" heeling, produced by flat-bottomed and multihull vessels, with the highly effective self-righting action of a deep keel sailboat, whose increasing angle of heel correspondingly increases the righting arm and righting moment of the vessel's heavy keel. Thus, a slight angle of heel immersing the right shelf **31** immediately moves the boom's center of buoyancy to the right, tending to reerect the heeling boom; simultaneously, the relative stiffness of the fin **22** against bending about horizontal axes serves to produce an increasing "deep-keel" righting moment of ballast **33**, further tending to reerect the boom **20** with highly efficient self-righting action.

A barrier design developed by *P. Preus; U.S. Patent 3,849,989; November 26, 1974;* provides an inflatable barrier for substances floating on water which avoids the disadvantage of the prior art by furnishing a flotation chamber which is fabricated from a tough, puncture resistant material and is provided with independent subchambers, each capable of supporting the barrier in the event of loss of pressure in the neighboring subchamber.

In a preferred embodiment, this barrier has plural elongated inflatable chambers along the upper edge thereof; a liquid impervious flexible skirt depending from the chambers to provide a barrier below the water line of the skirt; tension members disposed on each side of the skirt surjacent the chamber and substantially coextensive therewith and a ballasting tension member disposed along the lower edge of the skirt.

A device developed by *L. Ballu; U.S. Patent, 3,852,964; December 10, 1974; assigned to Kleber-Colombes, France* is a float-and-skirt barrier design in which means are provided to permit a change in the height of the skirt in such a manner as to vary the draught of the device.

A boom design developed by *C.H. Rudd; U.S. Patent 3,852,965; December 10, 1974* is a boom wherein a curtain extends downwardly in the water from a floating surface barrier with the upper portion of the curtain being impervious and the lower portion being open for the passage of water therethrough. The boom is towed by two lines, one extending through the floating surface barrier and the other connected along the lower extremity of the curtain, and the lower line is pulled in advance of the upper line.

A boom design developed by *R.A. Fossberg, U.S. Patent 3,852,978; December 10, 1974* is a flexible oil boom which combines the important features of strength and light weight as well as stability in choppy wave action. The boom comprises a barrier wall of sheet material for deploying in a substantially vertical position in the water and having an upper portion and a lower portion. These upper and lower portions are sewn together by an overlapping connection at a location below the water line.

At spaced locations along the length of the boom are positioned vertical stiffeners and immediately adjacent each stiffener there is provided a flexible float connecting strap which surrounds the lower edge of the upper portion and passes through the overlapping connection between the upper and lower portions so that free ends of the strap extend out of each side of the barrier wall. Each strap free end is detachably connected to an individual float member. Individual weights are connected to the barrier wall at the lower end of each stiffener rod pocket. This arrangement permits limited independent movement of each stiffener and adjacent portions of the barrier wall both horizontally and vertically relative to the remainder of the boom whereby the boom can respond to choppy wave action.

A boom developed by *R.A. Benson; U.S. Patent 3,859,796; January 14,*

1975; assigned to Submarine Engineering Associates, Inc. is a submersible oil barrier or boom capable of enclosing an area of water surface, said barrier being made up of sections each comprising a solid vertical dam and substantially continuous flotation elements extending laterally from opposite sides of the dam, the barrier being submersible, that is, selectively flotatable and sinkable so that the barrier may be sunk for passage of an oil tanker or the like thereover and subsequently again floated to enclose an area including such tanker for confining oil spills.

It uses a barrier design similar to that described in U.S. Patent 3,818,708 earlier in this section, now however, with added tubes integral with the barrier which may be filled with air, filled with water or deflated to provide varying degrees of buoyancy to the barrier.

A boom design developed by *R.R. Ayers; U.S. Patent 3,859,797; January 14, 1975; assigned to Shell Oil Co.* is a boom which is highly resistant to either splashover or underflow of floating pollutant. This has been achieved through provision of a boom composed of an upright corrugated skirt supported in the water by floats attached to outriggers which separate the floats from the skirt. Such a boom design is shown in Figure 70.

FIGURE 70: SHELL OIL CO. BOOM DESIGN

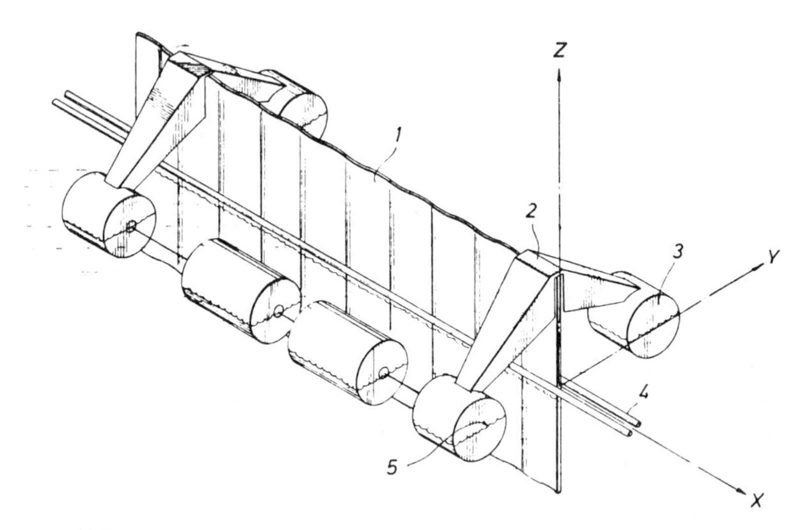

Source: U.S. Patent 3,859,797

The major components of the boom include a flexible corrugated skirt **1** which is in a substantially upright position, preferably vertical; outriggers **2**, which are attached to and support skirt **1**; floats **3**, which are attached to the outriggers and spaced from the skirt; and cables **4**, which provide reinforcement for the skirt.

Skirt **1** is preferably made of corrugated material with the axes of the corrugations running vertically (Z direction). Such construction makes the skirt particularly adaptable to use in moderate waves and currents. Thus, the corrugations give the skirt a vertical rigidity (rigidity in the Y Z plane) which resists deformation and consequent underflow caused by currents which would bend skirts not having such rigidity. On the other hand, the corrugations give the skirt longitudinal flexibility both in a direction normal to the plane of the skirt (flexible in the X Y plane), which helps to dissipate the effects of wave forces and prevent splashover, and in a direction within the plane of the skirt (flexible in the X direction), wich enables the skirt to conform to the surface of the water to resist submerging and consequent splashover. The skirt may be made of steel, aluminum, fiberglass or reinforced fabric, or whatever material is most suitable for either inshore or offshore service.

Outriggers **2** may take any of numerous suitable shapes provided the outriggers function to maintain the floats rigidly or somewhat flexibly connected to and substantially separated from the skirt. This construction is very important inasmuch as it makes the boom very stable and highly resistant to being overturned or tilted to any substantial degree by the action of wind, waves and currents without the use of weights. Most prior art booms utilize weights at the bottom of the skirt to overcome tendencies toward tilting and overturning. However, such weighting requirements are substantial, which makes the boom both cumbersome and expensive. Also, when flotation members are constructed integrally with the skirt as in conventional booms, waves tend to roll over the flotation element and then splash over the skirt. In the present design with the flotation element removed from the skirt area by outriggers, the waves meet a sheer vertical wall so that the wave action is reflected without appreciable splashover.

Preferred outriggers form a broad V which is inverted over the top edge of the skirt. The outrigger may be secured to the skirt by any suitable means, such as a bolt (not shown) which extends both through the outrigger and the skirt. At the extremity of either leg of the V there is secured a float. The float may be secured to the leg of the outrigger by any suitable means, either by temporary or permanent connection. While the form of outrigger shown is preferred, other forms of outriggers are suitable for use with the process. For example, the broad V may be attached to the bottom edge of the skirt; instead of a V shape, the outrigger may have a semicircular shape; the outrigger may be a simple rod or bar which attaches to the face of the skirt and has a float at one end or which pierces the skirt and has floats at either end; or, the outrigger may extend from only one side of the skirt at a given location and extend from the other side at another location. Manifestly, many outrigger designs are suitable so long as the outrigger performs the above stated functions.

Floats **3** may be of any buoyant material, for example, hollow spheres, foamed or expanded rubber or polymeric material such as foamed polyurethane, polyvinyl chloride or polyethylene and may be in any form which is compatible with the form of the outriggers. Such floats may be rigidly, semirigidly or loosely held to the outriggers.

Cables **4** function to carry boom tension. The cables are optional and may not be needed where the boom is of substantial strength or where only a short length of boom is being employed. Generally, it is preferred that the cables be spaced on both sides of the skirt at or near the water line of the skirt. However, as the situation may be, only one cable may be necessary if forces against the boom are substantially in only one direction.

As may be seen from the drawing, once the boom is deployed in the water, floats **3** maintain a substantial portion of the skirt above the water surface **5**. Generally, the cables **4** are arranged so that they are substantially at the center of forces applied to the skirt.

The boom of this design is easily stored and deployed. The skirts can be coiled and outriggers stored separately. In deployment, as the skirt is reeled out from a roll, the outriggers are slipped on prior to the boom reaching the water surface. The floats may be connected to the outriggers either before or after the outriggers are attached to the skirt. In addition, the cables may be attached to the skirt before or after the outriggers are attached.

A design developed by *K. Aramaki, Y. Kawaguchi and H. Kawakami; U.S. Patent, 3,867,817; February 25, 1975; assigned to Bridgestone Tire Co., Ltd., Japan* provides an oil fence having a unidirectional flexibility, comprising a plurality of rigid floats which are swingably connected by hinge means. Planar skirts are connected to the floats so as to define at least one continuous oil fence wall thereby. Such a fence design is shown in Figure 71.

FIGURE 71: BRIDGESTONE TIRE CO. DESIGN FOR TOWABLE OIL FENCE OF LIMITED FLEXIBILITY

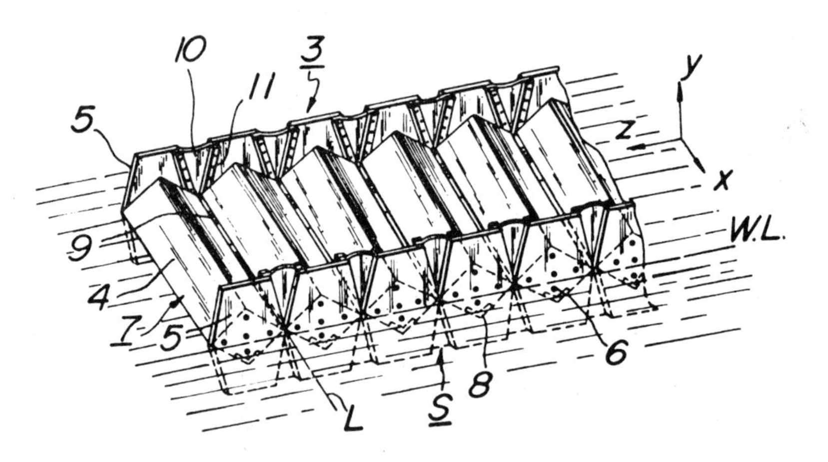

Source: U.S. Patent 3,867,817

The oil fence is formed by connecting a plurality of oil fence units **7**. Each unit **7** has a hollow prism-shaped float **4** and at least one planar skirt **5** secured to a base surface of the prism-shaped float by bolts **6** in a detachable manner. The hollow float **4** is made of a material having a high rigidity, such as iron sheet, fiber-reinforced synthetic resin sheet, wood board, and the like. The float shape is not resticted to the hollow prism, and any other suitable shape can be used as long as it provides a sufficient buoyancy to the oil fence. For instance, the float can be made of hard foamed plastic material in any desired shape. The skirt **5** is made of rigid sheet material, such as aluminum sheet, iron sheet, fiber-reinforced synthetic resin sheet, or the like. Preferably, two planar skirts **5** are used in each unit **7**, one at each of the opposing ends of the prism-shaped float **4**.

The buoyancy of each oil fence unit **7** is preferably so selected that, when it is placed in water, the longitudinal center line of the prism-shaped float **4** lies substantially on the plane of the water surface. To this end, a suitable weight member **8** may be attached to the float **4**. The weight member **8** is an angled steel member which is attached along the bottom edge of the prism-shaped float **4**.

In assembling the units **7** into an oil fence, the floats **4** are juxtaposed in such a manner that the planar skirts **5** which are secured to one end of the units **7** lie substantially on a common plane. More particularly, when each unit **7** has two planar skirts **5** secured to the opposite ends thereof, such skirts of the assembled units define two parallel planes extending along opposite ends of the floats **4**. The adjacent floats **4** in the oil fence **3** are connected by one or more hinge means **9** in such a manner that the adjacent floats **4** can swing relative to each other about a hinge line **L** which is substantially perpendicular to the plane of the skirts **5**. On the other hand, the relative movement of the adjacent floats **4** in parallel to the aforesaid hinge line **L** is restricted by the hinge means **9**.

To allow the relative swinging of the adjacent oil fence units **7**, a suitable spacing **S** should be provided between the skirts **5** of the units **7**, as shown. To close such a gap **S** against leakage of the floating matter e.g., spilt oil, a flexible membrane **10** is attached to the adjacent skirts **5** by fixtures **11**. The flexible membrane **10** may be made of a rubber coated cloth or a water-repellent asbestos cloth. Thus, a continuous wall is formed on the water surface by the skirts **5** and the flexible membranes **10**, which wall inhibits the floating matter from moving thereacross.

In operation, the oil fence **3** is pulled in the direction of the **Z** axis of the three-dimensional orthogonal coordinate system by a ship **1** for collecting or recovering floating matter. In this case, the oil fence **3** can flex on the **YZ** plane, so as to follow any variation of the wave contour on the water surface. On the other hand, the oil fence line **3** is restricted from flexing on the **XZ** plane. Thereby, the oil fence line **3** prevents the floating matter from moving in a direction normal to the line of the oil fence. Thus, undesirable leakage of the floating matter across the oil fence **3** can be effectively prevented.

A boom design developed by *R.K. Thurman; U.S. Patent 3,868,824; March 4, 1975; assigned to Merritt Division of Murphy Pacific Marine Salvage Co.* is a floating oil containment boom module which, while afloat, can be connected easily and rapidly end to end with other like modules to form a boom. The boom can be rapidly assembled and disassembled, and can be opened to any point along its length to allow removal or insertion of additional modules or open a passage for waterborne traffic. The flexible waterproof panels which provide oil containment continuity between adjacent float units are secured to the float units in such a manner to assure that they will not pull free, and are connected to adjacent panels with a seal that provides excellent watertight integrity between adjacent panels.

A boom design developed by *H.R. Appelblom and F.E. de Bourguignon; U.S. Patent 3,882,682; May 13, 1975* is a design rather similar to that described earlier in this section in connection with U.S. Patent 3,807,178.

An oil containment apparatus developed by *R.R. Ayers and E.V. Seymour; U.S. Patent 3,886,750; June 3, 1975; assigned to Shell Oil Co.* is an apparatus for use in water experiencing high current velocity which includes a barrier and a boom and skirt means upstream of the barrier defining therewith a capture area where the flow velocity is locally reduced. Oil or other floating pollutant liquid enters the capture area and accumulates therein because of the inability of the low current velocity to remove the oil. Oil thus collected may be skimmed or otherwise removed from the capture area.

In Figure 72, this apparatus **10** is illustrated as positioned in a body of water **12** on which there is floating a layer of liquid pollutant **14**. Although the liquid pollutant **14** typically comprises liquid hydrocarbons, it will be apparent that the apparatus **10** may be used to control floating liquid pollutants of any type. The apparatus **10** comprises as major components upstream and downstream spaced apart buoyant float means **16, 18** defining therebetween a capture area **20** for the liquid pollutant **14**, means **22** defining a path of water movement generally horizontally into the capture area and generally downwardly out of the capture area and means **24** for reducing the flow velocity in the capture area **20** below the relative current velocity to which the apparatus **10** is subjected. Means **26** are provided for positioning the apparatus **10** in an area of high current velocity or for towing the apparatus **10** through the water **12**.

The float means **16, 18** may comprise oil containment booms of the type commercially available. The float means **16** is illustrated as comprising a plurality of discrete spaced apart floats **28** connected together by suitable flexible lines or cables **30**. It will be apparent that the upstream float means **16** acts to retard wave movement in the capture area **20**. Waves breaking over the upstream float means **16** act to deposit some liquid pollutant in the capture area **20**. The float means **18** may comprise a single elongate float **32** or a plurality of discrete floats forming a barrier.

The flow path defining means **22** comprises a flexible impervious skirt **34** depending from adjacent the float **32** and may conveniently be attached thereto in a conventional manner. The lower end of the impervious skirt **34** merges with a flexible foraminous member **36** having a plurality of apertures **38** which function as the velocity releasing means **24** as will be more fully explained hereinafter. The upstream end of the foraminous member **36** is conveniently connected to the upstream float means **16** in any suitable manner, as by the use of a foraminous member **40** which is illustrated as a screen or net. Suitable weights (not shown) or other means may be provided at the upstream end of the foraminous member **36** to tension the member **40** and thereby maintain the member **36** at a desired location. It will be apparent that the means **22** defines a path of water movement generally horizontally into the capture area **20** and generally downwardly out of the capture area **20**.

The positioning or towing means **26** is provided to position the apparatus **10** in an area of high current velocity or to tow the apparatus **10** through the water **12**. The term relative velocity or relative current velocity is used to define the situation where water moves past the apparatus **10** regardless of whether

FIGURE 72: SHELL OIL CO. BOOM DESIGN FOR USE IN FAST FLOWING WATER

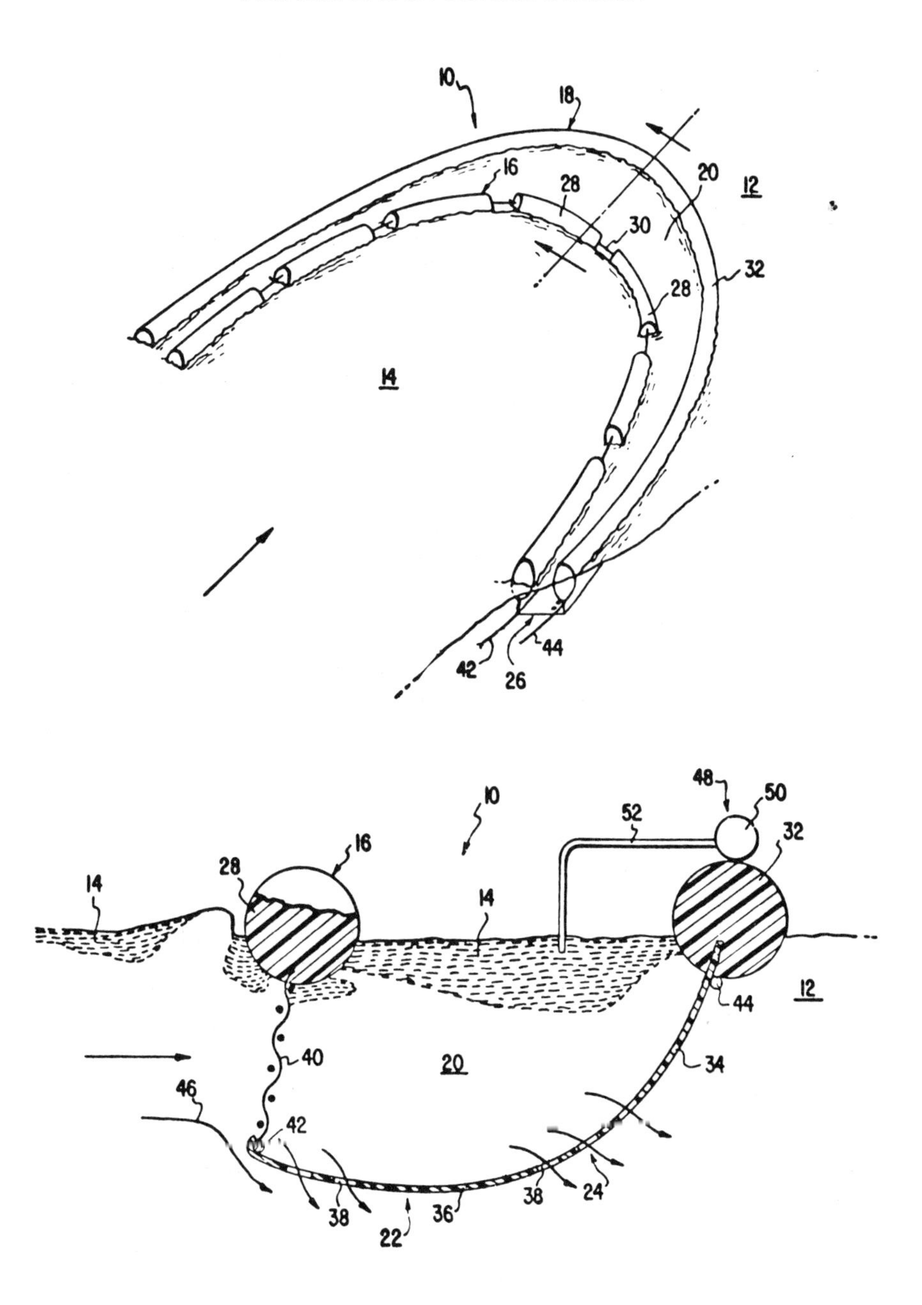

Source: U.S. Patent 3,886,750

the apparatus **10** is placed in an area of high current velocity, the apparatus **10** is towed through the water, or combinations thereof. The positioning or towing means **26** may be of any suitable type but is illustrated as comprising a first flexible line or cable **42** connected to the upstream end of the foraminous member **36** and a second flexible line or cable **44** connected to the float means **18** in a conventional manner. The towing means may be unnecessary since the skimmer may be self-propelled.

It is well known that conventional oil containment booms comprise a single elongate float means and have a maximum containment relative velocity of 1.2 to 1.3 feet per second. At greater relative velocities, oil bubbles break off of the headwave at the oil water interface and escape under the boom. Since current velocities in excess of 1.2 and 1.3 feet per second are not uncommon, it will be apparent that the use of conventional oil containment booms is somewhat limited. Even more important, oil containment booms are often towed to collect the liquid pollutant **14** and the low maximum containment velocity of containment booms substantially prolongs a collection effort.

A prototype of this device which could be considered a finite width skimmer, or alternatively as a length increment of a boom was tested in a current tank and subjected to relative current velocities up to 2.54 feet per second before failure occurred. The typical mode of failure of the prototype is that the buoyancy of the float means is insufficient at maximum velocity and the float means submerges thereby allowing escape of the liquid pollutant **14** over the top of the rearmost float. It is apparent that additional floats may be added to the float means of the prototype to increase the maximum containment relative velocity. Further tests with the prototype without any additional float means were conducted at a relative velocity of 2.20 feet per second without loss of the liquid pollutant **14**.

The velocity reducing means **24** acts to reduce the flow velocity in the capture area **20** sufficient to allow the liquid pollutant **14** to gravitate to the surface. The apertures **38** comprise the velocity reducing means **24** and act as a flow restriction in the outlet of the flow path through the capture area **20**. The size and number of the apertures **28** are selected to provide a flow capacity therethrough substantially less than the volume of water approaching the inlet to the capture area **20**. For all reasonable values of relative velocity, the apertures **38** constitute a flow restriction in the flow path tending to deflect water approaching the inlet to the capture area **20** downwardly under the foraminous member **36** as shown by the arrow **46**. As the relative velocity increases, a greater quantity of water is so deflected to provide a greater difference between the existing relative velocity and the flow velocity in the capture area **20**. It will thus be seen that the flow velocity in the capture area **20** is controlled by the flow capacity of the apertures **38**. In the prototype, the apertures **38** comprise ¾ inch diameter holes staggered on three rows 1¾ inches apart.

The capacity of the capture area **20** to collect the liquid pollutant **14** is necessarily finite. There may accordingly be provided a suitable pollutant removal means **48** which is illustrated as comprising a pump **50** having an inlet **52** extending into the liquid pollutant collected in the capture area **20**. When the oil is removed from the oil capture area of the boom-type aparatus, the apparatus could be called a mechanical skimmer or alternatively (since the apparatus as shown draped in the water has the appearance of a boom) a "boom-skimmer." It follows then that if the apparatus is relatively rigid along its length so that it does not appreciably drape as a boom and if it is perhaps relatively short in length, oil removal would then distinguish the apparatus as a "skimmer." A suitable outlet conduit (not shown) may be connected to the pump **50** for delivering the liquid pollutant to a suitable barge or the like. Other suitable means for removing the liquid pollutant **14** in the capture area **20** may be provided as, for example, scattering oil sorbtive particles in the capture area **20** and then collecting the same.

A boom design developed by *G.W. Robertson and T. Sturgeon; U.S. Patent 3,888,086; June 10, 1975; assigned to Uniroyal Inc.* is a floating boom comprising a barrier for containing pollutant-debris along the surface of a body of water. The barrier is provided on each side thereof with oppositely directed buoyant members, and is ballasted along the lower longitudinally extending edge by means of an appropriate medium so that it maintains an upright attitude. Figure 73 is a transverse section of such a boom.

FIGURE 73: UNIROYAL FLOATING BOOM DESIGN

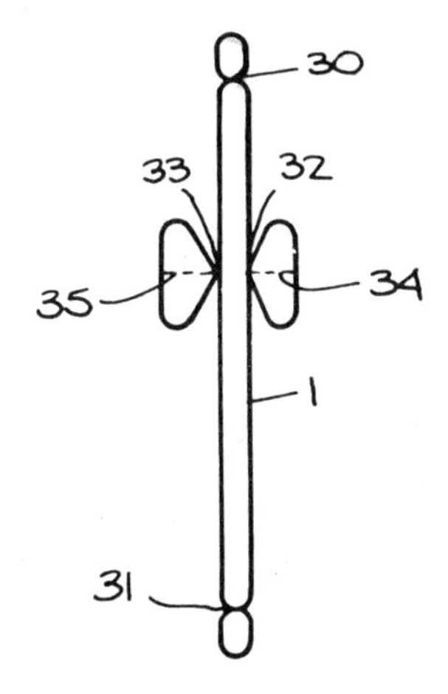

Source: U.S. Patent 3,888,086

Preferred materials for the boom are polyvinyl chloride or polyethylene, or woven or nonwoven fabric coated with polyvinyl chloride or polyethylene. In one method of fabricating the boom shown, a suitable length of plastic or plastics-coated fabric tube is taken, and lines of those portions to be pinched and sealed by preferably heat-welding are marked thereupon. The first seams made are seams **30, 31** to define the tensile member-receiving compartments. The material is then folded and pinched as shown and welded longitudinally along seams **32, 33** to form the buoyant compartments. The ends of these compartments are then sealed at a desired length along lines **34, 35**, and excess material is trimmed away.

A boom design developed by *G.J. Gauch; U.S. Patent 3,903,701; September 9, 1975; assigned to Uniroyal, Inc.* is shown in transverse section in Figure 74.

As shown the boom includes a sheet member, designated generally at **18**, stiffener means **20**, a float member, depicted generally at **22** and weight means. The boom floats on the surface of a liquid **26**, for example, a lake, river, or ocean. Sheet member **18** is supported in liquid **26** by float member **22** secured thereto at an intermediate height so as to divide the sheet member **18** into upper and lower portions **28** and **30**, respectively. Upper portion **28** is adapted to extend substantially in a vertically upwardly direction from float member **22**, thereby forming a dam, or fence, for preventing floating materials confined by boom **10** from being washed over. Lower portion **30** is adapted to extend substantially vertically downwardly from float member **22** into liquid **26**, for preventing materials floating thereon from passing beneath the boom.

As shown, upper marginal edge portion **32** of sheet member **18** is folded over a first reinforcing strip **34** and is stitched or otherwise suitably connected to itself so as to form a sleeve, or overlap, through which reinforcing strip **34** extends. Reinforcing strip **34** is folded back onto itself so as to provide additional stiffness and tensile strength at upper marginal edge portion **32** of sheet member **18**. Similarly, lower marginal edge portion **36** of sheet member **18** is folded over a second reinforcing strip **38** and is stitched or otherwise suitably connected to itself so as to form a sleeve, or overlap, through which reinforcing strip **38** extends. Reinforcing strip **38** also is folded back onto itself so as to provide additional stiffness and tensile strength at lower marginal edge portion **36** of sheet member **18**. Sheet member **18** and reinforcing strips **34** and **38** are made from flexible waterproof material which may be a polymer coated fabric, e.g., rubber or plastic coated nylon.

The upper marginal edge portion **42** of the cover piece **40** is illustrated as being folded back over the upper end of strut **20** and against strut **20**. Similarly, lower marginal edge portion **44** of cover piece **40** is also folded back over the lower end portion of strut **20** and against strut **20**. Strut **20** and the corresponding cover piece **40** thereof are secured to sheet member **18** such that both strut **20** and cover piece **40** have their corresponding upper and lower portions interposed between the folded over portions of reinforcing strips **34** and **38**, respectively. The aforementioned folded over portions **42** and **44** of cover piece **40**, in conjunction with reinforcing strips **34** and **38** strengthen those regions of sheet member **18** which are subjected to large shear stresses from strut **20**. The struts **20** reinforce the upper and lower portions **28** and **30**, respectively, of sheet member **18**.

As a result, the portions **28** and **30** of sheet member **18** are adapted to pivot substantially as a unit about float member **22** so as to cooperate with one another in resisting external moments and forces applied thereto by waves on the surface of liquid **26**. Hence, when waves, currents or other activity on or near the surface of liquid **26** applies a force to sheet member **18**, it will be resisted by the respective upper and lower portions **28** and **30** thereof cooperating with one another via the interconnection of struts **20** which act substantially as a rigid element. As a consequence thereof, sheet member **18** will tend to remain substantially in a stable upright position even though subjected to external moments and forces by liquid **26**. In this manner a relatively inflexible barrier is formed so as to thereby entrap pollutants floating therein.

FIGURE 74: ANOTHER UNIROYAL FLOATING BOOM DESIGN

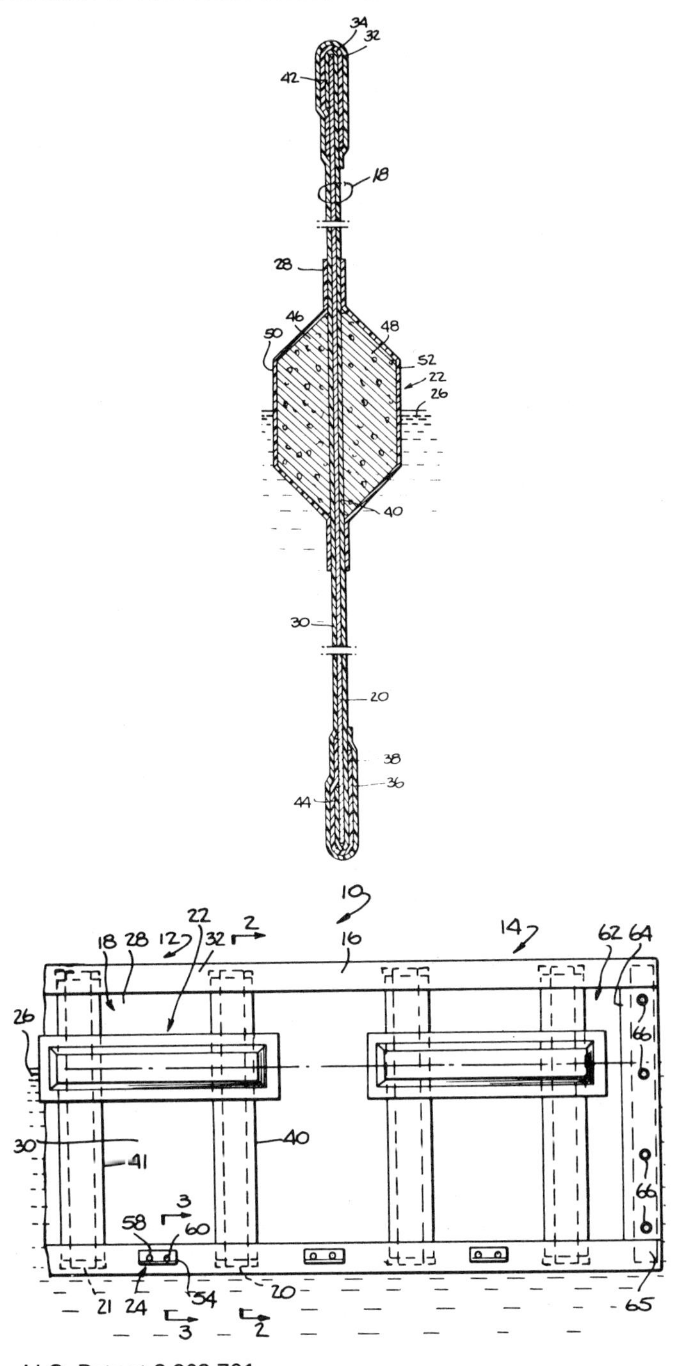

Source: U.S. Patent 3,903,701

As shown, float member **22** may include a pair of buoyant members e.g., a first buoyant member **46** and a second buoyant member **48**. Buoyant members **46** and **48** overlap one another throughout at least a common longitudinal portion of float member **22**. Sheet member **18** is interposed between the pair of buoyant members **46** and **48** in at least the common longitudinal portion of float member **22**. Moreover, struts **20** are also interposed between buoyant members **46** and **48** within the common longitudinal portion of float member **22**. Buoyant members **46** and **48** are made preferably substantially of a solid material, such as any suitable type of foam plastic. Although buoyant members **46** and **48** may be made from various types of foam, it is desirable to use a suitable polyethylene foam. The foam utilized for buoyant members **46** and **48** is preferably of the closed cell variety. This insures that the foam will not absorb any liquid and will remain buoyant even if surrounded by liquid.

Casings **50** and **52** secure corresponding buoyant members **46** and **48** to sheet member **18**. The casings **50** and **52** are also made preferably from a polymer coated fabric, such as rubber or plastic coated nylon, and are vulcanized or otherwise suitably fastened to sheet member **18** so as to form a liquid-tight repository for respective buoyant members **46** and **48**. Thus, casings **50** and **52** secure corresponding buoyant members **46** and **48** to sheet member **18** and also provide for protection of the buoyant members, i.e., protection from any deterioration caused by the environment in which they function. The location of buoyant members **46** and **48** with respect to the height of sheet member **18** is chosen such as to provide an effective barrier to pollutants floating on the surface of liquid **26**.

By way of example, struts **20** are made preferably from thin strips of spring steel; the strips may be about 0.042 inch thick. Preferably, float section **12** extends about 19 inches lengthwise of boom **10**, whereas sheet portion **16** extends about 21 inches lengthwise of boom **10**. Buoyant members **46** and **48** support sheet member **18** such that lower portion **30** thereof depends into liquid **26** substantially in a vertically downwardly direction for a distance of approximately 24 inches while upper portion **28** of sheet member **18** extends substantially in a vertically upwardly direction for a distance of approximately 11 inches. Thus, it is evident that about two-thirds of sheet member **18** extends downwardly beneath the surface of liquid **26**, and about one-third thereof extends upwardly above the surface of liquid **26**. Hence, when waves on the surface of liquid **26** apply a moment to upper portion **28**, both upper and lower portions **28** and **30**, via the interconnection of struts **20** and **21**, pivot as a unit about float member **22**, its movement is impeded by liquid **26** and a counter moment is developed by the force of liquid **26** exerted thereon so as to restore sheet member **18** to a stable vertical position.

A boom design developed by *J.H. Neal; U.S. Patent 3,921,407; November 25, 1975* consists of an erect band of oil-containing mesh material attached to spaced, vertical spars. The spars comprise hollow, rigid tubes partially filled with buoyant foamed plastic. Apertures are provided in the wall of the weighted, hollow lower end of each spar to permit water to enter therein and displace the contained air. By virtue of this arrangement, the weighted end of this spar will right it when it is dropped in water. The boom is uniquely light and can be unreeled quickly from a drum for deployment around an oil spill.

As shown in Figure 75 the boom comprises a series of erect mesh bands **2** interconnected by spaced, vertical spars **3**. Each such spar extends longitudinally across the full width of the mesh band. Each mesh band **2** may be discrete and connected at each of its ends to a spar or a longer band could be used, attached to more than two spars.

FIGURE 75: NEAL DESIGN FOR SPILL-CONTAINING BOOM

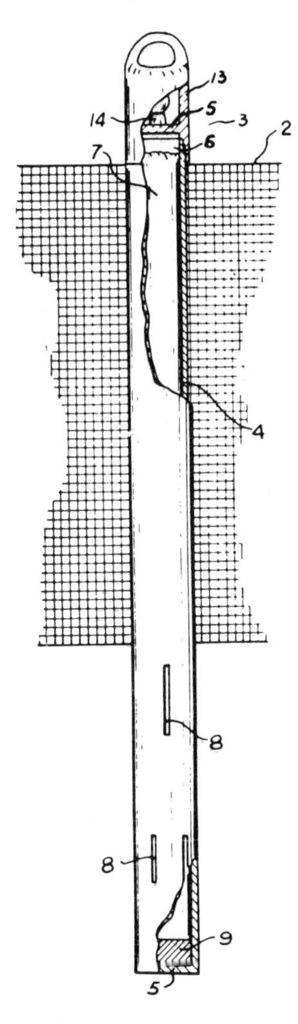

Source: U.S. Patent 3,921,407

The mesh band **2** is formed of longitudinally flexible, water-permeable, oil-impermeable material, which is stiff enough to hold itself erect between spars. Several such materials are commercially available. They often consist of a close weave which defines small apertures through which water will pass but which are too small to permit the passage of oil. A suitable material of this type is polypropylene fiber.

Each spar **3** preferably comprises a hollow, cylindrical casing **4** having end walls **5** and defining a chamber **6**. The casing is quite rigid, and may satisfactorily be made of polyvinyl chloride pipe. In its upper end, the casing **4** is filled with a buoyant, water and oil-impermeable material **7**, such as foamed urethane. Apertures **8** are formed in the lower end of the casing wall. The casing **4** is weighted at its lower end, as by providing a lead weight **9** in the chamber **6**. When the spar is dropped onto a body of water, the weight **9** forces the lower end of the casing **4** to submerge; water enters the casing chamber **6** through the apertures **8** and replaces the air therein. The spar **3** will then right itself to the vertical position shown. Buoyancy is assured by the presence of the foamed urethane.

A lamp **14** or other signalling devices may conveniently be mounted in the cap **13**, if desired, to assist in locating the boom.

The boom may quickly be deployed. Because of its lightness and integral construction, it may be carried on a drum by a launch or helicopter. It can be unreeled at the oil spill site. On contacting the water, the spars **3** automatically right themselves and hold the band **2** in a generally vertical, partially submerged position for containing the spill. The boom can be gathered in by reeling onto launch-carried reels.

A boom design developed by *N.D. Tanksley; U.S. Patent 3,922,860; December 2, 1975; assigned to Pacific Pollution Control* includes an elongated sheet-like partition and a plurality of float means mounted to the partition at periodic intervals over the length thereof. The float means include plate-like float elements mounted to the float means for rotation about a horizontal axis to and from a deployed position, for floating support of the boom, and a relatively rotated stored position, which enables a substantial reduction in the bulk of the boom. The float elements are biased by buoyancy and gravity forces to the deployed position, and the partition is formed as a flexible member to enable folding of the boom with the float elements thereon into a compact package. A method of folding the boom for compact storage and rapid deployment is also disclosed. The boom as deployed is shown in Figure 76.

The floating boom, generally designated **21**, is comprised of an elongated partition or barrier **22**, preferably having a sheet-like configuration, and a plurality of float means, generally designated **23**, mounted to partition **22** at periodic intervals over the length of the partition. Each float means **23** includes a pair of float elements **24** and **26** with one float element disposed on each side of partition or barrier **22**.

Mounted to partition **22** is a float means frame **27** having laterally outwardly extending arms **28** which terminate in an arm axle portion. Float elements **24** and **26** are preferably formed as hollow plate-like members having a foam material, such as polyurethane foam, disposed therein. A central opening or bore dimensioned for sliding receipt of the axle portion is provided in the float element. Float elements **24** and **26** are held on the axle portions by an end retainer **33**. Openings **44** can be advantageously used for securement of anchor lines to the boom or for towing or similar manipulation of the boom in the water.

FIGURE 76: FLOATING BOOM DESIGN HAVING ROTATABLE FLOAT ELEMENTS

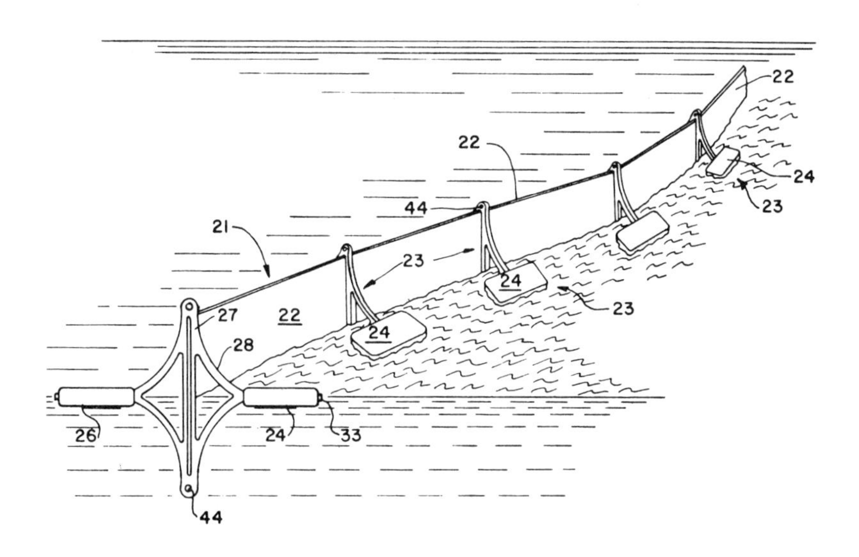

Source: U.S. Patent 3,922,860

The float means are shown in a form suitable for injection molding from a plastic material such as nylon. It will be understood, however, that frame **27** can be advantageously formed of other materials and other configurations. For example, frame **27** can be formed from stainless steel providing outwardly extending arms on which the float elements can be rotatably mounted. Similarly, the frame is shown with arms which space the float elements at a distance from the partition and provide a triangular opening in the arms for passage of oil and water between the float elements and the partition. This is a preferred form of construction; however, it is possible to employ a float means frame in which the float elements are mounted in immediate proximity to the partition. Moreover, the float elements themselves can be formed in a number of different configurations and from different materials.

It is preferable to rotationally mold the float elements from a plastic material such as nylon and then fill the float elements with a foam, such as polyurethane. The quantity of foam and size of the float elements depends upon the buoyancy desired and the intervals at which the frames are to be spaced along the sheet-like partition. Typically, the frames are spaced at about four foot intervals and the float elements are about three inches in thickness by 24 inches in length and 12 inches in width. Finally, the end sections of the belting are normally provided with connecting means (not shown) which allows the sections to be linked together to any desired overall length of the boom.

A boom design developed by *J.A. Bennett and I.R. McAllister; U.S. Patent 3,924,412; December 9, 1975 assigned to Bennett Pollution Controls, Ltd., Canada* is a floating oil containment boom especially adapted to remain continuously in the body of water in which it operates. The boom comprises an elongate, moderately flexible skirt, having an upper edge above the water surface and a lower weighted edge positioned below the water surface. Pairs of floats are mounted to the skirt at regularly spaced intervals along the length of the skirt. In a first embodiment, each float is mounted to the skirt by means of tongue and groove connecting device, made up of a longitudinal groove in the float and two tongue members connected to the skirt and positioned at opposite ends of the groove of the related float.

In a second embodiment the tongue and groove connection is made by a plurality of clips mounted to the skirt and defining a groove to receive upper and lower flanges of the float. Such tongue and groove mountings permit moderate relative longitudinal movement between the connecting portions of the skirt and float so that when the skirt is subjected to moderate elongation under tension loads, there is not undue wear of the connecting portions.

A fence design developed by *T. Tezuka, H. Kawakami, and K. Miura; U.S. Patent 3,939,663; February 24, 1976; assigned to Bridgestone Tire Co., Ltd., Japan* is one having a directional control device consisting of rudder plates or wind receiver plates secured to vertical plates or skirts of the fence at an angle thereto, which are subjected to a force due to relative movement of water or wind to cause the fence to position at an angle relative to the direction of the movement of the water or wind, whereby when the oil fence is being tugged by a towing boat the oil fence line is positioned at an angle without being subjected to any bending moment or any other undue force so as to collect or recover pollutants, e.g., spilt oil effectively from water surface.

A barrier design developed by *J.H. Milgram; U.S. Patent 3,943,720; March 16, 1976; assigned to Offshore Devices, Inc.* is one for use in the open sea to contain oil on the surface, comprising a flexible barrier provided with a plurality of flotation means only on the outside of the barrier, thereby avoiding churning of the oil and water on the inside of the barrier into an oil-water mixture that could pass under the barrier in rough sea conditions.

A barrier design developed by *K. Aramaki, Y. Kawaguchi and H. Kawakami; U.S. Patent 3,962,875; June 15, 1976; assigned to Bridgestone Tire Co., Ltd., Japan;* is very similar to that described earlier in this section in connection with U.S. Patent 3,867,817.

A fence design developed by *T. Kinase, I. Yano, K. Okubo, H. Kitakoga and H. Tayama; U.S. Patent 3,971,220; July 27, 1976; assigned to Mitsubishi Denki KK, Japan* features float units each divided into small individual floats. This results in several advantages. For example, the present construction provides oil fences easy in both manufacturing and setting up. Also those portions of the shielding screen member covered by the float units total a wide area resulting in a high resistance to a side wind. Further, the oil fence can readily be housed in place because it can be spirally wound on a drum or a shaft. In addition, the oil fence can be readily spread because it is required only to be unwound from the drum or shaft.

A boom design developed by *C. in 't Veld; U.S. Patent 3,979,291; September 7, 1976; assigned to National Marine Service, Inc.* includes a plurality of elongated, parallel rows of vertical, buoyant, barrier screen members connected together by flexible means and rigidly braced to hold the rows in parallel relationship in upright position when afloat in water. The basic elements of such a boom design are shown in Figure 77.

FIGURE 77: NATIONAL MARINE SERVICE INC. BOOM DESIGN

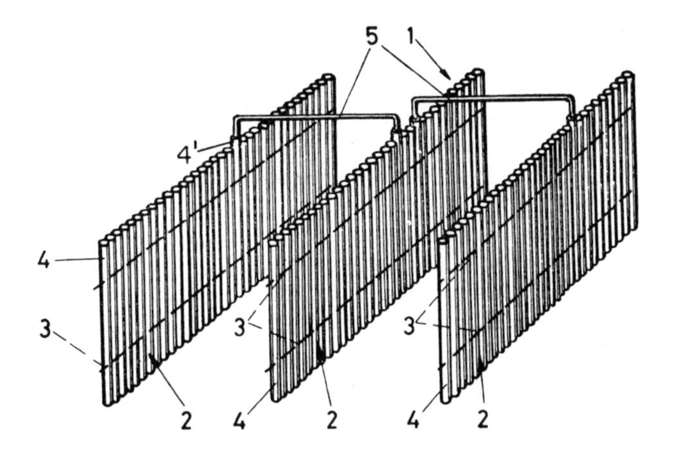

Source: U.S. Patent 3,979,291

The oil boom **1** shown consists of three parallel screen sections which are of the same construction. Obviously, the number of screens (variously referred to as partial booms) **2** may be chosen according to the circumstances.

Each screen **2** is built up of a large number of parallel vertical members **4** which are strung together by means of at least one strand **3**. In operation the members **4** in fact adjoin one another so that they form a screen for an oil patch floating on the water surface. The members **4** are in the form of hollow tubes or pipes of circular section of a material which floats in water, preferably plastic. According to one embodiment, the bars **4** are filled with foam plastic which is nonpermeable to water. The strand **3** by which the members **4** are strung together and of which the number may be chosen according to the circumstances extend through a transverse bore made in each bar extending perpendicularly to its axis. In the figure, each screen **2** is provided with two strands. Although these strands may be of any suitable material they preferably comprise nylon rope or cable.

Each two adjacent screens **2** are held in spaced parallel relationship by horizontal spacers **5** the number of which may be chosen according to the circumstances. Each spacer **5** is preferably made in the form of an inverted U-shaped brace. The material of the spacers is chosen with a view to the buoyancy of the oil boom. The legs of each are spacers **5** telescoped into a vertical member **4** of screens **2** which are opposite to each other in the direction perpendicular to the direction in which the oil boom extends. In order to assist in this telescoping, and for locating the positions of the members to receive the legs, the members **4'** have a greater upper length than the other bars.

The oil boom has a buoyancy of its own. In operation the screens **2** float in the water in a vertical position as shown, which vertical position is maintained by the employed spacers **5**. The vertical members **4** which have a length of approximately 40 cm have approximately half their length below the water surface. Due to the members being tightly strung together, in the event of a current in the water which is perpendicular to the direction into which the oil boom extends, the oil floating on the water cannot pass through the screen between the members. In the event of a strong current in the water, a small quantity of oil which lies against the outer or most forward screen may be entrained with the water flow passing beneath the bottom of the forward screen into the space between the latter and the central screen. This amount of oil in the space rises to the water surface and is displaced against the forward face of the screen.

Extremely strong oncoming currents in the water will have to set in to cause oil to underpass the central screen, but the thrid screen will catch such underpassing also. The oil boom thus effects a complete blocking of oil flow. The number of screens may be chosen in accordance with the circumstances (weather conditions, current). In this connection the small draught of the oil boom is pointed out which results in the flow of water below the boom due to current in the water being less than other booms, and therefore the chance that oil is entrained in the underpassing water is smaller than with a boom having a greater draught. To this advantage is added the advantage that the amount of oil which has possibly underpassed the first screen is arrested by the second or third etc. screens.

When storing the oil boom, the spacers **5** are removed while the screen comes into abutting relationship and the boom may be wound on a reel like a conventional oil boom.

A barrier design developed by *P. Preus; U.S. Patent 3,998,060; December 21, 1976* consists of a series of end to end connectable boom sections containing a floatable material which may comprise an oleophilic-hydrophobic-lighter than water composition which will selectively absorb hydrocarbons.

In a typical case, such an oleophilic material is "Sorbent C," an oleophilic-hydrophobic-lighter than water composition which will selectively absorb hydrocarbons floating on water. Such oleophilic-hydrophobic-lighter than water composition may comprise expanded pearlite 60-80% by weight; cellulose fibers 13-33% by weight; clays 4-8% by weight; and asphalt 1-5% by weight, all as more fully described in U.S. Patent 3,855,152.

The boom design then contains "Sorbent C" inside the boom structures whereas in U.S. Patent 3,795,315 discussed earlier in this section, "Sorbent C" was added to the slick to make the slick impenetrable to a barrier skirt.

A boom design developed by *F.J. Campbell and D.J. Graham; U.S. Patent 4,016,726; April 12, 1977; assigned to the U.S. Secretary of the Navy* utilizes an improved connector hinge as shown in Figure 78.

The female connector **10** and male connector **12** are each connected to an oil containment boom. Quick disconnect clevis pin **17** prevents sliding of male connector **12** within the female connector **10**. The connector hinge comprising female connector **10** and male connector **12** allows numerous oil containment booms, such as oil containment boom **15**, to be rapidly connected and disconnected to contain an oil spill on the open sea.

FIGURE 78: CONNECTOR HINGE DESIGN FOR OIL CONTAINMENT BOOMS

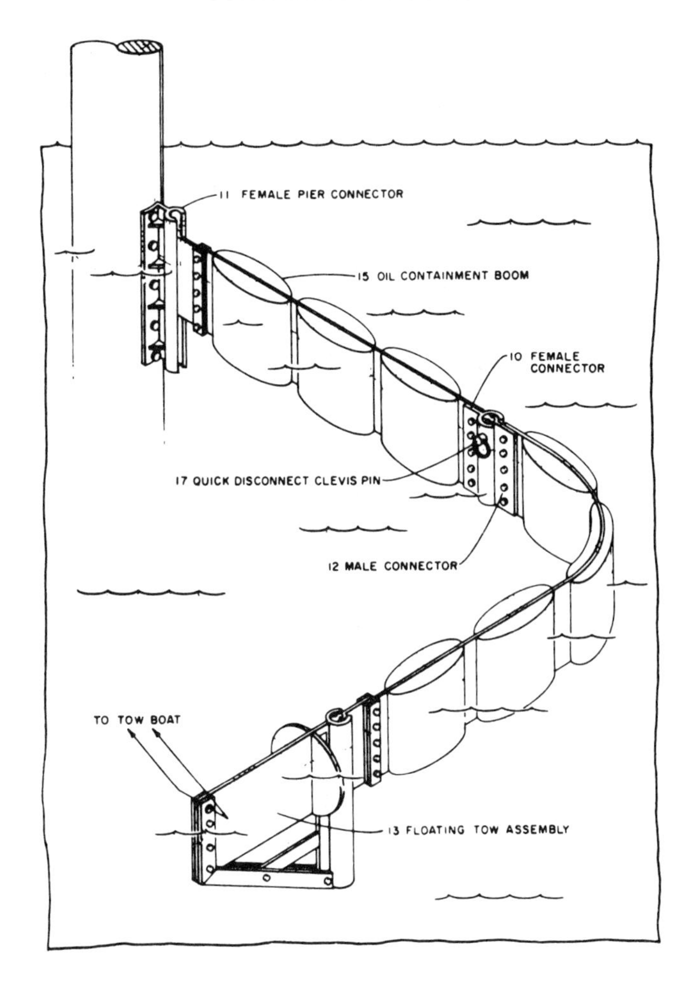

Source: U.S. Patent 4,016,726

Female pier connector **11** is an extended section of the female connector **10**, which can be attached to a stationary object such as a pier. The female pier connector **11** and male connector inserted therein provide a connection which allows adjustment of the oil containment boom **15** with tidal variations due to their loose assembly. Floating tow assembly **13** is connected to the last oil containment boom of the series of oil containment booms **15** and serves to guide the series of oil containment booms around the oil spill for proper confinement.

A boom design developed by *R.E. West; U.S. Patent 4,030,304; June 21, 1977; assigned to Cascade Industries, Inc.* is shown in cross section in Figure 79.

FIGURE 79: CASCADE INDUSTRIES INC. FLOATING BOOM DESIGN

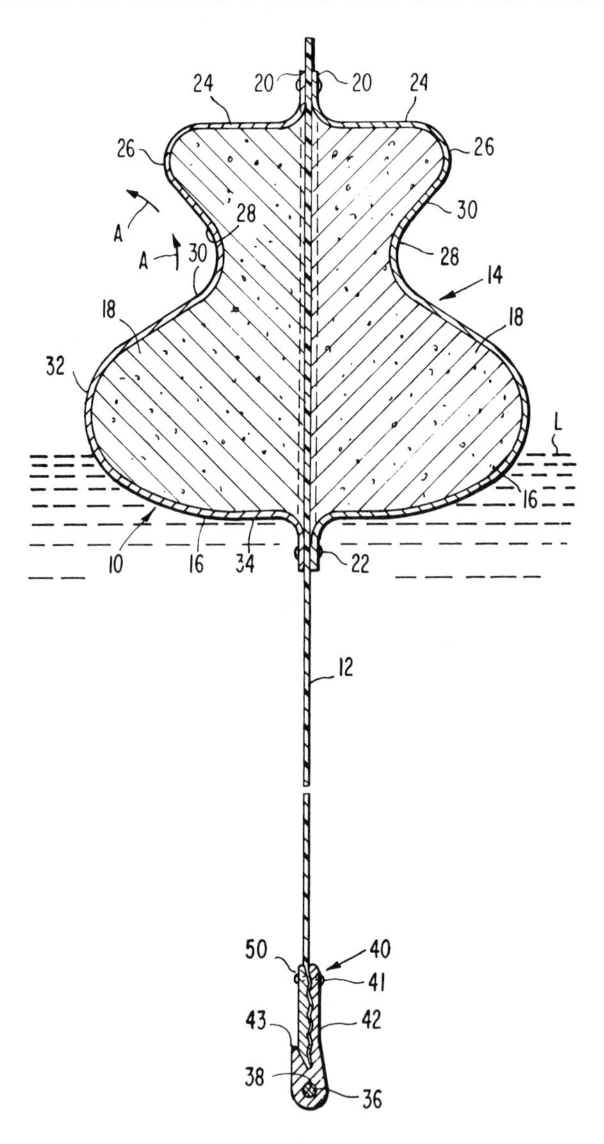

Source: U.S. Patent 4,030,304

Each section **10** of the floating boom comprises an elongated skirt **12** which is formed of a plastic material, as for example, a nylon fabric laminated with neoprene or some other type of flexible plastic that would be substantially unaffected by the action of oil, sea water or the like.

Spaced uniformly along the upper edge portion of the skirt are floats **14** each of which is of elongated configuration, each float in the illustrated example being of rigid formation, comprising, for example, a molded shell confining solid or particulate buoyant material **18**. Each float **14**, is symmetrically disposed in respect to the skirt **12**, and comprises a pair of identically formed, oppositely arranged float sections **16**, secured to opposite faces of the skirt **12** through the provision of peripheral flanges **20** formed upon the respective float sections, said flanges having registering apertures through which rivets or other fastening elements extend as at **22**, for the purpose of permanently and fixedly securing the float sections to the opposite faces of the upper edge portion of the skirt.

The particular cross-sectional shape of each float **14** is of importance. Each float, as seen, is bisected by a vertical plane coincident with the plane of the skirt **12**, and at opposite sides of the plane, each float is formed with horizontally disposed top wall portions **24**, merging along their outer edges along curving lines into outwardly curved, bulged top float portions **26**, which in turn merge along their lower edges into oppositely, inwardly curved reentrant midportions **28**.

It will be observed that the inwardly curved reentrant midportions of the float when viewed in cross section define recesses **30** that extend along the full length of each float, continuously up to the peripheral flange **20**, said recesses being provided at opposite sides of the skirt by reason of the symmetrical relationship of the float to the skirt. As seen by the directional arrows **A**, waves contacting the float **14** from either side thereof will be caused to deflect or travel upwardly along the outwardly curved lower, bulbous portion **32** into the inwardly curved reentrant recess **30**, and will be deflected by the wall of the recess reversely, instead of washing pollutants across the boom structure, thus returning to the side of the boom structure from whence they approached the same.

The bulbous lower portion is provided with a bottom wall **34** which is curved to a position approaching the horizontal. This arrangement or form of the lower portion **34** is calculated to impart a high degree of stability to the boom, tending to retain the boom in its proper operational posture within the water, despite the use of the same in rough water, in that it presents a "hard bilge" supported by the upper part of the wall of portion **32** where it merges into reentrant portion **28**.

A construction as illustrated causes the floats to ride quite high in the water as shown with the recesses **30** disposed above the normal water level **L** a distance such as to assure to the maximum the efficient discharge of their function of deflecting waves in normal use of the boom.

A tension member **36** is in the form of a length of strong wire cable or the like, capable of being tensioned in a longitudinal direction after passage through a longitudinal bore **38** formed in a clamp assembly generally designated **40**. A plurality of the assemblies would be provided in each boom section, the assemblies being generally coextensive in length, and being transversely aligned with, the several floats **14** of the section. Assemblies **40** are riveted at **41** to the skirt.

Each clamp assembly comprises a pair of cooperating, extruded metal elements, formed of aluminum or similar material. Each element is of elongated formation, and one of said elements has been designated **42**, being formed approximately to a J-shape when viewed in cross section. The other element is designated **50**. At the base of the J there is provided, in the clamp element **42**, the longitudinal bore **38** mentioned previously.

A barrier design developed by *J.J. Geist; U.S. Patent 4,033,137; July 5, 1977* consists of buoyant sections connected by a sliding seal for allowing sliding relative vertical motion and limited relative angular movement between the sections to accommodate extreme and abrupt variations in water surface level and turbulence resulting from wave action. The seal connection is made up of a vertical inner slide member, either integral with or attached to one end of one section, and a mating outer member integral with or attached to the opposite end of a similar section, the outer member having an undercut, open-ended slot for engaging the slide member with sufficient clearance to permit relative vertical sliding movement between the members while at the same time providing an effective seal between the sections. The inner slide member is preferably in the form of a cylindrical rod, and the opening of the mating slot in the outer member is preferably wide enough to permit limited relative angular movement between the two sections in the horizontal plane so that the connection serves as a combination sliding seal and hinge. Such a barrier design is shown in Figure 80.

FIGURE 80: ARTICULATED FLOATING BARRIER DESIGN

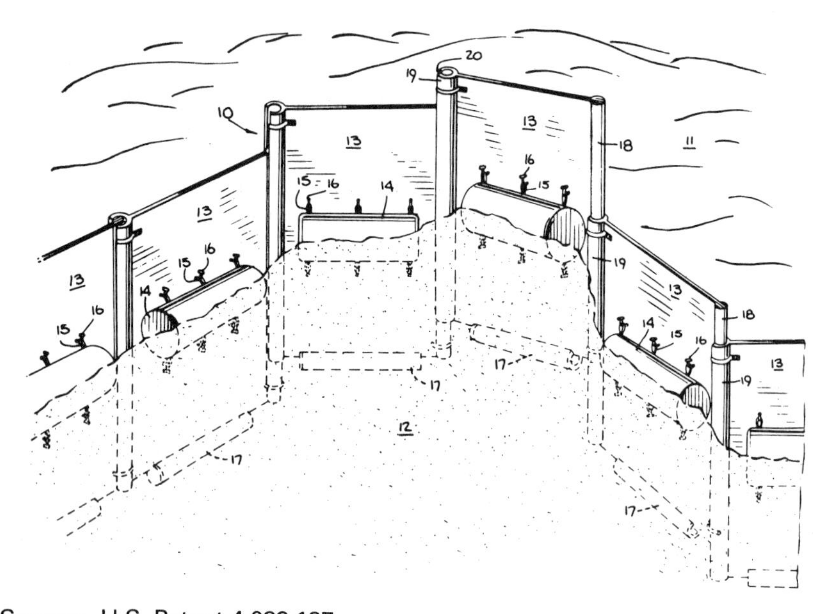

Source: U.S. Patent 4,033,137

As shown in the figure, a portion of an articulated segmented barrier **10** floats on a body of water **11** to contain an oil spill **12**. The barrier comprises a plurality of buoyant sections in the form of flat rectangular panels **13** supported in an upright position by flotation means in the form of floats **14** attached by loops **15** to hooks **16** on the sides of the panels approximately midway between the top and bottom. Ballasting means in the form of weights **17** stabilize the panels in the upright position by increasing their metacentric height (e.g., the vertical distance between their center of gravity and center of buoyancy).

The buoyant sections of the barrier are connected end-to-end by means that serve as a sliding seal to allow relative vertical movement and limited angular motion in the horizontal plane between adjacent sections while effectively preventing escape of the contained oil. In the illustrated embodiment, the connection means includes a slide member in the form of a vertically disposed cylindrical rod or hinge pin **18** extending along one end of each panel **13** and a member having a mating undercut slot in the form of a vertically disposed tubular hinge journal **19** extending along the opposite end of each panel. Hinge journals **19** have outward-facing, longitudinal openings or cuts **20** extending along their length, openings **20** being wide enough to accommodate the thickness of the adjacent panel when its hinge pin is slidingly engaged by the hinge journal, with sufficient additional clearance to permit limited angular motion between the adjacent panels.

A design developed by *B.J. Sessions; U.S. Patent 4,043,131; August 23, 1977* is an oil-retaining boom structure for disposition upon a body of water, preferably surrounding an offshore oil well. It has a plurality of buoyant float members movably connected together by a connecting cable which extends through the body structure of each float member. Each float member comprises an elongate housing having weights disposed in one end thereof adapted to retain the housing end downward in the water. The floats are interconnected with each other through the connecting cable with washer means surrounding the cable between the adjacent floats. The washer means preferably have grooved means surrounding the circumference of the washer to accommodate elongate rods which can be used to retain portions of the boom in parallel relationship if desired to form a channel for the oil within the boom. The floats have sheet means attached to at least one side thereof extending along the boom to retain the oil on one side of the sheet means as the boom rides at or near the surface of the water.

A device developed by *H. Kawakami, I. Nagaoka and Y. Tsukagawa; U.S. Patent 4,049,170; September 20, 1977; assigned to Bridgestone Tire Co., Ltd., Japan* is a device for pulling up and paying out an oil boom. It comprises a structural frame, a pair of end pulleys horizontally journaled on the frame, an endless belt engaged with the end pulleys. There are a plurality of engaging projections for an oil boom and a pusher member for slightly urging the oil boom against the endless belt, the pusher member being resiliently supported by a supporting member with a proper distance held between the bottom portion of the pusher member and the upper surface of the endless belt.

TREATMENT

Removal of spilled oil may be accomplished in several ways, including:

a. Allowing evaporation to take place (gasolines and JP-4).

b. Use of physical removal methods such as manual collection or collection by mechanical equipment such as skimmers.
c. Removal by fostering biodegradation.
d. Removal by burning.
e. Removal by dispersion (emulsification).
f. Pumping of oil in land spills.

Because of other effects detrimental to the environment, methods (d) and (e) are not recommended, or practiced, by the Navy unless there is a direct threat to human life and property. Because of the lengthy reaction time involved, and because of the possibility of toxic products, method (c) is not practiced and is not recommended as a desirable Navy practice. However, it does occur, and can constitute a final polishing action if all the oil is not removed by physical means.

In addition, gelling agents (chemicals which convert the spill to semisolid mass) or sorbent materials, such as straw, polyester plastic shavings or polyurethane foam, may be used to facilitate the subsequent manual or mechanical removal of the spill.

Dispersion

When crude petroleum or fractions obtained by the refining of crude petroleum, hereinafter referred to as oil, are transported over water by tank ships, there is a possibility that spillages can occur. One way of dealing with such spillages of oil is to spray a detergent composition onto the oil, and to agitate the spilt oil and water to form an emulsion of the oil and water. The emulsion formed is diluted in the water, and consequently the spilt material is dispersed as very fine droplets thus reducing its effect on the marine environment and assisting its biodegradation.

Dispersants or emulsifiers are surface active agents which foster the development of oil/water emulsion. They may be ionic or nonionic in nature and are typically mixed with stabilizers to preserve the emulsion formed and solvents for cold weather use which reduce surfactant viscosity. A typical dispersant is about 70-80% solvent, 10-15% surfactant and 10-15% stabilizer.

The use of dispersants will expose a greater surface area for microbiological attack. However, many of the surfactants are not degradable; and they, or the materials with which they are mixed, may be toxic to microorganisms and aquatic species. Further, by dispersing the oil phase, they distribute it throughout the water column, extend its area of influence considerably, and have a resultant adverse biological impact. Also, the dispersant may have a short effectiveness period; thus the oil is released and resurfaces.

It is U.S. (EPA) policy that use of dispersants is not an acceptable method except for fire and explosion emergencies. Dispersion is not really a removal method but rather one of spreading the spill and reducing its visibility.

Dispersants (emulsifiers) are available with either a petroleum or water solvent base. Petroleum-based dispersants are more effective in dissolving weathered oil from equipment, whereas water-based dispersants are more effective in dispersing oil slicks in open water. Petroleum-based dispersants are more toxic to aquatic life, and their use is discouraged. The primary recommended use of dispersants is in the cleanup of booms, skimmers and other equipment following a spill. Use of water-based dispersants (detergents) in portable, high-pressure, heated washing units (temperature of 120°F and 1,200 psi) is also effective in the cleanup of equipment. In any case, care must be exercised in the control and disposal of dispersant, oil, and water mixtures resulting from equipment cleanup to prevent recontaminating the environment. During actual spills it has been shown that application and mixing of dispersants directly on an oil slick is not effective inasmuch as the oil slick is likely to resurface in patches and result in a more difficult cleanup task. For this and environmental reasons, the use of dispersants directly on an oil slick is not recommended. Use of dispersants may be deemed necessary by the OSC to prevent loss of property or prevent bodily injury, in cases involving potentially explosive products like AVGAS; but even in these cases, concurrence of the EPA and state agencies is required.

Concurrence by EPA and state agencies is not required to avert an immediate fire and explosion hazard especially when checking with these agencies would delay the initiation of life and property saving actions. When adequate lead time is available or if the operation is longer than a "one shot" situation, the RRT and EPA should be consulted. This caveat for fire and explosion hazards should not be abused by operating units who suddenly decide that an increasing number of incidents are "fire hazards."

If the use of dispersants is contemplated, it should be noted that they are most efficient on low-viscosity products and are ineffectual on Bunker C and weathered Navy Special or Distillate Fuel. The use of dispersants on sandy beach areas is not recommended since they cause sand to lose its compaction and drive the oil deeper into the beach. The oil then "bleeds" back to the surface with each incoming tide.

A technique developed by *P.G. Osborn, P. F. Nicks, and M.G. Norton; U.S. Patent 3,996,134; December 7, 1976; assigned to Imperial Chemical Industries, Ltd., England* is a technique for dispersing oil in water which is particularly useful in oil slick dispersion, secondary oil recovery and tar sand oil recovery. The technique comprises contacting the oil and water with an alkyd resin in which one component of the resin is the residue of a water-soluble polyalkylene glycol such as a polyethylene glycol.

A composition developed by *P.M. Blanchard and D. G. Meeks; U.S. Patent 3,998,733; December 21, 1976; assigned to The British Petroleum Co., Ltd., England* is a detergent composition for dealing with oil spills on water which comprises an ester of a polyoxyethylene glycol ester of a C_{10}-C_{24} fatty acid, an organic sulfate sulfonate detergent, an alcohol or glycol of molecular weight 90-250.

A small amount e.g., 0.1 to 5% wt of the total composition, of a low molecular weight alcohol can optionally be added to the composition in order to increase the rate of dispersion in the spilt oil. Examples of suitable low molecular weight alcohols are isopropyl alcohol, sec-butyl alcohol and hexylene glycol.

In use on the open sea, water can be added to the dispersant composition just before it is applied to the spilt oil e.g., by having a nozzle with two entries, one for seawater and one for the dispersant composition. Preferably the dispersant composition is mixed with from 2 to 10 volumes of water before being applied to the spilt oil.

The dispersant composition dispersed in seawater, can be applied to the spilt oil by any conventional method. A usual method for applying dispersant compositions is simply by spraying the diluted composition onto the surface of the spilt oil. The agitation of the mixture formed by application of the dispersant can take place as a consequence of hosing or spraying the dispersant composition into the spilt oil, and at sea by natural wave motion. Alternatively ships can be driven through the mixture, and the disturbance caused by their passage and the action of their propellers providing agitation. In some cases a surface agitator can be towed through the spilt oil/dispersant mixture.

The amount of dispersant composition added to the spilt oil is usually from 5 to 100% by volume of the spilt oil, preferably from 20 to 50% by volume based on the amount of dispersant composition depending on the viscosity of the spilt oil.

When oil is spilt on beaches or washed up on beaches or other land areas, the dispersant composition can be diluted with a hydrocarbon solvent such as a de-aromatized kerosene e.g., kerosene with an aromatic content of below 3% by weight. The amount of hydrocarbon solvent is preferably 50 to 90% by weight of the composition.

Sorption

Sorbents use both adsorption and absorption processes to scavenge oil. They are widely used for removal of small land spills and near shore spills and to remove traces of oil left after skimming operations, or to mark and retard the spread of spills on water. They are also used as protective barriers for shorelines.

Sorbents may be naturally occurring or modified natural products such as straw and clays, or synthetic materials such as polymers, resins, and rubber. The polyurethane foams are the most effective sorbent material and are commonly used in "adsorbent booms" (packed in netting), in sheets or pads. The Navy Oil Spill Containment and Cleanup Kit, Mark I (Shipboard) uses these foam mats. However, polyurethane foam is not recommended in water under relatively strong wind and wave conditions. Under these conditions fibrous perlite has been found to be more effective because it holds together well and can be picked up easily.

Table 21 presents several sorbent materials and their pickup ratios. The pickup ratio is defined as the unit of weight of oil sorbed per unit weight of material. The amount of oil adsorbed, however, will vary with the nature of the oil involved. Straw, for example, is most effective with bunker fuel oil and least effective with gasoline or JP-4.

These materials are generally nontoxic and can improve the effectiveness of pickup or skimming operations. However, distribution systems are deficient, and high costs are associated with handling and ultimately disposing of large quantities of sorbent materials.

TABLE 21: TYPICAL SORBENT MATERIALS AND THEIR "PICKUP" RATIOS

Sorbent Material	Pickup Ratio*
Ground pine bark (dried)	1.3
Sawdust (dried)	1.2
Reclaimed paper fiber (dried and surface treated)	1.7
Ground corncobs	5
Straw	3-5
Chrome leather shavings	10
Asbestos (treated)	4
Fibrous perlite	5
Vermiculite (dried)	2
Polyester plastic (shavings)	3.5-5.5
Resin type	12
Polyurethane foams	15-80

*Weight of oil removed to weight of sorbent used.

Source: Reference (9)

Properly used, sorbents can be very effective in minimizing environmental damage and overall cost of cleaning up an oil spill. However, their use should be carefully controlled since all sorbents dispersed in the control and cleanup of a spill must be recovered, and they do place a load on transfer pumps and ultimate disposal requirements of the collected product. Sorbents are manufactured in three forms: granular, mat, and sorbent boom. It has been verified by controlled laboratory and field tests that the most effective sorbent is polymeric foam which can be produced in all three forms. Its oil-sorbing capacity is much greater than organic materials such as straw or other natural fibrous products, but it is much more expensive. However, polymeric foam can be squeezed out and reused; and if desired, it can become more cost-effective than other single-use products. It is available in sheets, rolls, or small chunks. The Federal Supply System has available 4′ x 4′ x 1″ sheets of sorbents (FSN 9G 9330-158-2353 or 9330-113-5856). The FSN foam sorbent mats are part of the U.S. Navy Oil Spill Containment and Cleanup Kit, Mark I (9).

Key uses of sorbents in a spill operation include covering work areas of skimmers, support boats, and docks, to minimize personnel injury due to falls; and covering land spills to minimize oil leaching to ground water areas. Sorbents can be spread along the upwind edge of a slick to provide wind sail area in order to move the slick more rapidly to a waiting skimmer. They may also be spread around the edges of a slick floating in open water to reduce spreading and provide for operator visibility during subsequent pickup, or they may be spread along beaches or rocky shorelines prior to arrival of an oil slick. They are also used in or near drainage outfalls to trap or coagulate oil for easier pickup. Additionally, sorbents are sometimes used as a final polishing method to remove the last traces of an oil spill. Oil-soaked sorbent should be disposed of in a manner approved by state and federal environmental agencies. Incineration may not be allowed in some areas for some sorbents.

A technique developed by *C.O. Bunn; U.S. Patent 3,783,129; January 1, 1974; assigned to Col-Mont Corp.* is one in which a matrix material is provided for recovering oil from water comprised of finely divided coal particles bonded in spaced relation by polyethylene. The oil sorption capability of the matrix is exceptionally high and the matrix is highly selective to oil in the presence of oil and water. A closed system is provided for forming the matrix material and for separating the sorbed oil from the matrix material for reuse of the latter. The matrix material can be in the form of a fixed or movable bed through which the oil and water pass for selective sorption of the oil, or the material can be dispersed on the water surface and collected following oil sorption. Figure 81 shows a ship adapted for oil spill removal using this technique.

FIGURE 81: VESSEL FOR OIL SPILL REMOVAL WITH COAL/POLYETHYLENE SORBENT MATERIALS

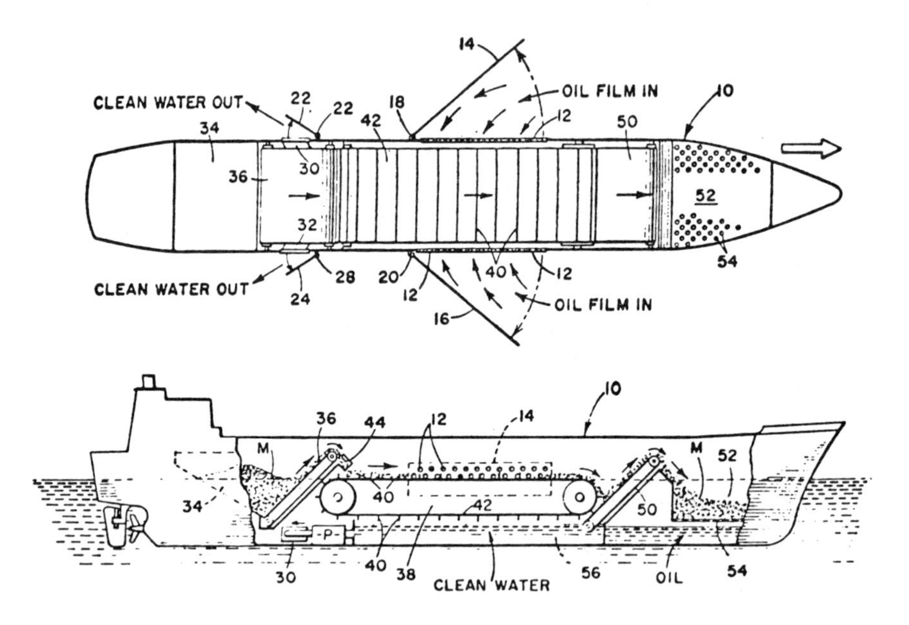

Source: U.S. Patent 3,783,129

The basic construction and powering for the recovery vessel **10** can be of conventional design, with the interior of the ship being specifically designed and equipped.

A plurality of openings, commonly designated at **12** are formed generally intermediate the vessel and are adapted to be closed when the vessel is inoperative by baffle members **14** and **16** mounted on the side of the vessel by pivotal connections **18** and **20**, respectively. The baffles **14** and **16** are shown in a fully open position in the figure and are adapted in such a position to guide the oil and water into the interior of the vessel through openings **12** when the vessel is traveling in the direction indicated by the arrow. When the vessel **10** is inoperative for the purpose presently intended, the baffles **14** and **16** are pivoted to a closed position contiguous the side of the vessel and retained in such position.

Outlet baffles **22** and **24** are pivotally connected on the vessel to the rear of baffles **14** and **16** by pivotal connections **26** and **28**, respectively. Water entering the vessel with the oil flows outwardly of the vessel through openings **30** and **32** formed in the vessel, as indicated by the arrows, with the assistance of pump **P**, if necessary. Although not illustrated, it will be understood that the discharge of the clean water is accomplished without the entrance of water into the vessel through the openings **30** and **32**.

Located in the interior of the ship is a lean matrix storage area **34** for storing the matrix material indicated at **M**. A matrix feed conveyor **36** is provided at the forward, inclined end of the storage area **34** for conveying the lean matrix material **M** to a matrix bed conveyor **38** extending longitudinally of the ship in the intermediate portion thereof. The conveyor **38** includes spaced transverse plates or structurally adequate screen members commonly designated at **40** which cause the generally forward motion of the matrix counter to the generally backward flow of the oil-polluted water to insure integrity of the matrix bed and to maintain the counterflow action between the matrix and the oil and water. A pair of transverse distributor mechanisms or baffles **44** are provided at the discharge end of the feed conveyor **36** for both directing the material **M** onto the matrix conveyor and regulating the bed thickness. If desired, the plates **44** may vibrate or reciprocate to transversely level the matrix fed by conveyor **36** from the storage area.

A discharge conveyor **50** is located at the discharge end of the matrix conveyor **38** for carrying the oil-enriched matrix material into the rich matrix storage area **52**, with the latter being provided with a perforated plate **54** which retains the rich matrix material while permitting gravity draining of the oil therefrom into the bottom of the storage area as illustrated.

In the operation of the recovery vessel as shown the vessel is directed through the area of oil spillage and the oil and water are directed into the vessel through the openings **12**, with the baffles **14** and **16** being open. The oil and water are directed over and through the bed of lean matrix material carried by the conveyor **38**, with the oil being sorbed by the matrix material and the water gravitating through the conveyor to the water sump **56** below the matrix conveyor **38**. Lean matrix material is continually delivered to the matrix conveyor **38** by the feed conveyor **36**, and the enriched matrix material **M** is conveyed by the discharge conveyor **50** to the rich matrix storage area **52**.

Depending upon the extent of the oil spillage, the rich matrix material can be stored in the storage area for continued draining of the oil from the matrix material or for subsequent processing of the material by other equipment. Alternatively, when the extent of the spillage makes reuse of the matrix material a desired objective, the separation of the oil from the matrix material can be accomplished by distillation equipment housed within the vessel by which substantially all of the distillable oil can be removed from the material and the latter conveyed to the lean matrix storage area **34** for subsequent conveying to the matrix conveyor. In either event, it will be understood that as the vessel passes diametrically through the oil spillage, the vessel will be turned around and the vessel again directed through the oil film.

A very similar technique is described by *C.O. Bunn; U.S. Patent 3,846,335; November 5, 1974; assigned to Col-Mont Corp.*

A technique developed by *J. Teng, J.M. Lucas and R.E. Pyler; U.S. Patent 3,788,984; January 29, 1974; assigned to Anheuser-Busch, Inc.* for removing spilled oil from water surfaces without further injuring the environment involves applying a carbohydrate fatty acid ester in powder, fibrous, or granule form to the oil covered surface. The preferred ingredient is cellulose acetate. The additive is nontoxic, biodegradable, water insoluble, and is not degraded by the acids in petroleum fuels. The additive is sprinkled on the surface of the oil coated water and after absorbing many times its weight in oil is easily removed. The picked up oil is easily removed from the cellulose acetate and up to 95% recovery of spilled oil is possible. The cellulose acetate also can be reused.

Cottonseed hulls and sawdust as the starting cellulose material are desirable because of low cost and the enhanced oil absorbing character of the resulting acetate.

A technique developed by *C.O. Bunn; U.S. Patent 3,798,158; March 19, 1974; assigned to Col-Mont Corp.* for removing oil from water is one in which the oil is sorbed by carbonaceous material either in the form of −200 mesh or finer carbon particles in the form of coal, fly ash or activated carbon, or a porous matrix comprised of such finely divided carbon particles and a powdered plastic such as polyethylene, or a combination of powdered carbon particles and matrix. The sorbent is stripped of the oil thereby conditioning the same for reuse in the system, or withdrawal from the system.

This closely follows the technique described earlier in this section for a ship-borne removal system as described in connection with U.S. Patent 3,783,129.

A scheme developed by *H.V. Hess and E.L. Cole; U.S. Patent 3,800,950; April 2, 1974* involves projecting open cellular particles of a highly oleophilic plastic foam such as polystyrene having a solvent affinity for petroleum substantially equivalent to that of polystyrene foam, to preferentially effect absorption into lumps, mechanically recovering the lumps by screening and thereafter completely burning the oil saturated lumps to final disposal thereof.

The apparatus for carrying out the method includes a heated foaming vessel supplied with foamable plastic particles and means for projecting the particles, after they have been foamed or expanded, upon oil floating on a water surface. These means include a barrel to which the foamed particles are conveyed as well as propulsive means which disintegrate the foamed particles into relatively small particulates, which are then cast onto the oil. Preferably the oil is surrounded by the foamed particles. The apparatus can be mounted on a barge or other vessel.

A process developed by *E. Muntzer and P. Muntzer; U.S. Patent 3,804,661; April 16, 1974; assigned to Westdeutsche Industrie-und Strassenbau-Maschinen GmbH* is one in which powdered comminuted or granular materials are made hydrophobic or water repellent and simultaneously oleophilic or oil attracting by first mixing the materials with a chromium salt solution and then adding to the mixture a bridging agent such as a thermoplastic hydrocarbon which envelopes the material particles. These materials are especially suitable for the clean-up of oil spills.

The following is one specific example of the application of such a process.

1,000 kg of perlite which is an expanded mineral substance having a volume of 12 cubic meters per ton is used as the carrier or base material. 0.200 to 1.200 kg, or 0.02 to 0.120% by weight of chromic sulfate are admixed to the base material together with 20 to 100 kg, or 20 to 100% by weight, of water. Further, 0.4 to 4.00 kg, or 0.04 to 0.4% by weight, of superstabilized bitumen emulsions in water are added to the mixture.

The addition of the emulsion in the above stated range is desirable in order to achieve a good adhesion or bond of the very fine perlite dusts. The mass is homogenized and then dried. The product has been found hydrophobic and is so oleophilic that it can bind in an adhesive manner the oil floating on the water to an extent or quantity corresponding to 5 times and more the weight of the mixture.

Activated perlite can be used as often as desired, whereby it is removed from the water as by skimming, burned-up and thereafter applied again to the water surface as an oil binding means. It is even possible to perform the burn-up step while the mixture is floating on the water because the water repellent perlite forms a wick only for the oil and does not provide a passage for the water into the combustion zone.

A composition developed by *L.E. Stern; U.S. Patent 3,812,973; May 28, 1974; assigned to Kritbruksbolaget I Malmo AB, Sweden* is a new floating-type oil pollution controlling agent which per unit of weight takes up a large amount of oil and which is characterized in that it consists of fibers of a mixture of polyethylene and paraffin.

This provides an inexpensive and effective oil pollution controlling agent which can easily be spread and collected. The agent is also suited for taking up firing oil No. 3 or thicker oil. It is characteristic of the fibers that they consist of a mixture of polyethylene and paraffin. It should be observed that the paraffin is not a surface coating on the polyethylene but forms a mixture with the polyethylene in connection with the actual manufacture of the fibers. From the polyethylene-paraffin mixture there are obtained fibers which are effective in taking up oil and also inexpensive to manufacture because of the admixture of paraffin. The mixing ratio of polyethylene to paraffin can vary; a high content of paraffin gives a relatively brittle fiber while a high content of polyethylene yields a stronger but simultaneously more expensive fiber. The quantity of oil taken up per quantity of fiber is not appreciably affected by the ratio of paraffin to polyethylene in the fiber. It has proved that a suitable ratio of paraffin to polyethylene is from about 50:50 to about 60:40.

At the manufacture the fibers are given a suitable cross sectional diameter. The smallest possible cross sectional diameter is desirable in view of the consumption of polyethylene per quantity of oil absorbed. A suitable cross sectional diameter is 0.02 mm or less.

At the manufacture the fibers are obtained in the form of monofilaments which are then preferably comminuted into staple fibers.

A composition developed by *B.H. Clampitt, K.E. Harwell and J.W. Jones, Jr., U.S. Patent 3,819,514; June 25, 1974; assigned to Gulf Research and Development Co.* consists essentially of a foam of an ethylene-alkyl acrylate copolymer having a melt index of at least 800, preferably an ethylene copolymer containing from about 10 up to about 45 weight percent methyl acrylate having a melt index in the range of from about 1,000 up to about 2,000. Such foams may be used in the selective removal of oil from water surfaces. The recovery of the absorbed oils from said foams is feasible. Figure 82 shows a technique for broadcasting such a foam and then recovering the oil-soaked foam.

FIGURE 82: SCHEME FOR BROADCASTING ETHYLENE/ACRYLATE POLYMER FOAM ONTO SOIL SPILL AND RECOVERING OIL-SOAKED FOAM

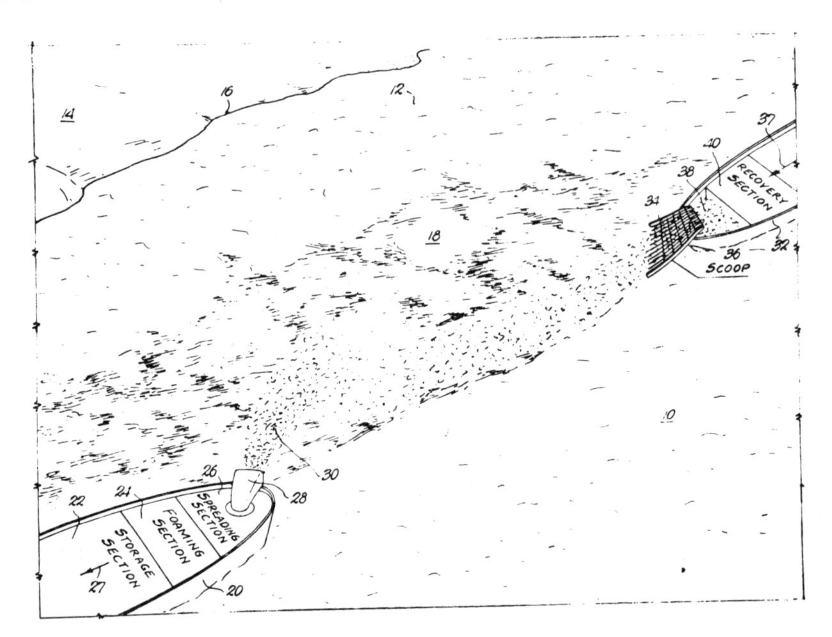

Source: U.S. Patent 3,819,514

As can be seen therein, a body of water **10** having a surface portion of area **12**, such as an ocean, bay, harbor, river, inland lake, or the like, exists adjacent a land area **14**. Body of water **10** and land area **14** meet and adjoin one another at a bank, beach, or the like **16**. The surface **12** of water body **10** has existing thereon an oil slick or spill **18** which it is desired to clean up or remove and generally recover. One method of accomplishing the selective recovery of oil spill or slick **18** from surface **12** of water body **10** is by employing ethylene copolymer foam absorber. In carrying out this particular method, a first ship, vessel or barge **20** is used and may be termed the spreading or casting ship.

Ship **20** typically has 3 compartments or sections therein being respectively identified as storage section or compartment **22**, foaming section or compartment **24** and spreading or casting section or compartment **26**.

In operation pellets or chunks, of conveniently transportable and storable size, of the blended mixture or material are stored within section **22** of spreader vessel **20** until the same are to be employed in the removal and recovery of oil. Then when spreader vessel **20** is in the vicinity of the oil slick **18**, the pellets or chunks in section **22** of vessel **20** are fed into foaming section **24** thereof. In foaming section **24** the pellets of the blended material are exposed to elevated temperatures in the range of about 150° to 180°C for periods of up to about 30 minutes for expansion into a free-rising open-cell ethylene copolymer foam composition. The resultant foam is then directed from section **24** into the casting or spreading section **26** for spreading or casting an oil slick **18** as vessel **20** moves in the direction of arrow **27**.

The foam composition is typically spread or cast over, on and about oil slick **18** by means of an air blower **28** in the form of small chunks or pieces **30**. It may be necessary to grind or chop the foam composition exiting section **24** or in its passage through section **26** before exiting through air blower **28** on to oil slick **18**. The foam particles or pieces **30** are allowed to remain on or within oil slick **18** for time periods sufficient for them to absorb their oil capacity before they and the absorbed oil are collected, gathered and picked up by a second or recovery ship, vessel or barge **32**. Vessel **32** is equipped with means such as scoop or rake **34** adjacent its bow **36** to generally pick up, collect and gather the foam chunks and chips and the oil absorbed therein as vessel **32** moves in the direction of arrow **37**.

Scoop or rake-like structure **34** delivers the foam particles and absorbed oil to a holding or storage section or compartment **38** and may employ belts or conveyors in so doing. From storage section **38** the foam particles and absorbed oil are delivered to recovery section or compartment **40** wherein the oil is recovered from the foam by expression and squeezing thereof. The recovered oil is then stored in the hold of vessel **32** and the particles of foam, after expression and recovery of the oil, may be immediately reused, stored for future use or disposed as may be desired.

A composition developed by *P. Preus; U.S. Patent 3,855,152; December 17, 1974* consists of a loose mass of expanded perlite mixed with clays and a fibrous filler formed by demoisturizing an aqueous suspension of the compound and comminuting the demoisturized mixture.

The compound is introduced into, or contacted by, an organic liquid-water system to selectively absorb the hydrocarbon whereupon the compound and the hydrocarbon absorbed therein are mechanically separated from the system. Similar ground is covered by *P. Preus; U.S. Patent 4,011,175; March 8, 1977.*

An adsorbent composition developed by *H. Hoshi, K. Murakami, I. Maeda and R. Susuki; U.S. Patent 3,862,963; January 28, 1975; assigned to Lion Fat and Oil Co. and Idemitsu Kosan Co., Ltd., Japan* is made of a foam substance consisting of 5-90% by weight of a synthetic resin insoluble in water as well as the oil and 10-95% by weight of an inorganic filler and having a bulk density of less than 1.

The following is one formulation of such a composition.

	Weight Percent
Polypropylene	10
Ethylene-vinyl acetate copolymer	30
Calcium sulfite	30
Calcium carbonate	30

100 parts of the above mixture is kneaded with 2 parts benzene sulfonyl hydrazide and 0.15 part tert-butyl peroxy isopropyl benzene.

A mixture having the above composition was foamed by heating at 180°C for 20 minutes, whereby a white sheet-type foam material having bulk density of 0.2 and a thickness of 2 mm was obtained. When a piece of said sheet-type foam material of a size of 1 square meter (weight: 0.4 kg) was thrown into a water tank where the surface of the water had been covered with heavy oil, the piece when recovered weighed 1.4 kg and was black as a result of adsorption of 1 kg of heavy oil. When this sheet was burned, it was easy to burn it completely, leaving but a small amount of ash.

A technique developed by *S.W. Miranda; U.S. Patent 3,886,067; May 27, 1975* is one in which oil slicks on surface waters are controlled by applying oleophilic foam material to the slick from a boat or airplane. In one case the material foams and binds the oil to form sponge-like clods which can be skimmed from the surface by another ship. In another embodiment chips of the foam material which are formed on board the craft are applied to the oil slick.

A process developed by *W.J. De Young; U.S. Patent 3,888,766; June 10, 1975; assigned to Uniroyal, Inc.* is one which cellular materials are impregnated with a hydrophobic and oleophilic sealant in an amount sufficient to coat the surface of the cells or capillary domains, for use in absorbing oil and like pollutants from the surfaces of either solids or liquids. Such a material is made by impregnating a sheet of open-celled polyurethane foam with a solution or latex of a sealant inert to water but capable of being swelled or dissolved by oil, drying the sheet by driving off all or part of the liquid medium, and compressing the sheet while the impregnated sealant is tacky, so that the sealant acts to retain the sheet in its compressed state.

The resultant article is characterized by the unique quality of being relatively inactive when in contact with water, but becoming activated by oil, so that a rapid absorption of large amounts of the oil is effected. The cellular material may also be a foam in chopped or powder form, or a fibrous nonwoven batt, and is particularly advantageous in selectively removing residual films of oil from the surface of a body of water.

Solid complexes for binding hydrocarbons have been developed by *G. Durand, A. Pareilleux, G. Goma and P. Monsan; U.S. Patent 3,901,818; August 26, 1975; assigned to Creusot-Loire, France.*

Said complexes are formed by an amine bound to a substrate (clay, glass, organic polymer) through a binding reagent which is an acid chloride. To prepare the complexes, the amine is reacted with the previously activated substrate and becomes covalently bound thereto. The hydrocarbons and other organic materials are bound to the amine by a hydrophobic linkage. Proteins may also be bound to the complex by means of a hydrophobic linkage. The complexes are useful in the purification of waters containing hydrocarbons.

A technique developed by *L.E. Bertram; U.S. Patent 3,902,998; September 2, 1975; assigned to The Standard Oil Co.* is one in which rice hulls are floated on water contaminated with oil to absorb the oil which is then removed by skimming the combined oil-rice hull material from the water.

Increasingly stricter standards for water purity in the waterways used by boat traffic have made it necessary to devise means for cleaning up smaller ecological accidents, such as spills of small amounts of oil, or accidental dumping of oil-like contaminants into the waterway from contiguous shores. It is impractical to use prior art devices on spills of, say, 25 or 50 gallons, since they are so large and in most instances require special equipment to position on and retrieve from the water. Moreover, it is prohibitively expensive for everyone who is in a position in which he might be responsible for a small spill to have readily available a large boom device, or to engage someone who has such a device for cleaning up the small spill.

A device developed by *C.H. Yocum; U.S. Patent 3,904,528; September 9, 1975* is an element which is relatively small and inexpensive and which can be used by individuals for cleaning up small oil spills and the like.

A number of materials in sheet form which are pervious to oily contaminants but not to water are known which are satisfactory for the container. In addition, the material should be resistant to deterioration by fresh and salt water to a degree that it will last from several months to several years in the water. One material which has been found to be very satisfactory is a multilayer sheet material formed of fiberglass in a polyethylene binder.

Such a material is used to form a container which is filled with spaced strips of an absorbent material such as cellulose sponge. Between the strips of such absorbent are placed particles of a second absorbent material. For example, diatomaceous earth or calcined or baked absorbent clay can be used. Attapulgite clay is effective as an absorbent. The amount of the granulated material is relatively small in comparison to the overall size of the element. For example, for an element which is about one foot by two feet and about three inches thick, a handful of granulated absorbent is sufficient.

The reason for this is that the primary means for absorbing the contaminant is the sponge . But because the sponge is normally dry when the element is first placed in the water in which the contaminant has been spilled, it does not absorb the contaminant very rapidly, just as a dry sponge does not absorb water very rapidly. It has been found that the granules of absorbent spread over the surfaces of the absorber increase the rate at which absorption of the contaminant takes place.

A composition developed by *J. Orban and C.E. Case; Patent 3,917,528; November 4, 1975; assigned to Sorbent Sciences Corp.* is a porous material treated with a drying oil.

The following is one example of the preparation and utility of such a material. 100 polyurethane sheets each of a dimension of 6′ x 1½″ are obtained and are cleansed in a tank of trichloroethane solvent. The trichloroethane solvent is squeezed out of the polyurethane sheet and the sheet is let dry. A treating solution having the following composition is prepared:

1,1,1-Trichloroethane	110 gal
Silicone solution 50% strength	15 gal
Safflower oil	6 lb
Safflower oil, conjugated	1 lb
Oil crude	3 lb
700X Chlorinated paraffin	150 lb

The reagents are mixed in the sequence stated above and are stirred at room temperature. The sheets are introduced at a residence time of one sheet for 2 minutes, i.e., each sheet is allowed to be suspended in the bath for a period of 2 minutes. The treated foam sheets are squeezed between mechanical rollers opposed to one another and the so-squeezed sheets are vacuum dried.

When tested, one pound of the foam so treated sorbs about 36 lb of spilled oil almost instantaneously. The so treated foam functions as a general purpose oleophilic adsorbent.

A composition developed by *E.C. Peterson; U.S. Patent 3,933,632; January 20, 1976; assigned to Electrolysis Pollution Control Inc.,* consists of an admixture comprising from between about 30% and 70% by weight of lead slag mineral wool, with the balance being a finely divided natural stone substance containing substantial quantities of iron, aluminum, and magnesium oxides, including such natural stones as trap rock, basalt and gabbro.

The lead slag mineral wool is treated with a hydrophobic-oil-soluble hydrocarbon chain substance, such as oleic acid to wet the surface of the mineral wool prior to mixing with stone flour. The lead slag mineral wool is preferably fragmented into nodules having a diameter of, for example, from ½" to 1". The composition may be also utilized for removing oil spills from water surfaces, lake beds or soil surfaces.

Very similar ground is covered by *E.C. Peterson; U.S. Patent 3,980,566; September 14, 1976; assigned to Electrolysis Pollution Control, Inc.*

A material developed by *M. Tomikawa, A. Tsunoda, K. Kaneda, H. Ohkawa and Y. Mugino; U.S. Patent 3,960,722; June 1, 1976; assigned to Idemitsu Kosan Co., Ltd., Japan* is an oil adsorbent plastic material for eliminating oils present in or on the surface of water through adsorption. Said adsorbent is produced by foaming by a physical means polyethylene containing 30-80% by weight of at least one inorganic calcium compound selected from the group consisting of calcium sulfite and calcium carbonate, to prepare a foamy substance having a density of 0.06-0.10 g/cc and then shaping the substance to have a network structure.

A composition developed by *A. Omori, I. Okamura, T. Imoto and T. Katoh; U.S. Patent 3,966,597; June 29, 1976; assigned to Teijin, Ltd., Japan* is an oil or organic solvent absorbent which is prepared by extruding a molten thermoplastic resinous polymer blend of polystyrene and polyethylene containing a foaming agent through a die having a slit aperture of 0.1-1.0 mm width, quenching the extrudate at the die exit to a temperature below the glass transition point of the resinous blend, drafting the extrudate at a draft ratio from the maximum draft ratio possible under the operating conditions to one-third the maximum draft ratio.

A structure is than prepared by laminating at least two sheets of the resulting unopened, sheet-like reticulated structure having numerous noncontinuous cracks along one direction so that the direction of the cracks of each such sheet is the same, pulling the laminate in a direction perpendicular to the direction of the cracks to separate the constituent fibers from each other, and crimping the opened, sheet-like laminate either alone or together with at least one other sheet-like material.

A process developed by *R.E. Langlois and C.R. Morrison; U.S. Patent 4,006,079; February 1, 1977; assigned to Owens-Corning Fiberglas Corp.* is one in which glass fibers are formed from melted glass, sprayed with a binder, and collected on a conveyor in a continuous process. A woven scrim of continuous glass filaments is fed from a roll to the conveyor along with the glass fibers. The scrim and fibers pass under a sizing roll which compresses the fibers into a mat, and then through a curing oven to cure the binder. The scrim-reinforced glass fiber mat is cut into lengths, such as one hundred feet or two hundred feet, and rolled into rolls for eventual use as an oil absorbent primarily to clean up oil spills from oil tankers.

A technique developed by *C. Stein and A. Marbach; U.S. Patent 4,011,159; March 8, 1977; assigned to Societe Chimique des Charbonnages, France* for the removal of petroleum products from solid or liquid surfaces is one in which the petroleum product is covered with a finely divided solid polymer, the polymer being selected from the group consisting of poly(bicyclo[2.2.1]-heptene-2) and poly(methyl-5 bicyclo[2.2.1]heptene-2) mainly obtained by opening the ring of bicyclo[2.2.1]heptene-2 and its methyl derivative. After the finely divided polymer is applied to the petroleum, there is obtained a strong solid rubbery film, the film being removable by mechanical means.

Sinking

Materials of high density and treated with lipophilic coating cause the oil to be sorbed and subsequently sink. The efficiency is limited. A major disadvantage of the material is that the sea bottom becomes fouled so that bottom dwelling marine life in the area may be killed.

Sinking agents are materials such as clay, fly ash, sand or crushed stone which when applied to spilled oil, will sink it. Sunken oil will cover and smother or taint bottom (benthic) organisms including shellfish. In addition, it will move and resurface as a result of turbulence or microbial degradation. The use of sinking agents is prohibited by federal regulations.

A technique developed by *J. Martineau and F.J. Biechler; U.S. Patent 3,886,070; May 27, 1975; assigned to Seppic, France* involves anchoring a trivalent chromium complex, in which the chromium is coordinated to an acyclic carboxylic acid, to solid particles having a grain size of 1 to 600 microns by mixing 5 to 20 weight parts of the particles with one weight part of an aqueous alcoholic solution of the complex and heating the mixture to dryness.

The following is one example of the use of such a complex. 1 kg of crude petroleum was poured on to the surface of seawater. 3 kg of a fine sand (average grain size 590 microns) treated with a Werner complex were added to the oil. A continuous pasty layer of petroleum product was formed, sinking to the bottom of the sea after about 1 hour. This continuous pasty layer retained its structure for a few months until the biodegradable products present in it were biologically degraded.

Gelling/Coagulation

Gelling agents absorb, congeal, entrap and fix the oil to form a semirigid or gelatinous mass, which may be more easily recovered, or will inhibit the spread of the spill. Gel agents include soap solution, wax, fatty acids, and various polymers.

A technique developed by *H.E. Alquist and A.C. Pitchford; U.S. Patent 3,785,972; January 15, 1974; assigned to Phillips Petroleum Co.* involves containing oil on the surface of water and removing the oil from the water surface by increasing the liquification temperature of the oil to 50-80°F above the temperature of the water on which it is floating by incorporating a wax into the oil to form a crust-like fused mass which will act as a boundary against extension of the oil mass and which can be easily skimmed from the water surface.

A technique developed by *C.E. Creamer; U.S. Patent 3,816,359; June 11, 1974; assigned to Union Carbide Corp.* is one in which organopolysiloxane carbamates can be employed to remove and/or recover undesired oil films from water, e.g., petroleum oil product spillage caused by shipping tankers, offshore oil-well leaks, and the like. The process merely involves contacting the organopolysiloxane carbamate with the oil film floating on the water surface by any suitable method, such as spraying, pouring, or otherwise mixing and the like. The organopolysiloxane carbamate then dissolves in the oil film and diffuses to the oil-water interface where the carbamate polymer and water react to form a crosslinked siloxane foam which absorbs and emmeshes the oil.

The consistency of the oil-foam product can be varied from rubbery gels to stiff rock-like foamed substances which float upon the water and which can be mechanically harvested from the water surface by any suitable method. Such provides a unique aide in recovering and/or removing such undesirable oil from the water. If desired the foam can be ignited and burned. Alternatively, it can be netted, scooped or skimmed off of the surface of the water by any suitable method. It can easily be removed from beaches where it has been dragged or washed ashore by scooping, shoveling, plowing, raking, etc.

A technique developed by *R.E. Gilchrist and J.C. Cox; U.S. Patent 3,821,109; June 28, 1974; assigned to Tenneco Oil Co.* is a method of treating an oil slick or spill on the surface of a body of water so as to render the slick amenable to mechanical or other such removal which comprises dispersing on the slick or spill a composition comprising a drying oil and a carrier for the drying oil. The carrier is of a type which is substantially nontoxic to marine life in the amounts employed.

The following is a report of a laboratory test of such a technique. To a 500 ml beaker approximately 4 inches in diameter and containing around 300 ml of tap water at about 72°F was added 1 cc of 30°API crude oil. The oil quickly spread over the surface of the water in the beaker forming a slick on the surface thereof which substantially covered the entire surface area of the water in the beaker. One cc of solution containing 23% of a drying oil known as FLOAT R-22 marketed by Tenneco Chemicals, Inc., 74% ethanol and 3% cobalt naphthenate was atomized onto the oil floating on the surface of the water in the beaker. The FLOAT R-22 which is a tall oil derivative had the following properties:

Acid value	174.2
Rosin acids	9.8%
Fatty acids (mainly oleic & linoleic)	79.0%
Esters and unsaponifiables	11.6%
Titre below −8°C	16°F

The beaker was agitated for 2-3 minutes and it was noted that a containing film soon covered the oil slick and formed a barrier around the edges of the slick. It was further observed that the area of the slick was substantially reduced. Upon vigorous shaking, the oil slick, upon cessation of the agitation gathered in a relatively confined area and remained contained therein.

A scheme developed by *J.A. King; U.S. Patent 3,835,049; September 10, 1974; assigned to Cities Service Oil Co.* is one by which hydrocarbon oil floating on the surface of water is recovered by admixing with the hydrocarbon oil in the presence of oxygen one to fifty parts by weight of a drying oil per one hundred parts by weight of the hydrocarbon oil. At least one of ten carbon-to-carbon bonds of the drying oil are double bonds and the dyring oil contains at least one carbonyl moiety per molecule.

The drying oil addition serves to coagulate the oil such that the mixture can be removed from the surface of the water.

A technique developed by *P.C. Stoddard; U.S. Patent 3,865,722; February 11, 1975* is one involving subjecting an oil slick freely floating on a water surface to the action of a corona discharge. The oil tends to conglomerate and become cohesive. As a result, removal of the oil from the water surface is facilitated.

A process developed by *D.J. O'Sullivan and B.J. Bolger; U.S. Patent 3,919,083; November 11, 1975; assigned to Loctite (Ireland) Ltd., Ireland* is one in which oil spills on water may be treated with monomers which polymerize in the presence of moisture, preferably, monomeric esters of 2-cyanoacrylic acid. Upon polymerization, significant portions of the pollutant are incorporated within a polymer matrix, thus reducing the danger to shore and marine ecology and aesthetics.

While the polymerizable monomer may be used in undiluted form, use of a solvent may aid in diffusion of the monomer into the pollutant mass. The composite mass which is formed may float or sink depending on its density relative to that of the fresh or salt water. If it floats, as is frequently preferred, it may be swept or raked up mechanically and disposed of. If it sinks, the preponderance of the pollution threat is also removed.

Compositions developed by *A. Winkler; U.S. Patent 3,929,631; December 30, 1975* may be used to effectively coagulate and recover oils from aqueous and solid surfaces, by application thereon of particulate expanded polystyrene and polystyrene-butadiene, the particles combined with meltable hydrocarbons such as paraffin, naphthalene, and mixtures thereof.

A technique developed by *D.M. Zall; U.S. Patent 3,977,969; August 31,*

1976; assigned to the U.S. Secretary of the Navy involves chemically treating the surface of the oil spill with a polymer of high molecular weight having jelling properties thereby causing the oil to coagulate. The oil is then easily raked off the surface of the water.

A technique developed by *S.L. Ross and O. Shuffman; U.S. Patent 4,031,707; June 28, 1977* involves controlling an oil spill by cryothermal means.

A possible application of the technique is illustrated in Figure 83.

FIGURE 83: CRYOTHERMAL MANIPULATION OF PETROLEUM SPILLS ON WATER

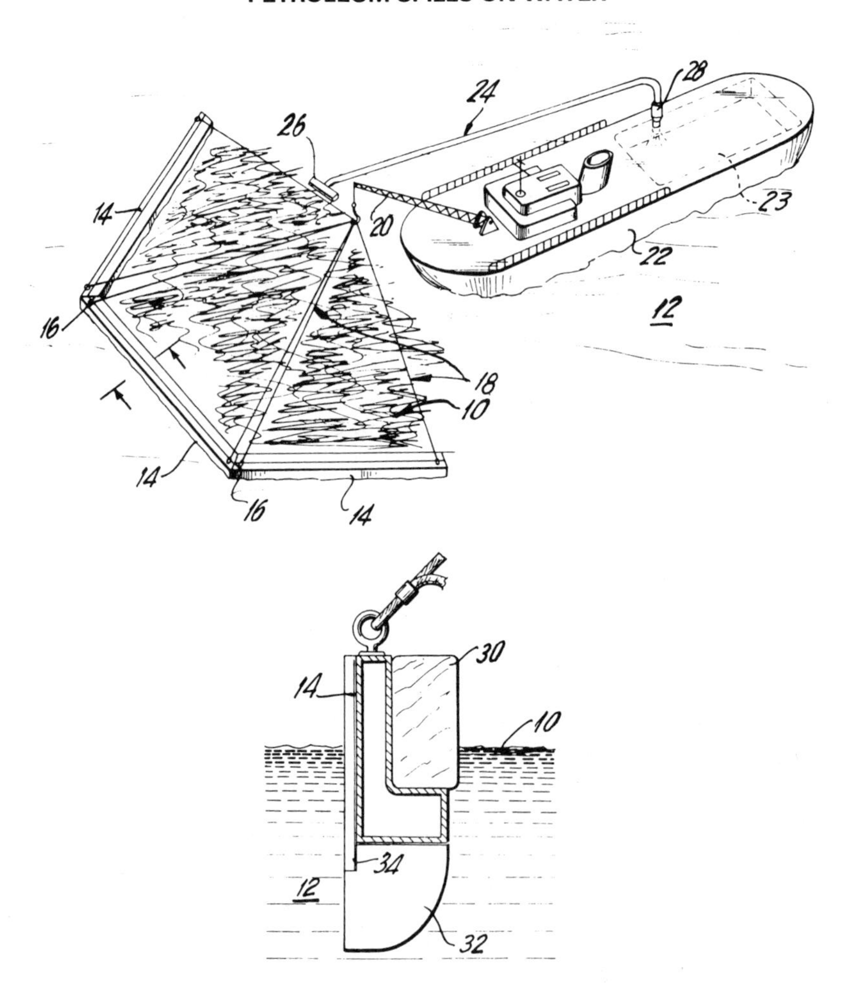

Source: U.S. Patent 4,031,707

The oil spill is designated by numeral **10** and is disposed in a relatively large area on a body of water **12**. In order to confine, collect, and remove the body of petroleum **10** from the water, an apparatus may be utilized comprising a plurality of booms **14**, which are pivotally interconnected as at **16**, and which are controlled by means of a system of guide wires **18** that are connected to a crane **20** of an oil collection vessel **22**. Also included in the apparatus is a suction apparatus **24** having a suction head **26** for collecting the body of oil, and a nozzle **28** which leads to an oil collection compartment **23** of the vessel **22**.

Each boom **14** is generally L-shaped in cross section, and is made of suitable material for flotation purposes, and includes a plurality of pivoted keels **32** which are interconnected to the boom by means of shafts **34**. Disposed in the L-shaped cut-out of each boom **14**, and facing in the direction of the oil spill is a cryothermal material **30**, such as dry ice. In order to herd an oil spill, the first step would be to position the booms in an arrangement to facilitate the collection of the oil spill, and with the cryothermal material **30** in place. The vessel **22** would be maneuvered in a manner such that the booms surround the periphery of the oil spill, and as the booms are positioned adjacent the oil spill, the dry ice immediately causes a lowering of the temperature of the adjacent water, and a simultaneous increase in the surface tension of the petroleum oil spill.

The increase in surface tension of the oil spill greatly increases the ability to manipulate and directionalize the movement of the oil spill, and more particularly, to directionalize the oil spill in a direction to enable the suction apparatus **24** to efficiently suck up the oil spill for collection into the compartment **23** of vessel **22**. It should be noted that the colder that water **12** becomes during the herding operation, its density increases, as does the density of the oil, and the accumulated densification of the oil and water results in a retardation of the wave action and cresting action of the water **12** which, of course, greatly aids in the herding and collecting operation. This is to be contrasted with presently available chemical herding processes which are generally inefficient when the water is rough.

In addition to the features of the technqiue as described above, there is a possiblity of forming an almost instantaneous mass about the periphery of the petroleum mass **12**, when the temperature differential between the cryothermal material and the water is significant. The formation of a peripheral solid mass about the oil spill **10** will further aid in confining the oil slick to a desired area for further manipulation and collection.

The provision of the movable keels **32** greatly aids in controlling the direction of the booms, and preferably the keels should be aligned with the current action of the water to minimize the amount of movement of the booms during the collection operation. Although the preferred technique has been described with respect to the use of dry ice as the mass of cryothermal material, it is readily apparent that other materials may also be used. For example, instead of using blocks of dry ice, each of the booms **14** may be provided with expansion nozzles and tanks of gaseous carbon dioxide or liquid nitrogen, which may be sprayed along the periphery of the petroleum mass **10**. As the liquid nitrogen or carbon dioxide gas is sprayed from the tank through the expansion nozzle, it loses heat, thereby forming a cryothermal solid which functions in the same manner as the dry ice.

Combustion

The loss of volatile components and the incorporation of water make oil spills difficult to ignite and sustain in the burning condition. The use of burning agents is essential if burning is to be pursued, and approved, as a disposal means. These agents contain combustion promoting and sustaining chemicals. Their use may be authorized when it will prevent or substantially reduce hazard to life or property. Such instances are rare in inland waters, and burning should be avoided (9).

A material developed by *J.M.C. Whittington, J.E. Meyer and G.D. Tingle; U.S. Patent 3,843,306; October 22, 1974; assigned to GAF Corp.* is a porous alkali metal silicate foam having oleophilic-hydrophobic properties for use in oil spill control and removal. The silicate foam is preferably formed from a blend comprising solid and liquid alkali metal silicates and an oil absorption-water repellent agent. The blend is pelletized, heated in an oven to expand the material into foam particles, and then shredded, graded and retreated with an oleophilic-hydrophobic agent to coat the internal and external surfaces and thereby further enhance the oil-absorption characteristics.

The foam particles float on water and can absorb about 3 times their own weight in oil without being wetted by the water. When spread over approximately 75% of an oil slick area, the oil is immediately wicked into the foam and can be ignited. Burning continues until the oil is completely consumed. The silicate foam is incombustible and can be reused. The foam also permits recovery of the oil, rather than burning. The oil soaked particles can be skimmed from the surface and the oil extracted by the use of solvents.

Magnetic Removal

A technique developed by *D.E. Weiss and H.A.J. Battaerd; U.S. Patent 3,890,224; June 17, 1975; assigned to Commonwealth Scientific and Industrial Research Organisation and Imperial Chemical Industries of Australia and New Zealand Ltd., Australia* involves removing oil slicks from aqueous media by: first treating the slick with sufficient fine particles or granules of ferromagnetic material so that the ferromagnetic material preferentially absorbs or adsorbs oil from aqueous media and also the particles or granules float on the aqueous media when associated with the oil; second removing the particles together with the associated oil by magnetic means.

One example of the preparation of a ferromagnetic polymeric particle of use in this process is now given. A dispersion of gamma-iron oxide was prepared as follows:

51 g gamma-iron oxide (Bayer S11) was added to a solution of an alkylene oxide condensate (Teric PE68) in 400 ml of water and stirred vigorously until the dispersion consisted of clusters of oxide particles smaller than 5 microns.

A solution of polyvinyl alcohol (491 ml of a 20% solution w/v) and 2 g of polyvinyl alcohol (Gelvatol 20-30) was added and the suspension stirred until it consisted of clusters of oxide particles smaller than 5 microns.

To the above suspension was added a 25% aqueous solution of glutaraldehyde (200 ml) and 2 N HCl (70 ml) with rapid stirring. The solution was immediately dispersed into 2 liters of kerosene to which has been added 40 g of sorbitan monooleate (Span 80) and 10 g of polyoxyethylene sorbitan monooleate (Tween 85).

Vigorous stirring was continued for one hour followed by gentle agitation for about 6 hours. The product was filtered off, washed with kerosene, hexane and finally acetone until the filtrate was clear. The particles so obtained were dried and cured for 1 hour at 100°C. 170 g of particles were obtained with an average size of 10 microns and containing 60% w/v of gamma-iron oxide.

The particles (26.7 g) were then added to 100 ml of styrene. The mixture was purged with nitrogen and irradiated in a nitrogen atmosphere with cobalt 60 gamma rays at a dose rate of 0.11 Mrad/hr to a total dose of 5.1 Mrads. The particles were removed and washed with benzene until free of homopolymer and finally washed with methanol and dried under reduced pressure at 65°C. 56.4 g of particles were obtained containing 52.7% polystyrene.

To a 12 inch dish containing 100 ml of water was added 1 ml of crude oil. The particles (approximately 200 mg) were dusted over the surface. The particles were wetted by the oil and when a magnet was moved close to the surface of the water the particles and associated oil were removed leaving an almost clean surface.

Biodegradation

The microbial degradation of oil is a process whereby oil is reduced to asphaltic or tar residue. However, this degradation does not remove the slick from the environment. It may remove the slick from view, but intermediate degradation products (which are still environmentally undesirable) may remain in the water column. It is clear that microorganisms that decompose petroleum components are widespread and that degradation of at least a percentage of the components of the spill is possible. Little information is available, however, on the rate of this degradation in the field. Biological agents (organism cultures and enzymes added to the spill to encourage its degradation) may be used only with the approval of EPA. Biological organisms operate at maximum efficiency over a very narrow temperature and food range, and because of this rarely operate at maximum efficiency in the open environment. EPA approval of any procedure fostering biodegradation is required because it is not a desirable process. Biodegradation can produce more toxic intermediate products. It also provides greater opportunity for aquatic life to come in contact with hydrocarbons. It is recognized that nonrecoverable oil biodegrades and this phenomenon will change the nature of the impact of oil, that is, reduce aesthetic effects.

A technique developed by *R.W. McKinney, A.L. Dixon and R.L. Jordan; U.S. Patent 3,843,517; October 22, 1974; assigned to W.R. Grace and Co.* is one in which cultures containing hydrocarbon and oil consuming microorganisms with or without nutrients are adsorbed in a carrier such as clays, vermiculites, silica gels, perlites and similar materials and freeze-dried to form useful compositions. These compositions are placed on oil films or layers which are on fresh or salt water, or on beaches or river banks, or in storage tanks, ships tanks and so on. The carrier adsorbs and concentrates the oil in the presence of the microorganisms and the microorganisms consume the oil. The microorganisms remain viable in these compositions even after long storage periods. In an alternative composition, the microorganisms may be freeze-dried and then admixed with a carrier such as clays, vermiculites, perlites, silica gels or so on.

A process developed by *R.L. Raymond; U.S. Patent 3,846,290; November 5, 1974; assigned to Sun Research and Development Co.* is a process for eliminating hydrocarbon contaminants from underground water sources by providing nutrients and oxygen for hydrocarbon-consuming microogranisms normally present in the underground waters, the nutrients and oxygen being introduced through wells within or adjacent the contaminated area, and removing water from the contaminated area until hydrocarbons are reduced to an acceptable level.

A process developed by *E.N. Azarowicz; U.S. Patent 3,856,667; December 24, 1974; assigned to Bioteknika International, Inc.* is a process for the microbial degradation of petroleum or oil waste materials which comprises treating the petroleum or oil waste with a strain of *Candida lipolytica* for a sufficient time until degradation has been achieved. The microorganism strains employed have a broad spectrum of degradation capability and are capable of degrading crude petroleum as well as a variety of organic molecules, including aliphatic, aromatic and heterocyclic compounds.

A similar process developed by *E.N. Azarowicz; U.S. Patent 3,870,599; March 11, 1975; assigned to Bioteknika International, Inc.* covers the use of a strain of *Candida parapsilosis, Candida tropicalis* or *Candida utilis*.

A similar process developed by *E.N. Azarowicz; U.S. Patent 3,871,956; March 18, 1975; assigned to Bioteknika International, Inc.* covers the use of a strain of *Nocardia corallina, Nocardia globerula, Nocardia opaca, Nocardia rubra* or *Nocardia paraffinae*.

A process developed by *R.R. Mohan; G.H. Byrd, Jr., J. Nixon and E.R. Bucker; U.S. Patent 3,871,957; March 18, 1975; assigned to Exxon Research and Engineering Co.* is one in which beaches, jetties, rocks, cement and the like can be pretreated with freeze-dried microorganisms to prevent the accumulation of oil thereon.

While the present process is applicable to a broad scope of operable microorganisms, there are a number of microorganisms which are especially suitable for dispersing and degrading oil spilled as well as preventing the accumulation of oil on beaches, rocks, jetties and the like. These species were specially selected by elective cultures and screening techniques upon a wide variety of hydrocarbons. The hydrocarbon dispersing and degrading microorganisms useful in this process include bacteria, yeasts, actinomyces and filamentous fungi.

A composition developed by *P.M. Townsley; U.S. Patent 3,883,397; May 13, 1975; assigned to John Dunn Agencies Ltd., Canada* is a material for promoting growth of petroleum-degrading bacteria to aid in the eradication of oil slicks of water. The material consists of a mixture of bacterial available nutrients, in particulate form, each nutrient particle having a coating of a salt of a fatty acid which is lipophilic, partially hydrophobic, and biodegradable, for retarding entry of the nutrient into the water at the oil water interface.

A technique developed by *P. Fusey; U.S. Patent 3,900,421; August 19, 1975; assigned to SA Banque pour l'Expansion Industrielle "Banexi" France* involves the use of a composition formed substantially by a phosphoamino-lipid mixed with a nontoxic and entirely biodegradable emulsifier, the phosphoaminolipid being preferably lecithin. This composition may be used in liquid form for spraying floating oil-slicks and for cleaning tanks for mineral oil products and it may be brought into the form of a paste or powder by addition of mineral charges.

Various compositions have been proposed for bringing mineral oil products into the form of a biodegradable emulsion, these compositions having a base of organic substances biodegradable by microorganisms, such as carbohydrates or hydrophilic polypeptides in the form of molasses, various residual wash liquors (vinasses), casein, etc., together with emulsifying agents or various chemical substances such as mineral or organic acids or alkali or alkaline earth metal salts thereof. These compositions give stable emulsions and are perfectly biodegradable, particularly when the biodegradation is effected in a closed medium.

In fact, in order to be biodegradable, an emulsion of mineral oil, especially petroleum products must not be toxic and it must contain sources of carbon, nitrogen and phosphorus in sufficient quantity to ensure, for the microorganisms, the elements necessary for the constitution of living matter. The biodegradation of mineral oil products results from the use by the microorganisms of the hydrocarbons as a source of carbon and it is necessary that the composition used for emulsifying should contain considerable sources of nitrogen and sources of phosphorus in a smaller quantity.

These sources of nitrogen and phosphorus exist in many previously proposed compositions, but the emulsion is made via the intermediary of oleophilic agents which are water-soluble or the nitrogenous or phosphorus elements are not closely bound to the hydrocarbon molecules. In every case the nitrogenous and phosphorus elements are dispersed in the dilute emulsion. If this dilution is contained as in the case of biodegradation effected in a tank or the like, for example for a biodegradation intended to transfrom the hydrocarbons into fertilizing materials or in the case of the use of the emulsion as a culture medium for microorganisms, said microorganisms find the nitrogenous and phosphated elements within their immediate reach.

On the other hand, in the case of very extensive dilution, for example, in the case of discarded material discharged into the river or sea, the content of nitrogen and phosphorus in the medium surrounding the hydrocarbon molecules is insufficient to permit the growth of microorganisms.

A technique developed by *P. Fusey; U.S. Patent 3,919,112; November 11, 1975; assigned to SA Banque pour l'Expansion Industrielle "Banexi," France* is one for the elimination of mineral oil products by biodegradation containing a source rich in amino acids. It is characterized in that at least one oleophilic element constituted by a fatty acid is combined with a source rich in amino acids and containing phosphorus, the pH then being adjusted between 7 and 7.5 by addition of a basic element. This composition may be used in liquid form for the atomization of floating oil slicks and for the cleaning of vats and tanks for mineral oil products and it may be brought into the form of paste or powder by the addition of mineral charges.

A process developed by *R. Bartha and R.M. Atlas; U.S. Patent 3,959,127; May 25, 1976; assigned to the U.S. Secretary of the Navy* is one in which free-floating oil slicks on bodies of sea and fresh water are disposed of by microbial degradation at a greatly enhanced rate by applying the essential microbial nutrients, nitrogen and phosphorus, to the oil slick in a form that dissolves in or adheres to the oil and thus selectively stimulates the activity of oil-metabolizing microorganisms.

To demonstrate this capability under controlled conditions, miniature, 1 ml, slicks of Sweden crude oil were floated on 113 liter seawater tanks. The individual miniature oil slicks were confined each to a 9.6 cm^2 surface area by floating glass frames. Fresh seawater was continuously pumped through these tanks at the rate of 450 liters per day. Some of the 1 ml oil slicks were left untreated, others were treated with Sun Oil CRNF, 62 mg, and pyrophosphoric acid dioctyl ester, 7 mg, and a third set was treated with KNO_3, 101 mg, plus Na_2HPO_4, 7.5 mg. The biodegradation of the oil slicks was monitored for 42 days. During this period, the seawater temperature gradually rose from an initial 15°C to a final 20°C. In weekly intervals some of the miniature slicks were retrieved and were analyzed for residual oil. The results were corrected for evaporation losses, and precautions were taken to assure that no oil was lost through escape from the containment.

Figure 84 depicts the time-course of the biodegradation of the miniature oil slicks if left untreated, **A**; if treated with phosphate and nitrate salts, **B**; and if treated with Sun Oil CRNF (a paraffinized urea slow-release garden fertilizer, nitrogen content 26.8% manufactured by the Sun Oil Co.) and pyrophosphoric acid dioctyl ester, **C**. While only 10% of **A** and 18% of **B** were biodegraded, 60% of **C** was eliminated during the same experimental period. This experiment clearly establishes that suitable oleophilic, i.e., oil-seeking sources of nitrogen and phosphorus accelerate the natural biodegradation rate of free-floating oil slicks, and demonstrates the usefulness of this process for the clean-up of free-floating oil.

FIGURE 84: BIODEGRADATION OF OIL ON WATER SURFACES

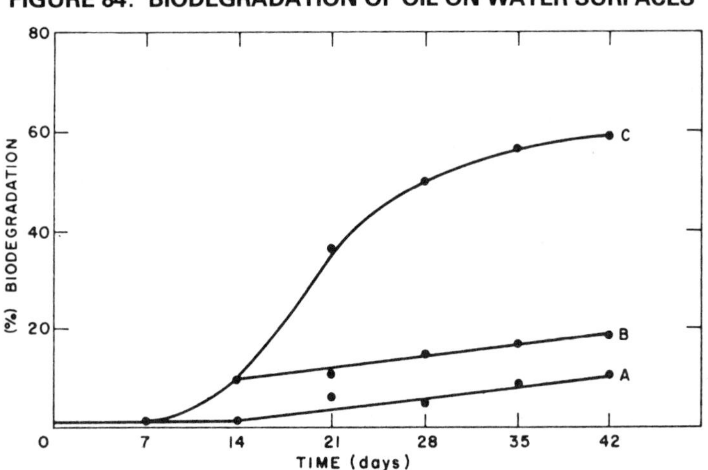

Source: U.S. Patent 3,959,127

A process developed by *W. Marconi, N. Oddo and L. Degen; U.S. Patent 4,042,495; August 16, 1977; assigned to Snam Progetti SpA, Italy* is one in which oily hydrocarbons which pollute the surface of water are removed therefrom by scattering over the polluted surface solid particles of nutrient salt containing nitrogen and phosphorus in a form that is readily assimilable by aqueous microorganisms capable of metabolizing hydrocarbons. The particles of nutrient salt are given a surface pretreatment with paraffin so that they are buoyant and only slowly soluble in water and lipophilic.

REMOVAL FROM OPEN WATER

Skimming Devices

Unlike containment booms, not all skimmers operate on the same physical principles or fail in the same ways. This appendix will describe the operating principles behind the various general types of skimmers available commercially and delineate criteria by which the effectiveness of any skimmer can be measured. Because of the wide variation in operating principles and effectiveness of skimmers, NAVFAC central procurement of skimmers is based upon the results of comparative tests involving all the major skimmer types in the same test environment and evaluations using the list of effectiveness criteria discussed in the following paragraph.

The rated effectiveness of a skimmer, regardless of its operating principle, is based upon the following criteria (9):

1) Oil Collection Criteria.
    a. Throughput efficiency (percent of oil entering skimmer which is picked up).
    b. Collection efficiency (percent of oil in the oil/water mixture picked up).
    c. Sensitivity to oil type (density and viscosity).
    d. Sensitivity to debris.
    e. Sensitivity to slick thickness.
    f. Sensitivity to currents.
    g. Sensitivity to waves.
2) Reliability Criteria.
    a. Seaworthiness (if manned).
    b. Complexity of machinery.
    c. Ease of field repair.
    d. Structural strength.
3) Storage and Deployment Criteria.
    a. Shipping and transportation requirements.
    b. Operational support required.
    c. Maximum transient speed to spill site when towed or if self-propelled.
    d. Deployment support required.
    e. Maintenance required.

The primary objective of a skimmer is to pick up as much oil and as little water as possible under the wide variations of oil type, slick and hydrographic conditions encountered. The degree of achievement of this objective is indicated by the values of the throughput efficiency and the collection efficiency when operating in various environments of oil type (density and viscosity), debris, slick thickness, currents and waves. Different skimmer types exhibit varying degrees of effectiveness when skimming oil of different density and viscosity. Generally, the less dense oils are easier for any skimmer to pick up. Some skimmers are better at picking up highly viscous oils, others at picking up less viscous oils. Paradoxically, debris may have both positive and negative effects. Coagulating oil slicks may be easier to pick up but could also result in the clogging of rotating components and pumping systems.

As may be expected, thin slicks are more difficult to remove than thick slicks without picking up a large amount of water. Based upon documented tests by NAVFAC and the Environmental Protection Agency, the maximum current in which commercially available skimmers can effectively operate is in the range of two-and-one-half to three knots. The maximum wave environment for effective operation is about three feet, a condition generally requiring large, vessel-type skimmers.

Reliability of a skimmer is largely based upon engineering judgment which must be rendered for each given type of skimmer. This judgment ranges from a simple one based upon the number of moving parts and ease of repair for smaller, unmanned skimmers, to a more complicated evaluation based upon these factors and naval-architecture calculations in the case of larger, manned units. Storage and deployment criteria have an impact upon the intended deployment and operational scenarios of a given naval activity. Depending upon the crane and rigging services available, larger, manned skimmers may require in-water storage between deployments, at a higher cost in maintenance and repair. Smaller skimmers can usually be stored ashore and deployed using only manpower or manpower together with trucks or forklifts.

Because of the low relative water speed in which skimmers are effective (three knots maximum), most of the self-propelled types are designed for nonskimming transit speeds not in excess of six or seven knots. To achieve greater transient skimmer speeds, these self-propelled types must be towed to the spill site using harbor tugs or other craft. NAVFAC central procurement of skimmers uses the previously listed criteria in defining skimmer requirements for the various naval activities.

In accordance with the operational needs of the various activities, NAVFAC has grouped all skimmers into three classes: small (unmanned, deployable by two men), medium (unmanned, trailer-mounted and deployable using an air-driven winch and jib boom), and large (manned, self-propelled). Each skimmer size has been designed for oil pickup with minimum water content and minimal required support equipment during deployment and operation.

Independent of the physical size of a skimmer (small, medium or large), each type may operate on a different physical principle. Figure 85 illustrates the principle of operation of the seven skimmer types considered to be the most effective of those available commercially. Figure 85(a) has been identified by laboratory and field tests to be the most effective operating principle of the types shown.

FIGURE 85: SKIMMER TYPES

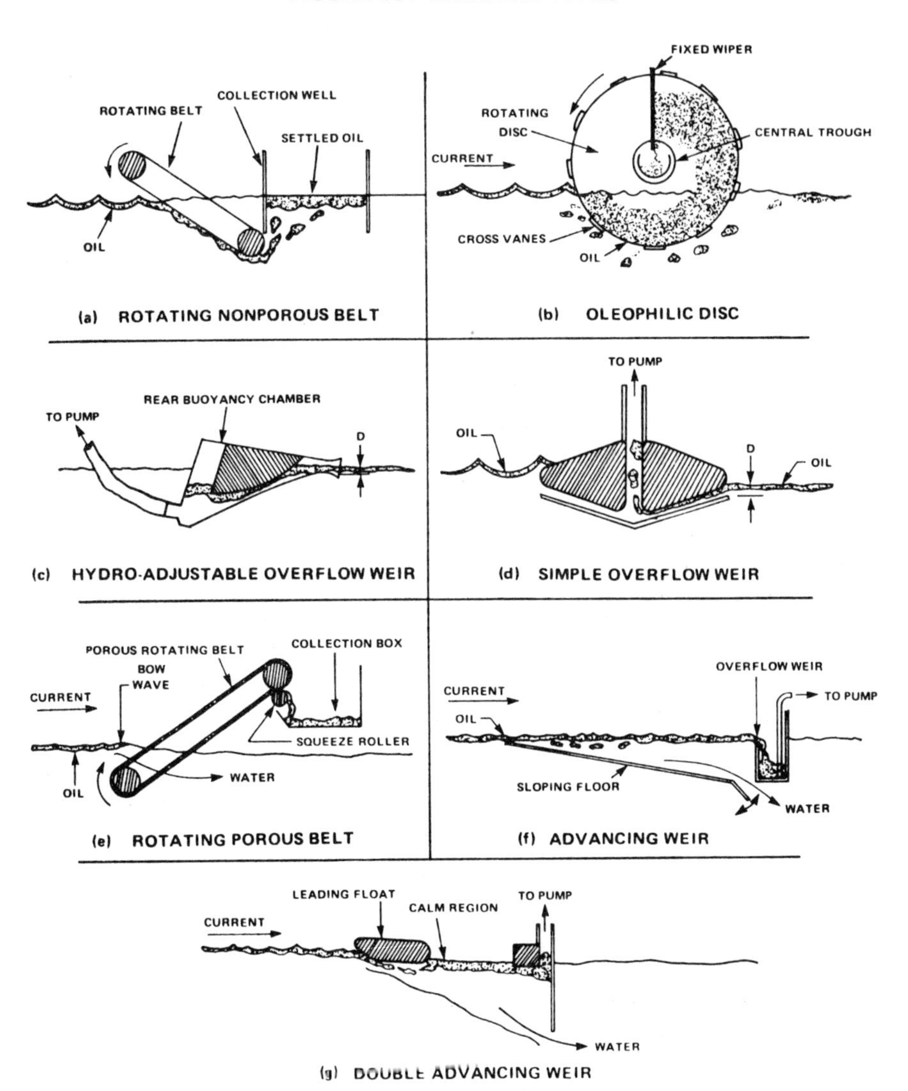

Source: Reference (9)

In operation, current, or the skimmer forward motion, brings the oil slick in contact with the rotating, smooth-surfaced, conveyor belt. The combined action of the rotating belt and the relative water velocity causes oil to be dragged below the surface of the water, out of the region of influence of surface waves. At the bottom roller, the oil leaves the belt, because of its buoyant force, and rises to the surface of the calm water region inside the collection well. This process is continued until the settled oil layer is thick enough to trip a signal light. At this time, the settled oil is pumped out.

As a result of the natural gravity separation of oil and water, the settled oil contains little or no water. The skimming principle is not affected by oil viscosity, which can vary widely for various oils and for a given oil during a spill due to weathering. The oleophillic disc principle of Figure 85(b) relies upon the adherence of floating oil to the rotating metal discs. Set side by side in a row, the discs rotate down into the slick from above causing oil to adhere to their surfaces. As they rotate past the stationary vertical wiper, any oil still adhering to the discs is scraped off and falls into a "U"-shaped trough located through a hole on the disc centerline. At this point, a screw conveyor transports the oil to one end of the row of discs where it is pumped to a suitable container.

This device is not very effective since rotation of the discs and their supporting cross vanes put mixing energy into the slick. This mixing energy causes oil droplets to be formed which do not come in contact with the discs and are swept under the device. The rotational speed of the discs must be set for each oil, depending upon its viscosity, so that the oil does not drain off the vertical surfaces of the discs by the time it reaches the scraper blade above the open trough. The device is only effective on very viscous oils occurring in thick slicks. Figure 85(c) shows an overflow weir device whose weir depth **(D)** can be controlled by varying the pumping rate out of the skimmer.

In thick slicks the pumping rate is increased, and the fluid inside the device does not have a chance to fill the rear buoyancy chamber. This results in a buoyancy force which lowers the front weir deeper into the water. Conversely, in thin slicks a slower pumping rate allows the rear buoyancy chamber to fill, weighting the rear of the device to lift the front weir closer to the water surface. A debris screen is placed over the frontal weir opening. The device is not very effective when skimming viscous oils, which tend to clog the front debris screen or the narrow sloping passage leading to the rear buoyancy chamber.

Figure 85(d) illustrates perhaps the oldest skimming principle, in which a simple overflow weir is floated on the surface and its weir depth adjusted to just skim the (usually) thin oil slick. The shallow weir depth must be manually changed by adding or removing weights, or moving attached floats up and down. The device is very inefficient in waves when the weir openings become flooded with water, carrying the floating oil above the weir opening.

Figure 85(e) shows a rotating, porous, oleophillic belt principle in which the floating oil adheres to the belt, constructed of open pore polymeric foam, allowing the water to pass through. The oil-laden belt rotates up and out of the water and through a pressure roller assembly where the oil is squeezed out, falling into a collection box. The device is very sensitive to high viscosity oils, which clog the pores of the foam and block the flow of water through the foam, causing a standing "bow wave" in front of the rotating belt. This bow wave prevents a large fraction of the incoming slick from reaching the front

Also, floating debris can easily tear the foam belt. Figure 85(f) is a schematic of the advancing weir principle in which the velocity of the oil and water flowing into the unit over the front weir is reduced during its flow over the sloping floor to the rear of the device, allowing oil to separate, by gravity, to the top. A hinged door in the floor of the device allows water to pass out while a mechanically operated weir is used to skim settled oil into the collection box.

In practice, the rear overflow weir is difficult to adjust, and the oil slick thickness at the weir is not sufficient to prevent large quantities of water from being taken into the collection box. Also, it is found that the waves coming into the mouth of the device create turbulence which floods the overflow weir and causes oil droplet formation. The oil droplets are then carried out the bottom door by the prevailing water current. Figure 85(g) illustrates the principle of the double advancing weir. This is similar to the advancing weir principle except that a leading float serves to absorb a large portion of the incoming wave energy, thereby providing a relatively calm water region for gravity separation of the oil. The inlet of a pump is placed near the waterline to skim the separated oil. Water flowing into the device exits through a fixed opening in the bottom of the device.

Based upon past results of controlled tests and field experience on actual spills, skimmers using the principles of Figures 85(a) and 85(c) have been determined as the most effective for Navy use (9). The rotating belt principle of Figure 85(a) is used in the NAVFAC central procurement of medium and large skimmers. The hydro-adjustable weir of Figure 85(c), together with a floating gravity separator, is used in the NAVFAC central procurement of small skimmers.

The Navy has expressed preference for physical-mechanical methods of removal, and has designated the types of skimmers for use with Navy spills in various locations (9).

(1)The small unit which is designated for use in congested harbor areas is based on the weir principle similar to that depicted in Figure 86. The weir depth of these skimmers is controlled by adjusting the flow rate of the attached pump. As the flow rate is increased, fluid is removed from the rear buoyancy chamber, tipping the unit clockwise, and thereby increasing the weir depth. Decreasing the flow rate allows the buoyancy chamber to fill, tipping the unit counterclockwise, and thereby reducing the weir depth. The unit is most effective in a stationary mode where it is positioned and oil directed to it.

FIGURE 86: SKIMMER WEIR PRINCIPLE

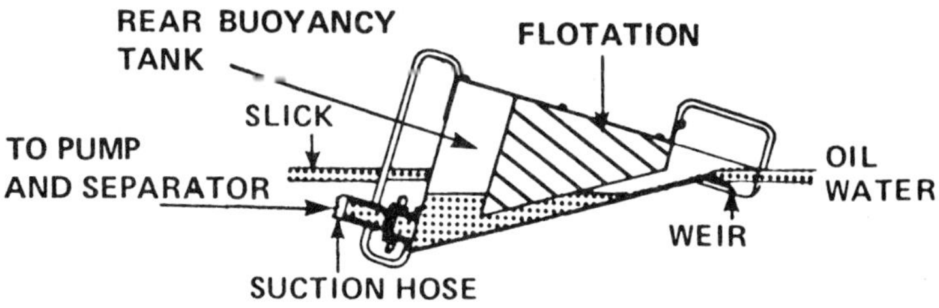

Source: Reference (9)

(2) The medium skimmer selected by the Navy is an "endless' belt unit. It is operable from a pier via hand-held controls. The principle of operation is shown in Figure 87. The rotating belt submerges the oil and directs it to the collection well where it concentrates and eventually from which it is pumped to temporary storage. This principle is entitled the dynamic inclined plane (DIP), in reference to the rotating and pitched belt. The unit may be used in the dynamic or stationary mode as described in Chapter 3.

FIGURE 87: PRINCIPLE OF OPERATION OF DYNAMIC INCLINED PLANE (DIP) SKIMMER

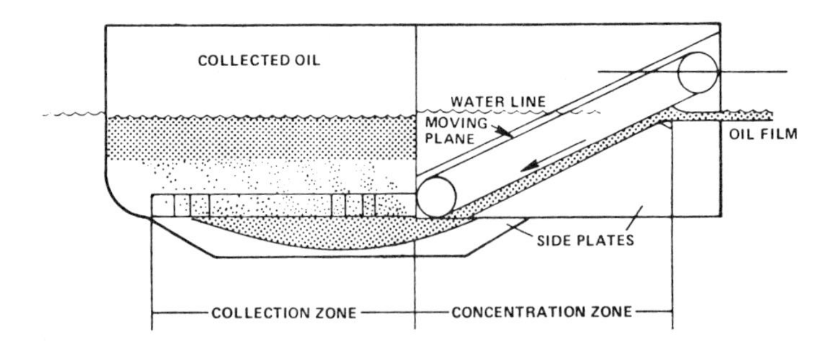

Source: Reference (9)

(3) The large skimmer selected for use by the Navy is a larger version of the medium skimmer (DIP). This unit is vessel-mounted for use in protected open waters in the moving mode. These units are quite effective even in choppy water in that they overrun and submerge the oil layer before collection. The rotating belt directs the oil to the collection well.

(4) Suction Based Skimmers. Other commercially available units for oil removal are based on suction, either taken directly off the surface of the water or by development of a submerged vortex. These units are highly susceptible to wave action and clogging; therefore, they work best in calm, debris-free waters and with thick oil layers. They are not extensively used for Navy spills.

(5) Sorbent Surface Skimmers. These units use an endless belt, hose or rotating drum, the surface of which will preferentially absorb the spilled oil from water surfaces. The concept is applied in large, craft-mounted units for large spills and in much smaller units using an endless, hose-width belt. The absorbed oil is conveyed to temporary storage tanks where it is squeezed from the belt or wiped from the drum or disc.

(6) Table 22 summarizes the principle of operation and advantages/disadvantages of the several skimmer systems.

(7) Manual removal methods are also used. Manual removal processes generally involve the physical pickup of the oil from shoreline areas with the use of sorbent materials, pitchforks and/or shovels. They also refer to "in water" removal operations such as that mounted for small shipside spills in which the Mark I Spill Control Kit is employed. In this instance, herder chemicals may be used to retard spreading of the spill, and hand-held polyurethane

absorbent pads or "mops" are used to "sorb" and remove the oil. The pads are squeezed out with conventional mop wringers.

TABLE 22: SUMMARY OF OIL REMOVAL DEVICES

Type of System	Principle of Operation	Advantages	Disadvantages
Weirs	Gravity	Simple devices. Good mobility.	Efficiency rapidly decreases in waves and debris.
Floating suction devices	Gravity	Compact. Shallow draft.	Work best in calm water. Easily clogged with debris.
Sorbent surface devices	Adhesion (exposed surface to which oil can stick)	Less vulnerable to wave action.	Expensive. Require training to operate. Poor throughput efficiency. Collect water with oil.
Vortex devices	Suction	High capacity.	Require mooring. High energy requirement. Sensitive to waves, debris.
Dynamic inclined plane (endless belt)	Depend on buoyancy of oil. Movement along inclined plane.	Less vulnerable to wave action. Work on thick/thin oil layers. Not easily clogged. High capacity. Rugged, simple concept.	–

Source: Reference (9)

Commercially available skimmers vary widely in type, size and operating efficiency. NAVFAC has defined harbor skimmer systems as either small, medium or large.

The small skimmer system is intended for use primarily on small spills in congested harbor waters around nested ships, piers or other structures. The system (Figure 00) consists of a small, lightweight weir skimmer; a floating, oil-water separator; a diesel-driven diaphragm pump; a collapsible, 300-gallon storage bag and interconnecting hoses. The pump is valved in such a way that oil and water can be pumped from the small skimmer head into the floating separator; and, after a suitable time has been allowed for separation of the oil and water in the separator, the separated oil can be pumped from the surface of the separator to the storage bag. Only two men are required for the deployment and operation of the system.

FIGURE 88: SMALL SKIMMER SYSTEM

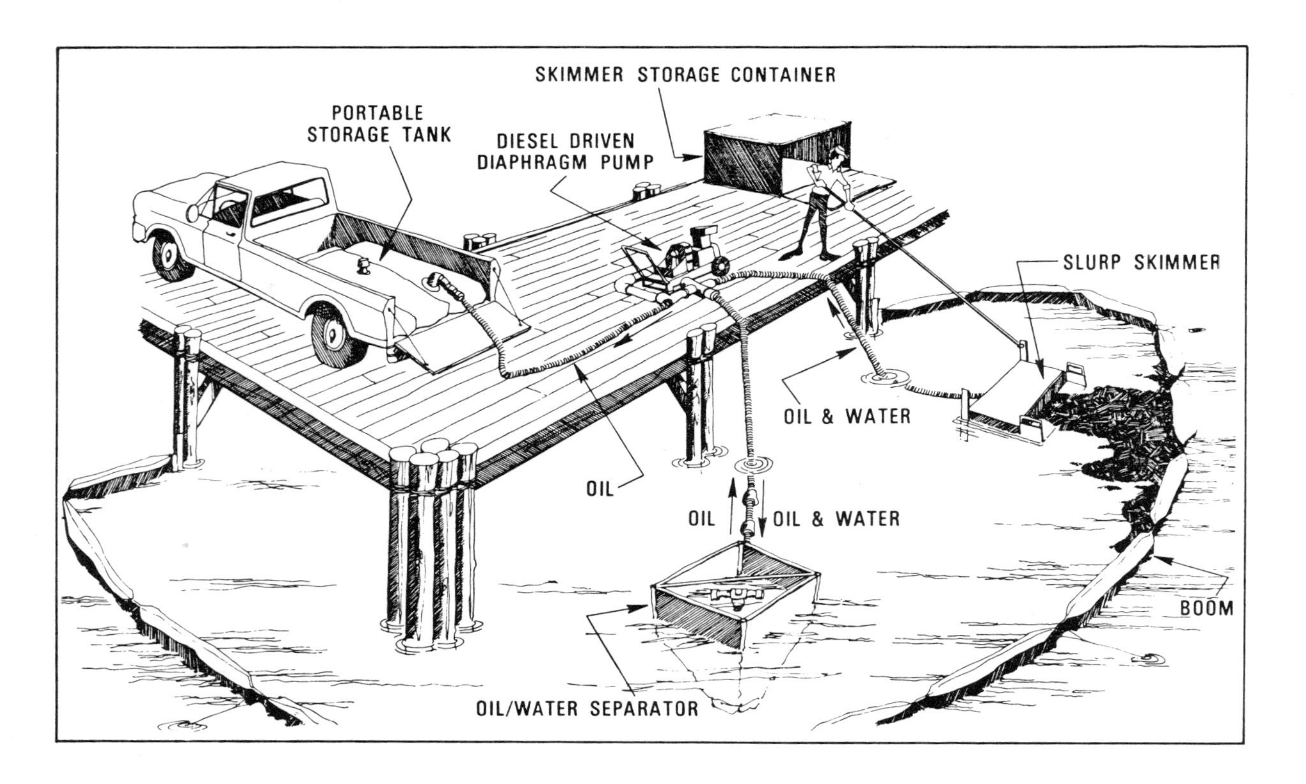

Source: Reference (9)

The small skimmer and floating separator are both designed and constructed to be dropped from a 20-foot high dock into the water without damage. Because of its shallow draft of only 10 inches, the small skimmer can be useful in spill situations near rock jetties or rip-rap walls, shallow rivers or ponds. Multiple small skimming systems, properly located, with boom and with the use of water jets or prop wash, can be very effective in cleaning up a spill.

The medium skimmer (DIP 1002) is a trailer-mounted system designed to recover small and medium oil spills at pier side. The system consists of a DIP 1001 skimmer, a diesel-driven air compressor which provides power for the skimmer, a jib boom with air hoist to launch and retrieve the skimmer, a 22-foot skimmer control wand, a 400-gallon oil storage tank, and 200 feet of Class I oil containment boom.

The oil recovery device, DIP 1001 skimmer, operates on the inclined, rotating, nonporous belt principle for recovering spilled oil. The DIP 1001 is launched from its support trailer using the jib boom and air hoist. In-water operational control of the skimmer is maintained by the 22-foot wand. The wand contains compressed air lines, to power the skimmer propulsion and the belt drive and transfer pump components. The wand is also used to transfer collected oil from the skimmer to the storage tank on the trailer.

The DIP 1002 system can operate effectively in both the maneuvering and stationary modes. The preferred method of operation for the DIP 1001 system is the stationary mode at the apex of a boom configuration. The skimmer is equipped with Navy standard boom end connectors to allow easy mating with containment boom. When operating in the stationary mode, natural forces of wind and/or current, or artificial forces generated by fire hoses or propeller wash, are used to push the oil to the skimmer. To improve the performance of the DIP 1002 system in the stationary mode, a combination air and water jet surface enhancement subsystem has been developed.

As shown in Figure 89, the air and water jets are deployed in a trapezoidal configuration. When activated, the jets induce an artificial surface current to push the contained oil toward the skimmer. The primary advantages of the enhancement subsystem are:

1) The active zone of influence of the skimmer is greatly increased.
2) The oil collection rate in the stationary mode is improved.

A minimum of two people is required to deploy and operate the DIP 1002 system.

The large skimmer system (DIP 3001) is designed to recover small, medium, and large spills in open harbor waters. Like the DIP 1002, the DIP 3001 skimmer system uses the rotating, nonporous belt concept for oil recovery.

As shown in Figure 90, the DIP 3001 is 25 feet long, 10 feet wide and has a normal draft of three feet. The large skimmer is powered by a diesel engine. It is equipped with twin screw propulsion, for improved maneuverability around ships and piers, with a 50-gallon per minute progressive cavity oil transfer pump, and 1500 gallons of on-board storage capacity. The skimmer maximum transiting speed is approximately five knots; however, oil collection is most efficient in the one-to-two-knot range. The greatest advantage of the DIP 3001 is its versatility.

The skimmer can operate independently in open harbor waters. It can also operate effectively around ships and piers and in conjunction with booms, utility boats, and other support equipment in both the maneuvering and stationary modes. Debris handling and water jet surface enhancement components are also available for the DIP 3001. Floating debris is often a problem in harbor areas used by the Navy. As oily debris is encountered by the large skimmer, it is harvested along with the oil. The debris handling system consists of two primary components. A one-half-ton capacity electric hoist, with an air-actuated expanded metal bucket, is provided to remove large debris from the collection well. The debris bucket is lowered into the collection well, scoops up debris, and is raised by the electric hoist. The collected debris can then be placed in a debris container on the skimmer or on a support vessel. Any debris remaining in the well can then be processed through a counter-rotating grinder mechanism on the intake side of the pump.

FIGURE 89: SURFACE ENHANCEMENT SUBSYSTEM FOR THE DIP 1002 SYSTEM

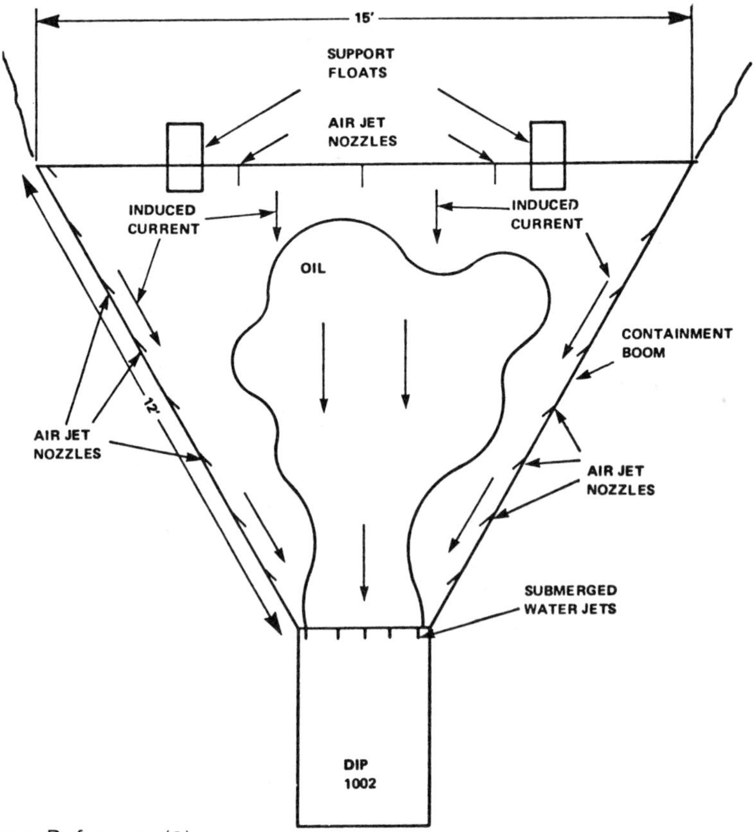

Source: Reference (9)

FIGURE 90: DIP 3001 OIL SKIMMER SYSTEM

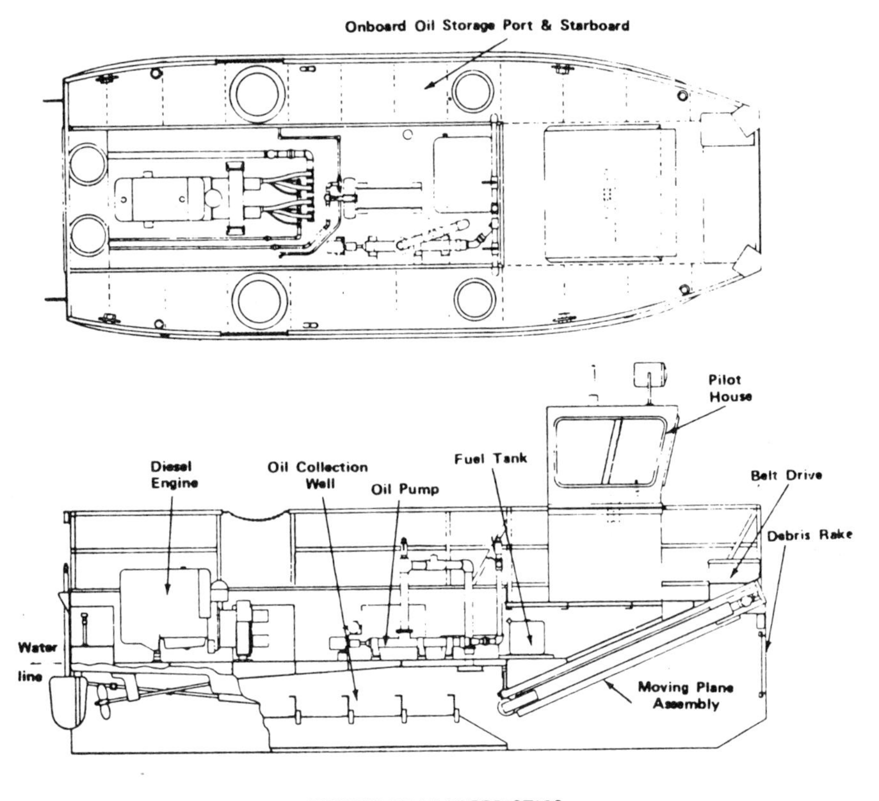

SYSTEM CHARACTERISTICS

Hull	Marine Grade Aluminum	Recovery Rate (2 mm slick)	60 gpm @ 1 knot
Length	25′	% Recovered in one pass	90 @ 1-2 knots
Beam	10′	% Water in Recovered Oil	< 1%
Draft	3′	Effective Oil Collection Speed	0-3 knots
Weight	6.5 LT		
Oil Storage	1500 U.S. Gallons		
Pump	50 GPM		

Source: Reference (9)

The grinder mechanism macerates the debris into small particles which can be handled by the oil transfer pump. The effluent from the grinder-pump combination is recycled back into the collection well through the debris bucket. This procedure reduces the volume of debris which has to be handled and minimizes the amount of ground debris mixed with the collected oil. Procurement of the debris handling system is limited to shore activities with significant debris problems. The surface enhancement system for the DIP

3001 consists of submerged water jets across the mouth of the skimmer and along the faces of the concentration sweeps.

Like the DIP 1002 system, the water jets induce an artificial surface current into the mouth of the skimmer. This increases the flow of oil into the skimmer and improves collection rates in both the maneuvering and stationary modes.

A variety of proprietary skimming devices will be discussed in the pages which follow.

A device developed by *J.K. Stewart; U.S. Patent 3,794,175; February 26, 1974* is a jet device in which floating oil is picked up by water jets and carried over a vertical wall into a receiving chamber. The water jet forming nozzles and the receiving chamber are parts of a floating vessel. The water jet forming nozzles are positioned to discharge upwardly through the floating oil. Float controlled mechanisms automatically maintain the jet nozzles properly oriented with respect to the receiving wall.

A technique developed by *R.L. Avey; U.S. Patent 3,796,656; March 12, 1974* involves removing an oil slick by dragging the open mouth of an initially empty collecting bag in partly immersed position along the surface of a body of water which is covered for instance by an oil slick, to cause the latter to enter into the bag, and to weight and release the open mouth of the bag to allow the same to sink and become suspended from the remainder of the bag. The system can be attached to one or two boats to so drag the mouth of the bag and has releasing and supporting elements adapted to release and allow sinking the mouth of the bag. The oil collecting bag is to float while holding collected oil, having a skin constructed of either open mesh material, fabric or film plastic such as to collect oil, with or without being water pervious.

An apparatus developed by *J.-C.J. Mourlon and E.M.R. Dubois; U.S. Patent 3,800,951; April 2, 1974; assigned to Bertin and Cie and Entreprise de Recherches et d'Activites Petrolieres-Elf, France* is a vortex device for removing from the surface of a body of liquid, an immiscible substance overfloating the same and spread thereover as a layer, the body of liquid and the layer of substance forming distinct horizontal strata, comprising:

a. means for locally whirling the liquid in the vicinity of the surface about a substantially vertical axis to generate a whirlpool, producing in the surface a depression of downwardly tapering closed-bottom cup-shaped outline into which the immiscible overfloating substance is centripetally drawn from the environment of the depression and accumulates in extra thickness compared with the layer; and
b. means, distinct from the whirling means, for discriminatingly extracting from the depression the immiscible substance accumulated therein, the extracting means including a portion adapted to engage a region of the depression located above the bottom of the depression.

A skimming device developed by *J.P. Oxenham; U.S. Patent 3,810,546; May 14, 1974; assigned to Shell Oil Company* is a skimmer utilizing liquid movement theretoward to facilitate separation of the polluting liquid from the water. The skimmer is arranged to remove the polluting liquid from all directions. The device further uses means for circulating water adjacent the polluting liquid/water interface to present moving liquid to the skimmer approaching from all directions.

Such a device is shown in Figure 91. The apparatus **10** comprises as major components a skimmer **16** having the capability of skimming the polluting liquid **12** from a plurality of different directions and a circulating device **18** for moving water toward the skimmer from a plurality of directions. It should be apparent that the device **10** may be used to skim any liquid floating on the surface of a second liquid, even though the polluting liquid **12** is generally a hydrocarbon while the liquid body **14** is normally water.

FIGURE 91: SHELL OIL CO. SKIMMER DEVICE

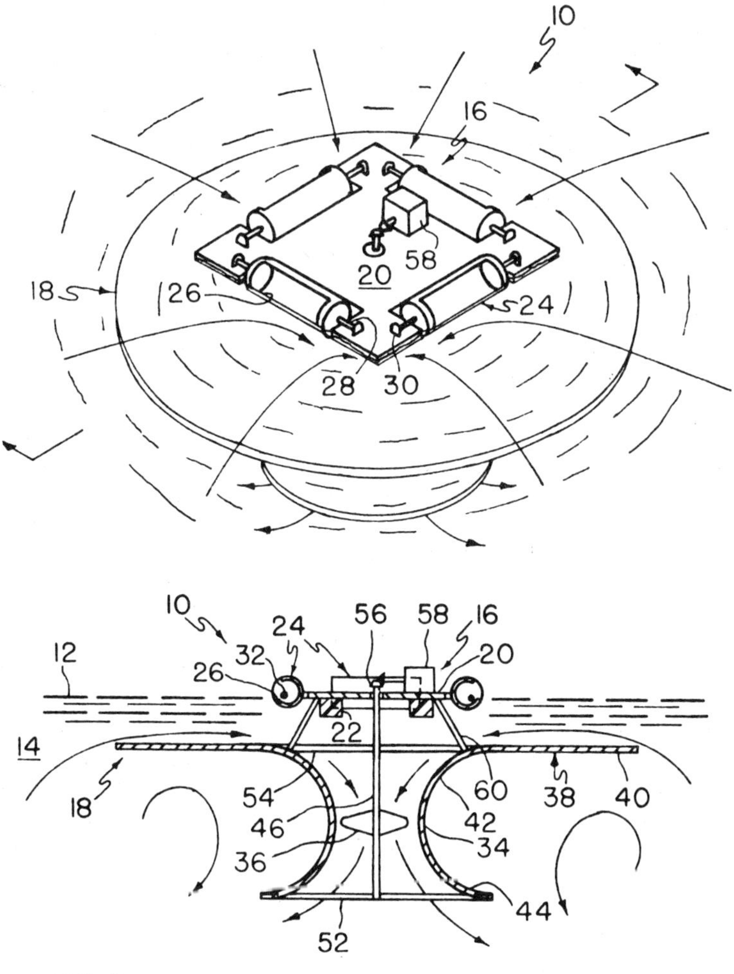

Source: U.S. Patent 3,810,546

The skimmer may be of any suitable type, the only requirement being the capability of skimming a floating liquid moving theretoward from at least two directions. Preferably, the skimmer is omnidirectional, i.e., has the capability of removing the polluting liquid from at least four mutually perpendicular directions.

Any suitable type of skimmer having this capability may be used. The skimmer which is described here is comprised of a platform **20** having a plurality of floats **22** thereon and a plurality of rotatable drums **24** spaced about the periphery of the platform **20**. The drums face in mutually perpendicular directions and each comprise a weir **26** positioned adjacent the interface of the liquids **12, 14** to allow a substantial amount of the polluting liquid to pass into the drums. Each of the drums is mounted for rotation about the axis thereof by a suitable shaft **28** and supports **30** on the platform **20**.

Inside each drum is a buoyant member **32** which acts to adjust the level of the weir **26** to maintain the weir adjacent the interface of the liquids **12, 14**. Suitable means (not shown), such as pipes and pumps, are provided to remove the liquid accumulating in the drums to a storage location (not shown) for ultimate disposal. It will be seen that the skimmer comprises means for removing the polluting liquid **12** from the water **14** upon movement thereof toward the skimmer from at least two directions.

The circulating means **18** comprises an important feature of this device and is adapted to circulate the water **14** to present moving liquid to the skimmer from at least two directions corresponding to the directionality of the skimmer. Preferably, the circulating device **18** comprises means for moving water toward the skimmer from the four directions corresponding to the directionality thereof or for moving water radially toward the skimmer from all directions.

To this end, the circulating device comprises a vertical duct **34** positioned below the skimmer having an upper end submerged in the water **14**. The circulating device also comprises an impeller **36** for moving water downwardly through the duct. The circulating device futher preferably comprises a generally horizontal flow directing member **38** comprised of a generally planar annular section **40** and a transition section **42** merging with the duct adjacent the upper end thereof. It will be apparent that the flow directing member directs water horizontally toward the skimmer. The flow directing member preferably extends substantially beyond the periphery of the skimmer as shown best in the section view at the lower part of the figure. The duct **34** may comprise an outwardly flared lower end **44** so that the circulating device comprises a streamlined shroud providing horizontal water flow from all directions and downward water flow immediately beneath the skimmer.

The impeller is secured to a suitable shaft **46** journalled in spaced bearings (not shown) carried by a plurality of suitable bracing elements **52, 54**. The shaft **46** may be driven in any suitable manner, as by extending the shaft through the platform **20** and providing a pinion gear arrangement **56** thereon driven by a suitable motor **58**.

The circulating means may be connected to the skimmer by suitable braces **60** to utilize the buoyancy of the floats **22** to support the circulating means **18**. In the alternative, the circulating device may be moored to the bottom of the water body **14** and provided with ballast tanks (not shown) or the like which may be emptied of water by the use of compressed air to buoy the circulating device adjacent the water surface. In this relationship, the skimmer may be moored to the circulating device or separately moored to the water bottom.

In the operation of this device, the apparatus is transported to the contaminated area and put into the water where it is desired to skim polluting liquids therefrom. The polluting liquid **12** passing over the weirs into the drums is removed in any convenient manner.

The motor is started to drive the shaft and consequently drive the impeller. The design of the impeller and the rotational speed of the shaft are selected to maintain a volume flow rate through the vertical duct sufficiently low to prevent drawing the polluting liquid **12** downwardly through the duct. As water is moved downwardly through the duct, a low pressure area is generated immediately beneath the skimmer. Consequently, water flows horizontally above the annular member into the top of the duct. As water adjacent the annular member is moved horizontally toward the center of the skimmer, flow of the polluting liquid in the same direction is thereby induced. The drums consequently act to skim a substantial quantity of polluting liquid from the water during the approach of the polluting liquid toward the predetermined central area of the skimmer.

The circulating means consequently presents a substantially greater amount of polluting liquid to the skimmer for removal. The skimmer is accordingly capable of removing substantially greater quantities of the polluting liquid without handling a disproportionately greater amount of liquid through the skimmer. This is particularly important when the liquid removed from the skimmer must be stored and subsequently rehandled as is normally the case.

A device developed by *A.J. Crisafulli; U.S. Patent 3,822,789; July 9, 1974* is a skimmer for removal of a layer of oil or other floating pollutants from the surface of a body of water incorporating a sump box having a pump disposed therein for conveying material from the sump box into a floating barge, vessel or other area. A free floating weir forming means is disposed forwardly of the sump box and connected and communicated therewith in such a manner that the weir forming means may vary in elevational relation to the sump box without the sump box being elevationally varied in relation to the surface of the water. The weir forming means is supported by float means which orients the weir of the weir forming means in desired relationship to the surface of the water.

A somewhat similar device is described by *A.J. Crisafulli; U.S. Patent 3,923,661; Dec. 2, 1975.*

An apparatus developed by *P.G. Bhuta, R.L. Johnson and D.J. Graham; U.S. Patent 3,831,756; Aug. 27, 1974; assigned to TRW Inc.* is one for separating oil from water by a surface tension action utilizing a hollow liquid surface tension separator having a surface tension screen wall. The separator is filled with oil and the outer side of its surface tension screen is placed in contact with the body of water to be separated, such that each screen pore exposed to the water contains a liquid-liquid interface whose interfacial surface tension resists passage of the water through the pore.

A pressure differential, less than the critical pressure differential necessary to overcome the interfacial surface tension force acting across the pore, is established across the screen to drive the oil only through the screen into the separator. The primary application of the devices is in an oil recovery apparatus which floats on and in some cases is propelled along the water suface and is equipped with one or more surface tension liquid separators for extracting the oil from the water surface.

A device developed by *U. Favret; U.S. Patent 3,836,004; Sept. 17, 1974* is a buoyant float disposed in a body of water which carries oil inlet ports at a level to withdraw primarily oily substance floating on the surface. The ports are sized relative to the size of a communicating oil collection suction chamber so as to permit impeded flow into the chamber and to aid in creation of a vortex like effect tending to withdraw fluid with a heavy concentration of oil. A remote pump connected to a portion of the float provides suction sufficient to withdraw the oily substance from the collection chamber. Where wind may tend to displace the oily substance, a sail-like wind reaction surface which is attached to the float, carries the float, and its oil inlet ports, with the displaced oily substance. The oil inlet ports are disposed so as to be carried preferably above the water level so that in the absence of an upper layer of oily substances, only air is withdrawn.

A device developed by *A.A.R. Larsson; U.S. Patent 3,838,775; Oct. 1, 1974* is a device for collecting loose material from a surface, especially oil floating on water and is particularly useful for such purposes where viscous oil is involved which is to be removed from a water surface, even where the oil is frozen into or mixed with ice and the like. However, it may also be useful in many cases for other purposes, as for example removing oil from sand beaches, etc., and removing sludge from sedimentation basins and the like. It employs a construction somewhat generally resembling a rotary snowplow.

The device utilizes a rotor which is rotatable on a horizontal axis and disposed within a housing which substantially completely encloses the rotor structure, with the housing being more or less complementally shaped to the generally cylindrical configuration of the rotor. The latter is rotatably supported in the housing for rotation about a substantially horizontal axis and thus parallel to the surface of the water from which the material is to be collected, such material being adapted to be received in a horizontally extending intake opening extending generally parallel to the rotor axis and formed in the lower portion of the housing at the leading side thereof with such opening being so arranged that it may be disposed at or adjacent to the water surface for receipt of material therein. The housing is also provided with a discharge duct which is so located with respect to blades carried by the rotor, which are provided with a helical configuration, that material entering the intake opening is transported, and in effect "scooped up" and discharged in the duct provided therefor.

A device developed by *G.D. Aulisa; U.S. Patent 3,844,950; Oct. 29, 1974; assigned to Sun Oil Co. of Pennsylvania* is one in which an elongated skimming blade is mounted for rotation, about a substantially vertical axis, across the surface of a liquid, and is rotated about this axis by means of a suitable propulsion device. A collection trough attached to the blade serves as an accumulation means for the skimmed material. A float whose buoyancy is adjustable maintains the blade in a predetermined position relative to the liquid surface, the float being anchored to the bed of the body of water being skimmed. The whole constitutes a portable rotary skimmer.

A device developed by *L.Chastan-Bagnis; U.S. Patent 3,847,815; Nov. 12, 1974* is a device for collecting hydrocarbons, granular materials, absorbing bodies and various polluting agents which float as a layer on the surface of water such as sea, lakes, rivers, harbors, estuaries, pools, etc. The device operates by producing a sufficiently thick layer of polluting material so that it can be collected. It comprises a scoop and a header with diffusing tubes circulating water under pressure towards the rear of the scoop where it is allowed to escape through a suitable opening. The thickened layer of polluted material which floats on top of water is removed through a separate duct.

The device as shown in Figure 92 comprises a scoop **1** having a bottom, a rear opening **3**, and journals **4** to serve as fixation and articulation of the apparatus. Water is admitted at **5** in a header **6** which is provided with diffusing tubes **7** the latter being directed towards the rear of the scoop **1** to define so-called hydroejectors. Pumping of the hydrocarbons is carried out above the opening provided for the removal water, by means of a tube **8**. The layer of hydrocarbons is referred to by reference numeral **9** and water by reference numeral **10**. Finally, the device comprises a blower **11** which creates a current of air at the water surface.

The operation of the apparatus is quite simple. After having immersed the scoop and its support at a suitable depth, the device is allowed to advance slowly and water is thereafter admitted under pressure in the duct **5** after which it escapes at **7** thereby causing a current which attracts a layer of water and the film of hydrocarbons floating thereon. Water escapes through the opening **3** and the hydrocarbons are accumulated at the rear of the scoop where they can easily be sucked by means of suction tube **8**. It should be noted that the free surface of hydrocarbons in the scoop is at a higher level than the layer of water and that the current of air created by blower **11** which is directed from the front to the rear at the surface of the layer of hydrocarbons will help **9** to substantially increase this difference.

FIGURE 92: SCOOP FOR COLLECTING A LAYER OF POLLUTING MATERIAL ON WATER SURFACES

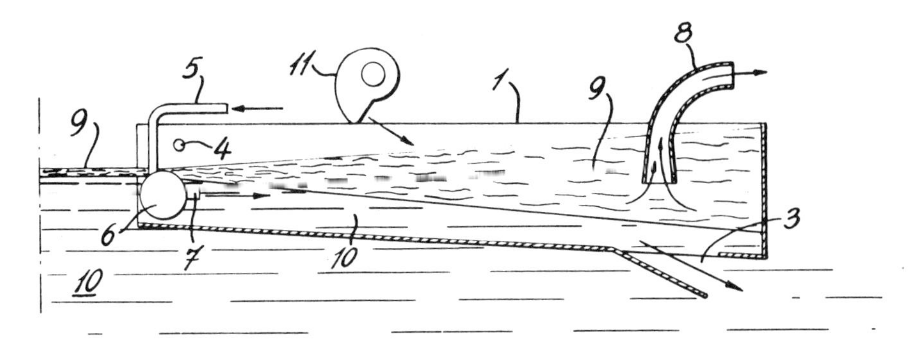

Source: U.S. Patent 3,847,815

A device developed by *F. Mohn; U.S. Patent 3,853,767; Dec. 10, 1974; assigned to Patents and Developments A/S, Norway* consists of an oil pump surrounded by a bowl or disc, the periphery of which is adapted to be positioned exactly at the plane of separation between water and oil. This precise adjustment can be done by means of the adjustable floats. The periphery of the disc will then, when correctly positioned, effect the first separation of oil from water. The oil pump is disposed in the center of the bowl or disc and, in order further to separate the oil from the water also at the center portion of the bowl, a chamber may be provided having inlets which are aligned with or somewhat lower than the periphery of the bowl, the oil concentration over the central portion of the bowl being substantially greater than that outside.

In order that the oil-covered water can be brought to the bowl in large amounts, a plate is arranged beneath the bowl which plate is provided at its center with a powerful pump for circulation of the water. When this pump is in operation, large amounts of surface water will be drawn toward the bowl and the plate and, on flowing over the plate to the pump thereof, the oil will trickle over the periphery of the bowl together with very small amounts of water. The water is drained from the central portion of the bowl through channels which open toward the plate.

A skimming device developed by *L. Bagnulo; U.S. Patent 3,853,768; Dec. 10, 1974* consists of a chamber adapted to be partly submerged in a body of liquid and having an upper and lower zone. Admitting means for admitting liquid into the lower zone of the chamber is provided and sufficient liquid is maintained in the chamber so as to allow substances floating on the admitted liquid to rise and float inside the upper zone of the chamber. Discharging means discharge the skimmed liquid at a lower portion of the lower zone of the chamber and conveying means conveys the floating substances from the chamber.

A device developed by *F. Galicia; U.S. Patent 3,880,758; Apr. 29, 1975* is a floatable collection device where contaminated liquid such as water polluted by oil is collected into a central drum by means of an elongated spiralled inverted V-shaped trough which is circumferentially mounted on the drum. The device includes expulsion means by which the contaminated liquid is expelled from the drum to separation apparatus located elsewhere for separation of the oil from the water.

A similar device is described by *F. Galicia; U.S. Patent 3,907,684; Sept. 23, 1975.*

A device developed by *R.G. Hoegberg and W.S. Tyler; U.S. Patent 3,905,902; Sept. 16, 1975* is one in which water pickup is minimized in the recovery of oil from water, including seawater, by negatively electrostatically charging hydrophobic rotating discs which are immersed in the water. Oil can be applied to the discs to help maintain the charge in the oil phase. The application of oil to the discs can also be used to remove thin films of oily material which contain oil-soluble contaminants from the surface of a body of water.

A device developed by *P. Degobert, F. Kermarrec and Y. Nadaud; U.S. Patent 3,912,635; Oct. 14, 1975; assigned to Institut Francais du Petrole, des Carburants et Lubrifiants, France* is a device for recovering polluting liquids spread over a water surface, comprising at least one floatable container having a porous hydrophobic wall which is easily wettable by these polluting liquids. The container includes a recess limited by an impervious wall, in combination with means for transferring into this recess the polluting liquids which have selectively traversed the porous wall.

Figure 93 shows such a device. In the embodiment illustrated, the container used for recovering polluting products forming a slick **1** on the water surface comprises a capacity **2** with an impervious wall, located above a capacity **3** having a porous wall which is hydrophobic and easily wettable by the polluting liquid (oleophilic material if the polluting liquid is an oily liquid, such as an oil product).

FIGURE 93: I.F.P. SKIMMER DEVICE UTILIZING A POROUS HYDROPHOBIC WALL

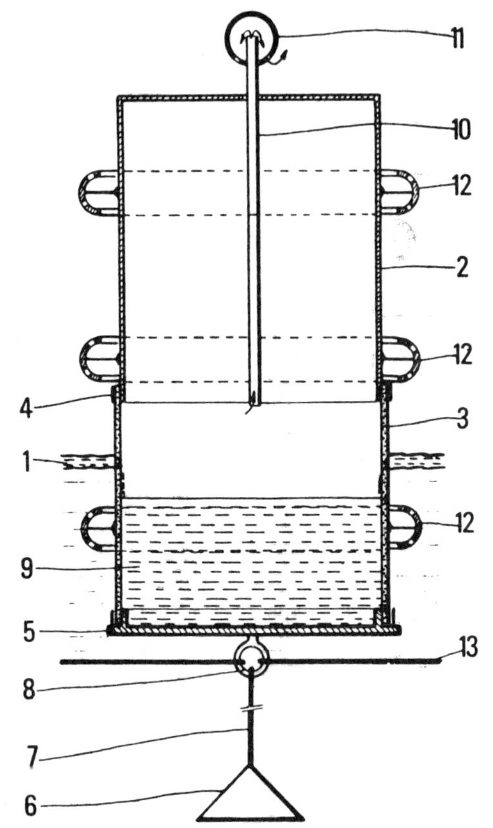

Source: U.S. Patent 3,912,635

These two capacities are connected at **4**. The bottom **5** of the capacity **3** is impervious. The assembly is provided with ballasting means **6** supported through a cable **7** from a ring **8** secured to the bottom of the capacity **3** and floats at the water surface.

The polluting liquid selectively penetrates into the capacity **3** through the porous lateral wall thereof and gathers at **9**. As this capacity is being filled, the air within the container can escape through tube **10** (arrows), overtopped by a protecting bell-shaped element **11**.

The container is provided with shock absorbing means **12**, capable of withstanding shocks between containers of the same type floating on the water surface. The container sinks progressively as the capacity **3** is being filled with polluting liquid. This filling is stopped when, upon sinking of the container, the lower edge of the capacity **2** with an impervious wall reaches the water surface. At this time the polluting liquid in the capacity **3** reaches the level of the lower opening of tube **10**.

The container is collected by turning it over, so that the polluting liquid contained in the capacity **3** with a porous wall passes into the capacity **2** with an impervious wall whose volume has been selected at least equal to that of the capacity **3**. A not-illustrated valve, associated with the vent **11**, prevents any ingress of water in this position. Under these conditions, the polluting liquid can no longer escape from the container when the latter is raised from the water surface, since this liquid is housed in a capacity having an impervious wall.

The turning over or inverting of the container is facilitated if a cable **13** connects to one another the rings **8** of a plurality of such recovering containers which float on the water surface, since it is then sufficient to exert a pull on the cable **13**, as is done with a fishing line, to cause the turning over of the various containers.

The recovery of a pollutant spill with an endless belt is particularly attractive because of its efficiency. Thus, the pollutant is continually sorbed by the belt and removed therefrom by a relatively simple operation. In addition, the belt can be employed under a wide range of weather conditions, the only limiting factor being weather too rough for the use of vessels transporting the endless belt. However, the endless belt system does suffer some drawbacks. Thus, conventional endless belts generally have a short service life and a low ratio of exposed surface area per unit mass of belt.

A technique developed by *E.V. Seymour and R.R. Ayers; U.S. Patent 3,928,205; Dec. 23, 1975; assigned to Shell Oil Company* utilizes a series of discrete sorbent bodies connected by a tension member to form an endless belt.

The devices are superior to conventional belts since (1) deformation of a discrete sorbent pad caused by wringing or squeezing to extract pollutant is not accompanied by the high stress intensity found in continuous sorbents of conventional design, and the durability of the belt is thus enhanced; (2) structural failure of one or more sorbent pads does not constitute failure of the complete belt; (3) the sorbent is not forced to transmit belt tension; (4) the belt design allows utilization of different types of sorbent pads for different applications, for example, different pollutants or different grades of the same pollutant; and, (5) the pollutant is imbibed in a shorter residence time because the oil reaches the interior of the belt by free flow, in which surface forces do not predominate, to increase the sorption rate. Free flow, or open flow, is possible where the belt structure, e.g., discrete bodies, allows natural forces such as wind, current and wave forces to predominate over surface tension forces which favor porous flow. This device is an improvement over the "sorption only" continuous belt of the prior art inasmuch as the belt of this process allows a combination of open flow and porous flow to increase sorption rate.

The device is shown in its simplest form in Figure 94. There is shown a sorbent belt composed of a tension member **20** to which is attached discrete sorbent pads **21** by means of lines **22**. While two pads are shown attached to the tension member at a single location, it is of course evident that one, three or more pads could be used similarly. Also, while the pads are shown as being cylindrical, other shapes are suitable, for example, rectangular or spherical.

FIGURE 94: SHELL OIL CO. DEVICE USING SORBENT BODIES ON AN ENDLESS BELT

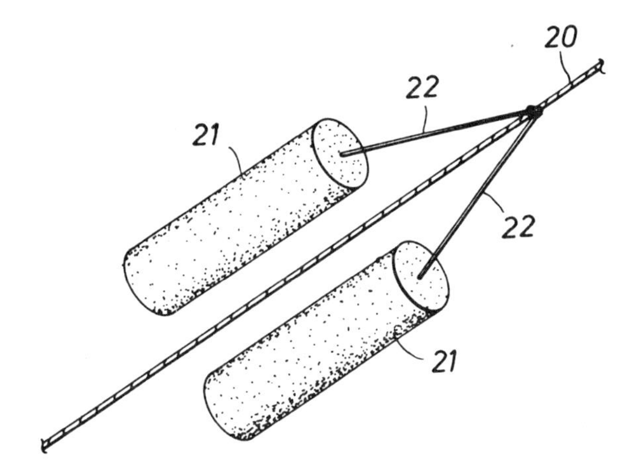

Source: U.S. Patent 3,928,205

A device developed by *H.E. Bagot and S.P. Funkhouser; U.S. Patent 3,986,959; Oct. 19, 1976* is one in which a floating sponge remains essentially fixed on top of the water and bouyant squeezing means are moved over the belt to squeeze out oil picked up thereby without having to lift the sponge from the water.

The essential elements of such a device and its mode of use are shown in Figure 95. Two buoyant vessels **13** and**15** respectively are spaced apart with a sponge like belt **17** extended therebetween lying on the surface of the water. Also floating on the water with the belt passing therethrough is buoyant squeezing device shuttle **18**, an essentially hollow member having an entrance opening **19** and exit opening **20** and containing in its inside a pair of squeezing rollers **21** and **23**.

At each of the entrances **19** and **20** seals **25** such as rubber seals are provided to prevent the entrance of excessive amounts of the water beneath the sponge belt **17**. Attached to the top of the buoyant squeezing apparatus **18** is an eye or the like **27** to which is rigidly coupled a cable **29**. The cable passes between winch drums **31** on the vessels **13** and **15**. At least one of the drums **31** is powered in conventional fashion to cause the buoyant squeezing apparatus **18** to be moved back and forth across the belt **17** squeezing the oil therefrom.

FIGURE 95: FLOATING SPONGE DEVICE FOR OIL SPILL REMOVAL

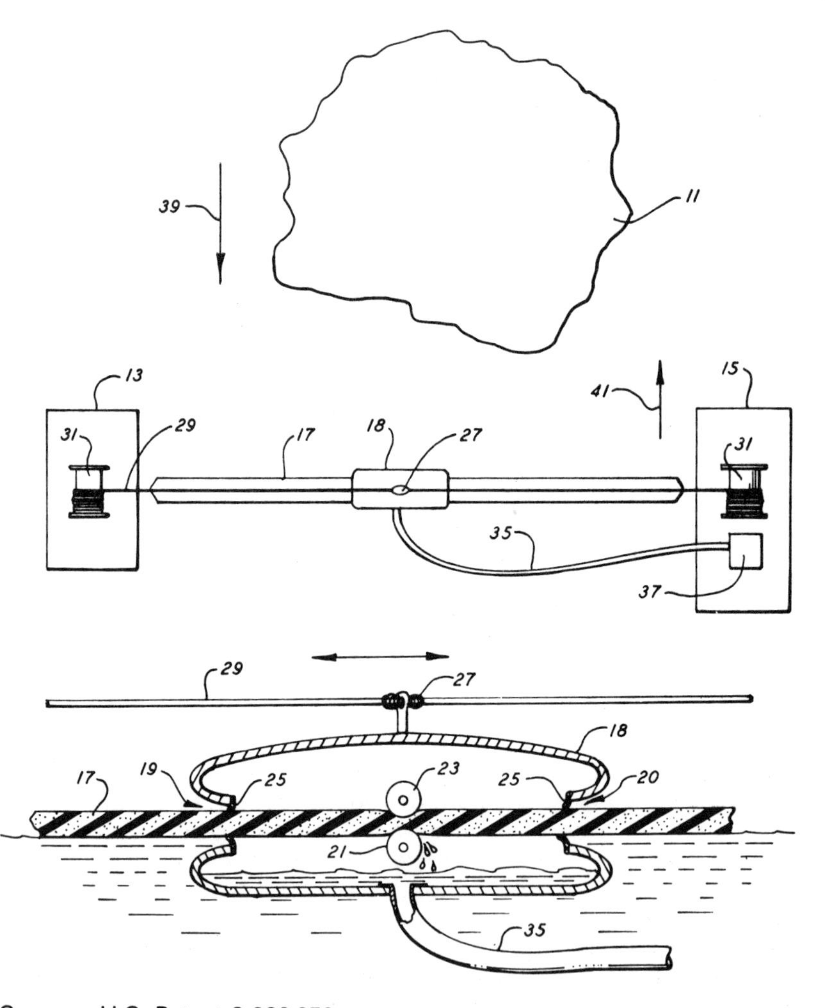

Source: U.S. Patent 3,986,959

The oil removed will drip down into the bottom of the buoyant squeezing apparatus **18** from which it is removed through a line **35** coupled to a pump **37** on one of the vessels **13** or **15**, in this case on the vessel **15**. From the pump it can be directed to suitable storage tanks within the vessel **15**. The vessels **13** and **15** can simply be buoys on which the necessary equipment is mounted anchored in a river for example. Should the river or the current in a bay be moving in the direction of the arrow **39**, so placing the apparatus would result in the water currents moving the oil spill **11** past the belt **17** whereby the oil would be picked up and transferred to the barge **15**. Similarly, the vessels **13** and **15** can be barges which are towed or self-propelled barges which move in the direction of the arrows **41** to sweep over the oil spill to pick it up.

An apparatus developed by *C. McLellan; U.S. Patent 3,990,975; Nov. 9, 1976; assigned to Oil Mop Inc.* is a rigging system for causing an endless oil mop driven through an engine or motor driven wringer mechanism to make multiple passes over and through an oil contaminated body of water from the deck of a vessel. This is accomplished by using sampson posts on the deck of a vessel and connecting spars to the posts and buoyantly supporting the free ends of the spars and mounting mop pulleys at the end of each spar and one pulley between the spars on the side of the vessel at the water line.

The mop is of the type described in U.S. Patent 3,668,118 and is made up of narrow thin gauge strips of polypropylene secured to a center pull multi-strand polypropylene line. The mop floats in water even when laden with oil.

A device developed by *E.L. Grimes and D.W. Lerch; U.S. Patent 3,992,292; Nov. 16, 1976; assigned to Marine Construction & Design Co.* is an oil spill recovery apparatus utilizing a low-resistance flow-through endless belt of reticular oleophilic, hydrophobic material and forced flow of water with oil through the belt through induction effected by propulsion means operating at a position immediately behind and beneath the submerged active portion of the belt.

A scheme for emergency small spill control which is described by *D.E. Irons; U.S. Patent 4,006,082; Feb. 1, 1977; assigned to Murphy Pacific Marine Salvage Co.* is shown in Figure 96. A ship's boat **A** is shown on the surface of harbor water **B** having boom **C** surrounding an oil spill **D**. The stern section of the ship's boat at **14** is broken away so that a pollutant receiving container **E** can be seen within the boat. This container **E** has a pump **F** connected through hosing **G** to a sump **H**. By the expedient of operating sump **H** and collecting boom **C** through rollers **J**, the oil spill **D** can be collected from the surface of the water **B** and thereafter pumped interior of the container **E** in the ship's boat **A**.

Boom **C** is a flexible curtain-type boom of approximately twelve inches overall in depth. It includes a circular polyvinyl chloride material **60**, electronically welded into typical 50 foot lengths. This material has disposed in the upper portion thereof an ethyfoam float material **62** preferably one-half inch thick.

The lower end of the curtain contains one-quarter inch chain **64**. As is typical, the respective ends of the boom are sealed as by a seal **66**. Preferably, over half the material of the floating boom is disposed as a curtain below the surface of water **B**. This underwater curtain penetrates the water/oil interface and enables the isolation of the oil on one side of the boom from unpolluted water on the other side of the boom.

Preferably, boom **C** is dead ended to one side of lip **26** at a bracket **70**. The boom is disposed in its partially immersed condition around the periphery of the oil spill **D** and threaded through a roller assembly **72**.

FIGURE 96: PROCESS FOR EMERGENCY SMALL SPILL CONTROL

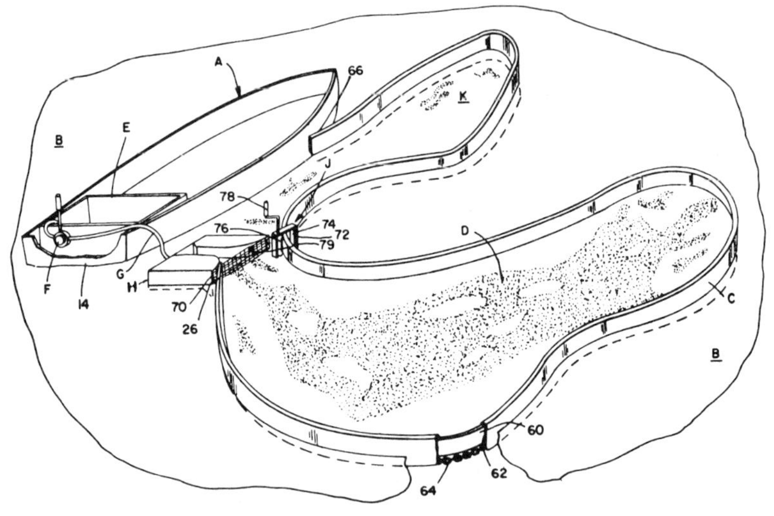

Source: U.S. Patent 4,006,082

Roller assembly **72** consists of paired upright rollers **74, 76**. Typically, roller **76** is rotated by an attached handle **78**. By confronting the floating boom **C** between the paired rollers, the boom is collected with clockwise rotation of handle **78** so as to collect the pollutants **D** interior of the sump.

Preferably, a screen **79** is placed across the opening to the sump. This screen serves to block the entrance of clogging debris into the funnel, hose, and pump assembly. Those experienced with shipboard life will appreciate that many harbors include a measure of surface debris far above that normally found in open waters. As such harbors are the scene of most spills, provision for debris separation is made.

It can be seen that a hose **G** connects sump **H** and pump **F**. Typically, hose **G** at the pump and at the sump is fitted with quick release clamps such as Kamlock clamps.

The pump here shown is a 20 gallon-per-minute, hand operated, positive displacement, bilge pump. This pump is of the type wherein a cover can be removed for the cleaning of debris from the pump.

It will be appreciated by those skilled in the art that a preferred location for the placement of the pump will be on the inside of the box. Thus, any cleaning of the pump which occurs will not cause oil to be discharged interior of the ship's boat **A**. The pump is shown here at the outside for ease of understanding. The discharge line to the pump is passed interior of the pollutant receiving container **E**.

In operation, boom **C** is moved to surround the oil spill **D**. Thereafter, the end of the boom at **66** is threaded between the vertical take-up rollers **74, 76** and collection of the boom commenced by the rotation of handle **78**. When a sufficient section of the boom has been passed through the vertical take-up rollers **74, 76**, the boom in turn is dead ended to the side of boat **A** working the spill.

It is important to note that the boom defines two containment areas. The first is the containment of the gross spill at **D**. The second is the secondary containment area of the spill at **K**.

This second area may be subsequently cleaned by using towels on the surface of the water. By the process of manipulating the boom back and forth to define respective decreasing and increasing pollutant collection areas, collection by pumping or sponging can occur at leisure until complete abatement of the spill occurs.

An apparatus developed by *I. Tsunoi; U.S. Patent 4,006,086; Feb. 1, 1977; assigned to Kyoei Senpaku Kogyo KK, Japan* is a floating oil recovery device which prevents the agitation and mixing of the oil with water at the time when the oil layer floating on the surface of the water flows into the oil dam, and also makes it possible to recover the oil in an extremely high purity.

Such an apparatus is shown in cross section in Figure 97. The device may be supported between two hulls of a twin-hull ship, one hull indicated at **4**. To the support rod **5** is hung down a dam section **8** via arms **7** capable of swinging freely. Spacer boards made of a rigid rubber, project both sides of the dam section respectively and fill up the slits formed respctively between both sides of the dam section and the two hulls, whereby the passage of the oil behind the dam section past thereover is prevented. A float **11** having adjustable buoyancy is provided to the lower surface at the lower end section of the dam section and balance weights **13** are fitted to the upper edge of each of the arms. A bottom plate **14** is placed between the lower edge section of the arms, and also connects the lower edge of the dam section so as to thereby define a chamber, that is, a water tank **15**.

A weir **16** is connected to the bottom plate **14** foldably by means of a hinge **17**, and is operable aboard the ship through a link **18** and a lever **19**. Though the weir may as well be made of a wood, it has advantageously a float-shape as shown in the figures, because buoyancy takes part in floating the weir upwards automatically. In this case, manipulation of the lever is necessary only when the weir is required to be pushed down.

An oil guide plate **20** is adapted at an upward position of the weir so as to project ahead of the dam section, and functions to guide an oil layer **21** passing over the weir into a chamber **22** defined at the innermost part of the dam section along the lower surface of the weir. Oil suction ports **24** of an oil suction passage **23** are open downwards at the upper section of the chamber **22**, and the oil is sucked up through the passage by an oil suction pump which is not shown in the figures. The chamber as an oil basin is formed at a position a little higher than the level of the water in order to allow the elevation of the oil having a specific gravity smaller than that of water, but prevent the elevation of water, and enhances the purity of oil to be recovered.

A number of vertical holes **26** are bored in between the dam section and the weir. Water passing over the weir easily falls downward through these vertical holes into the chamber formed therebelow, to wit, into the water tank, since specific gravity of water is greater than that of the oil. The oil flowing into the vertical holes, on the other hand, floats upward in the holes, hardly any falling downward because it is lighter than water. Therefore, hardly any oil gets mixed with the water which is down in the water tank, even when the velocity of the liquid passing over the weir is quite high. Thus the oil floats upward, with the vertical holes as its boundary, while the water remains below thereby effecting separation between the oil and water.

Inside the water tank, there are provided water suction ports **28** at the lowest possible portion of the tank. The openings of these ports are connected to a main passage **30** for a pressure connection with a water suction pump of a large capacity (not shown) to suck up water flowing into the water tank.

FIGURE 97: VACUUM CLEANER-LIKE DEVICE FOR OIL SPILL REMOVAL FROM WATER

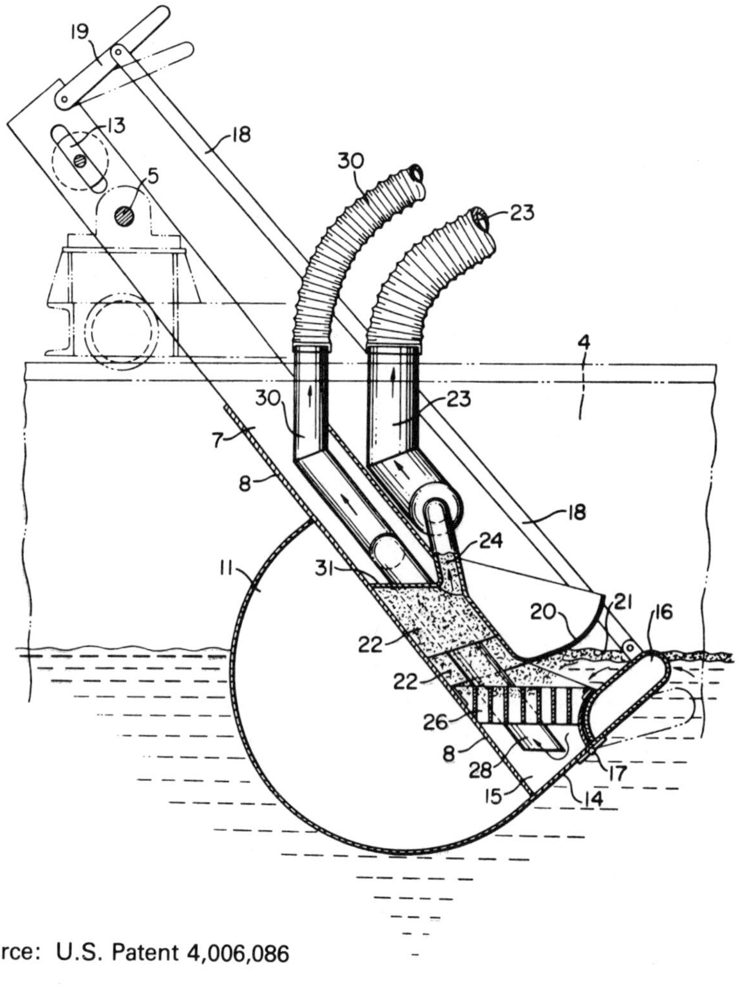

Source: U.S. Patent 4,006,086

In the embodiment shown in the figure, an oblique plate **31** to form the chamber as the oil basin is shown as a fixed type. However, the oblique plate **31** may also be made of a material such as a rubber which can be displaceable in a vertical direction so that the positioning thereof can be adjusted from the above. At the initial state of the oil-suction operation or when the oil layer is very thin, the oblique plate is maintained at a level that is not quite as high. As the oil layer becomes thicker, the oblique plate is pulled upward gradually so as to enlarge the capacity of the chamber as the oil basin. By this arrangement, occurrence of cavitation is eliminated when the oil layer is thin, and invasion of the oil layer into the water tank is prohibited when the oil layer is thick. Thus, smooth suction of only the oil layer is ensured.

The mode of operation of this device is as follows. The dam section is first launched down to the level of the water by proper means whereupon the lower section of the dam section is floated up by the action of the float. Next, the level of the dam section is adjusted by controlling the balance weights as well as the buoyancy of the float by means of an electromagnetic valve (not shown) so that the upper periphery of the weir appears and disappears from the level of the water. Subsequently, the lever is pushed forward so as to thereby push down the weir via the link whereupon water flows into a number of vertical holes, and the oil layer likewise flows thereinto along with the inflow of water.

As the water tank is empty at the beginning, the oil and water drop into the tank in the state of admixture for a while, but the oil soon floats up into the upper layer and separates from water. By driving then the water suction pump as well as the oil suction pump (not shown), negative pressure is applied respectively to the water suction ports and the oil suction ports, and a large amount of water is sucked from the former while the oil is sucked up by the latter.

While adjusting the weir vertically by the lever, the oil and water are supplied continuously, and after a short while, they are separated from each other and sucked up separately. In other words, the oil layer is guided into the dam section by the inflow of water as a conveyor.

As heretofore noted, the device has a large number of vertical holes provided inside the dam section in order to prevent the occurrence of a turbulent flow inside the dam section as much as possible, and to separate the oil from water into the upper and lower layers respectively. Meanwhile the oil suction ports are located at a position relatively higher than the level of the water so as to prevent the suction of water which is heavier than the oil. By these combined arrangements, the very high purity oil can be recovered both efficiently and continuously.

A device developed by *M.G. Webb, U.S. Patent 4,021,344; May 3, 1977; assigned to The British Petroleum Company Limited, England* is an oil pickup device for removing oil off the surface of water which consists of a series of vertical plates driven from a central drive member. The plates pass through the oil/water surface and oil is picked up. Scrapers remove the oil and lead it away. There is also incorporated a pneumatic device for maintaining the pickup device at the right level in the water.

An apparatus developed by *J.N. Koblanski; U.S. Patent 4,032,438; June*

28, 1977; assigned to Ocean Ecology Ltd., Canada is one employing an ultrasonic focusing transducer for removing a contaminant such as oil which is floating on the surface of a body of water. The transducer is supported beneath the water surface with its focal region aimed at the underside of the contaminant oil. A source of alternating current is connected to the transducer to generate ultrasonic waves which travel through the water and converge at the focal region. The apparatus includes a collecting arrangement which catches most of the oil bounced upwardly as the result of the focused ultrasonic waves before that upwardly discharged oil can fall back onto the water surface.

A skimmer device developed by *E.V.M. De Visser and K.I. Ghyselen; U.S. Patent 4,032,449; June 28, 1977; assigned to SA Texaco Belgium NV, Belgium* is suitable for use on a pond or reservoir where the surface layer contains frangible solid matter such as ice. It employs a weir structure with a curved edge, and a pair of arms having interlocking teeth extending radially from the edge. One arm is stationary and one rotates so that the interlocking positions of the teeth will break up the solid matter.

A device developed by *R.S. Jenkins; U.S. Patent 4,038,182; July 26, 1977* operates by forming a vortex in the water mass to attract the oily film substance and water in the vicinity of the vortex to flow into the vortex and flowing the attracted oily film substance and water from the vortex to a quiescent zone to enable separation of the oily film substance and water. The separated oily film substance is retained in a receiver and recovered from the receiver as desired.

A scheme developed by *E. Irons; U.S. Patent 4,046,691; Sept. 6, 1977; assigned to Ballast-Nedam Groep, NV, Netherlands* is one for collecting oil floating on water, the oil being carried into a collecting reservoir with the outer side of a wall of the collecting reservoir extending below the water surface. Away from the water surface, a downstream is produced by which the oil and water are passed beneath an edge of the wall located beneath the water surface and into the collecting reservoir. After this the oil rises up in the reservoir to the water surface, where it is collected, while the water is conducted away through at least one outlet near the bottom side of the collecting reservoir.

A device developed by *M.G. Schwartz, A.P. Lorentzen and D.J. Bucheck; U.S. Patent 4,052,306; October 4, 1977; assigned to Minnesota Mining and Manufacturing Company* is a floatable oil sweep useful in controlling an oil spill on a moving body of water. It comprises an elongated web of oil sorbent adapted to float on the body of water with its large-area faces parallel to the waterline, and a weighted open-mesh netting attached to the web and adapted to be suspended below the floating web when the oil sweep is deployed. Use of the netting has been found to significantly extend the period of time before oil droplets are carried under the oil sweep by movement of the body of water.

A device developed by *G.H. Rolls; U.S. Patent 4,052,313; Oct. 4, 1977* is an oil recovery system which utilizes a rope of an adsorbent material arranged for floating on the surface of a liquid contaminated by a contaminating material preferentially adsorbed by the rope. The rope is in the form of a continuous loop extending between a desorption station through which the rope is advanced to remove adsorbed material and a rope-guide station or each of a plurality of rope-guide stations at which the rope is guided round guide means. Hold-off means controlled by or from the desorption station are provided for holding off the rope-guide station or each of the rope-guide stations from the desorption station by a predetermined distance without the use of an anchor.

Skimming Vehicles

Skimming vehicles are generally self-propelled vessels incorporating skimming devices.

A technique developed by *A.L. deAngelis; U.S. Patent 3,788,481; Jan. 29, 1974* employs a boat in which:

a. the polluted liquid surface is successively, fully or in part, withdrawn from the action of the wind and, at least in part, from the natural wave motion, and segregated;
b. the segregated floating polluting substance is skimmed together with a minimum quantity of the underlying water;
c. the liquid skimmed is collected and decanted while the floating substance is separated.

Such a vessel and its fittings are shown in Figure 98. The chambers **1** are supplied with openings **2** pierced in the boat. These openings may be opened or shut by a guillotine-like member **3** or the like. The chambers are equipped with funnel devices **4**. These devices are formed by a funnel **5**, a float **6** from which the funnel is suspended at an adjustable distance, and by a flexible pipe **7**. The upper end of the flexible pipe is connected with a pump **8**. From the pump two deliveries are branched off, on conveying the liquid into the floating vessels **B** and the other into the tanks **9**. The sucking end of the pump **8** is connected with the bottom of the funnel **5** and also—even if not represented—with the bottom of the tank **9**.

The funnel-device **4** is represented, for the sake of simplicity, in two chambers only, in its highest and lowest positions; it is to be understood that each chamber may be supplied with one or more funnel devices. Though not shown, each funnel is preferably guided in its upward and downward motion by rails preventing side movements.

The elimination of the floating oil, considering the boat empty, starts upon lifting of the guillotine **3** with the funnel in its lowest position. The pump **8** sucks the liquid and conveys it in the tanks **9** or in the vessels **B**.

In the first case, when the filling of the tanks is over, the liquid is left at rest, in order to separate it into two layers; the lower aqueous one is discharged off and the oily one recovered, for example, by conveying it into the vessel **B**.

The boat is equipped with retractable or unlockable flaps **10**, to facilitate the flow of the oily stratus into the chambers. In the figure, the flaps **10** are shown in correspondence with one chamber only, but it is understood that also the other chambers may be furnished with them.

FIGURE 98: DE ANGELIS DESIGN FOR SKIMMER VEHICLE

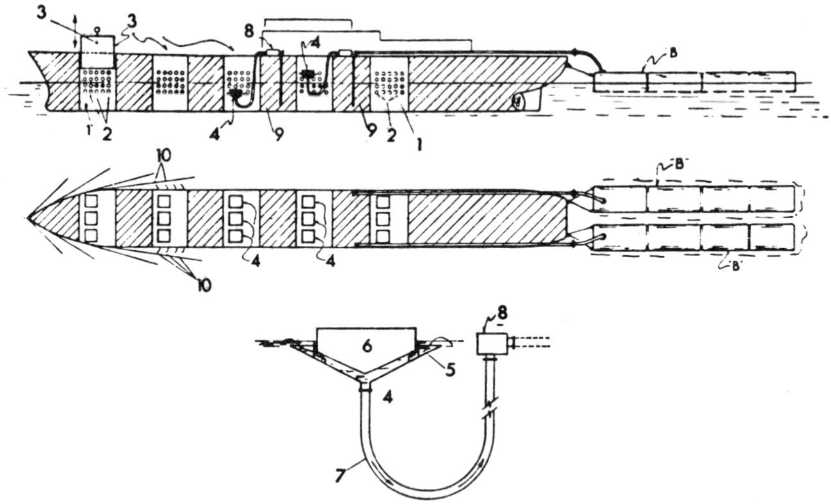

Source: U.S. Patent 3,788,481

A skimmer vehicle developed by *A.Y. Derzhavets, P.G. Kogan, V.N. Semenov and V.I. Tabachnikov; U.S. Patent 3,823,828; July 16, 1974* is a self-propelled oil skimmer craft having a collecting receptacle. It comprises a water jet and a duct for taking in water and delivering it to a water jet. The intake hole of the duct is located below the means for water inflow to the collecting receptacle of the skimmer craft, due to which a water stream entering the duct draws in the upper layer of water together with floating impurities towards the means for water inflow to the collecting receptacle.

A scheme developed by *J. DiPerna; U.S. Patent 3,847,816; Nov. 12, 1974* comprises an oil tanker ship containing a series of oil collection tanks within its hold, a floating oil harvesting ring being towed in front of the ship by director tugs that are electronically controlled from the ship, the harvesting ring including a skimmer section into which gathered floating oil is sucked and conveyed from the skimmer through intake tubes to the interior of the ship, where it is passed into the series of tanks.

Such a scheme is shown in Figure 99. Referring to the drawing in detail, the reference numeral **10** represents a pollution suction water sweeper, wherein there is an oil tanker ship **11** having an oil collection tank system within its hold.

In front of the ship, a harvesting ring **14** is pulled so as to gather up an oil slick **15** from a surface of a sea **16**. The ring **14** consists of a flexible empty accordian-like hose **17** at each end with an apron **18** attached thereto, the aprons having weights to hold down the bottom of the apron. Between the ends of the ring, a central portion forms a skimmer **19** and consists of a wide hose having a series of slotted vertical openings **20** on a front side so that the oil slick can enter inside. The skimmer is connected to two flexible hoses **21** that pass through openings in the hull **22** and into the ship hold. Gravitation may be employed to move the collected mixture of water and oil slick to the tank system which is below a sea level of the seawater.

As shown, the ring **14** is pulled in front of the ship by a director tug **30** at each end of the ring, the tub being controlled electronically by signals from a control station within the ship, so to guide the director speed and direction of steering. A guide line **31** from the ship bowsprit physically engages the ship with the ring additionally.

FIGURE 99: DI PERNA DESIGN FOR SKIMMER VEHICLE SYSTEM

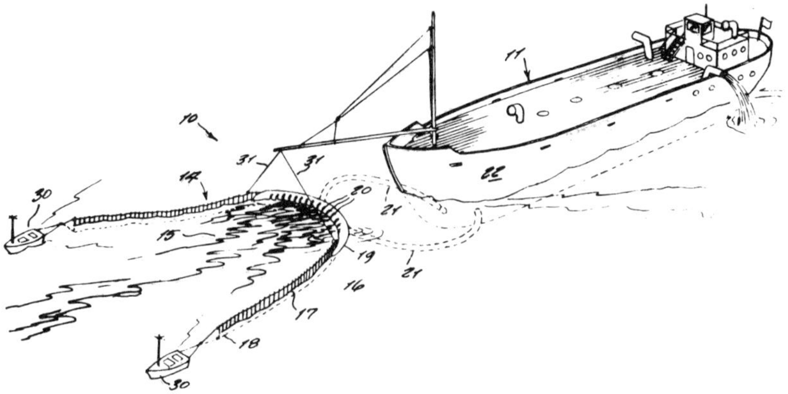

Source: U.S. Patent 3,847,816

A skimmer vehicle developed by *D.J. Weatherford; U.S. Patent 3,860,519; Jan. 14, 1975* is shown in Figure 100.

The vessel is driven by a suitable engine **12** which is connected to rotate propeller **13** in the usual fashion for moving the boat. Steering of the boat is through a rudder **14** mounted at the stern of the boat and controlled from the operator's console **15**, where all other control and measurement devices are positioned for easy reading.

The appended hull **16** has a separate front panel **17** mounted by a water sealed hinge arrangement **18** at the bottom **19** of the boat. The panel can be rotated around hinge arrangement **18** to a plurality of positions from a closed position where hull **16** is closed from water entering therein to one of a plurality of open positions which bring the top edge **20** of panel **17** below the water level **21**, thereby allowing a controlled amount of surface water and the top floating oil slick to enter the boat's isolated lower compartment **22**.

Two flaps **23** are hingedly attached to the hull so as to form a funnel or scoop with panel **17** at the bow of the boat to maintain a seal between themselves and the front panel, as shown. These side panels extend considerably beyond the front panel **17**, thereby serving both as a funneling device to gather in surface slick and as stabilizers for the boat. The front panel is controlled by one or more hydraulic devices **24** mounted between the bottom **25** of the hull **26** and the movable panel **17**.

FIGURE 100: WEATHERFORD SKIMMER VEHICLE DESIGN

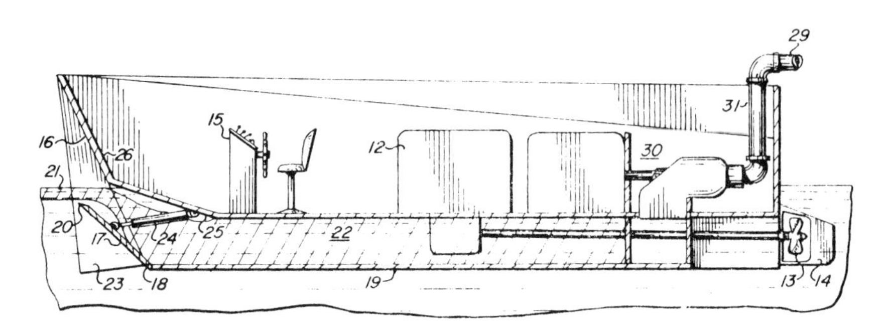

Source: U.S. Patent 3,860,519

Similarly, the flaps **23** are also held against the edges of the front panel by hydraulic devices mounted between the flaps and the outside surfaces of appended hull **16**. The operator at the console **15** can remotely control the hydraulically actuated front panel, reading its position below the sea level on a console gauge by sensing the position of the rim of the hydraulic device **24** and also can measure and read the rate of water flow pumped out of a pipe or hose **29** at the stern of the boat.

As the front panel is lowered below the water level **20**, the oil slick and minimal adjacent water are skimmed off continuously into the lower isolated hull compartment **22**. As noted from the drawings, this compartment may take up the entire lower portion of the boat. The amount of water and oil skimmed from the body of water is controlled by the operator, who adjusts the front panel position and the boat speed, taking into consideration the condition of the sea.

At the rear of the boat within the hull **26** are mounted one or more jet pumps **30**. As shown, two pumps of a type equivalent to a Jacuzzi Marine Jet Propulsion Pump, series 12WJ, each handling up to 36 gallons per second or more at 2,000 rpm, may be used. Any other pumping arrangement to accomplish the handling of a large quantity of water may be used, such as electric pumps or their equivalent used singly or in multiple combination, depending on the volume of oil and water handled and the size of the boat. When the water-oil mixture scooped into the lower compartment **22** reaches a given volume as indicated by a float-type gauge on the console **15**, the operator starts one or more of the jet pumps, causing liquid in a surging fashion to be pumped out of the lower hull of the boat through pipe **31**, hose **29** and into a towed boat or barge.

A skimming vehicle developed by *R.R. Ayers and D.P. Hemphill; U.S. Patent 3,865,730; Feb. 11, 1975; assigned to Shell Oil Company* is one with which an oil spill, in particular a small volume spill, can be skimmed from the water in an efficient and economical manner.

Figure 101 shows such a vehicle in plan view and in two modes of operation in vertical sectional views. The figure shows a skimmer **1** located between the pontoons **2** of a catamaran vessel. This vessel may be provided with independent propulsion means such as outboard motors **3**, a hydraulic power system **4** for adjusting the position of the skimmer as shown, a positive displacement pump **5** for removing collected oil from the skimmer, and oil storage tanks **6** for storing collected oil.

FIGURE 101: SHELL OIL CO. SKIMMER VEHICLE

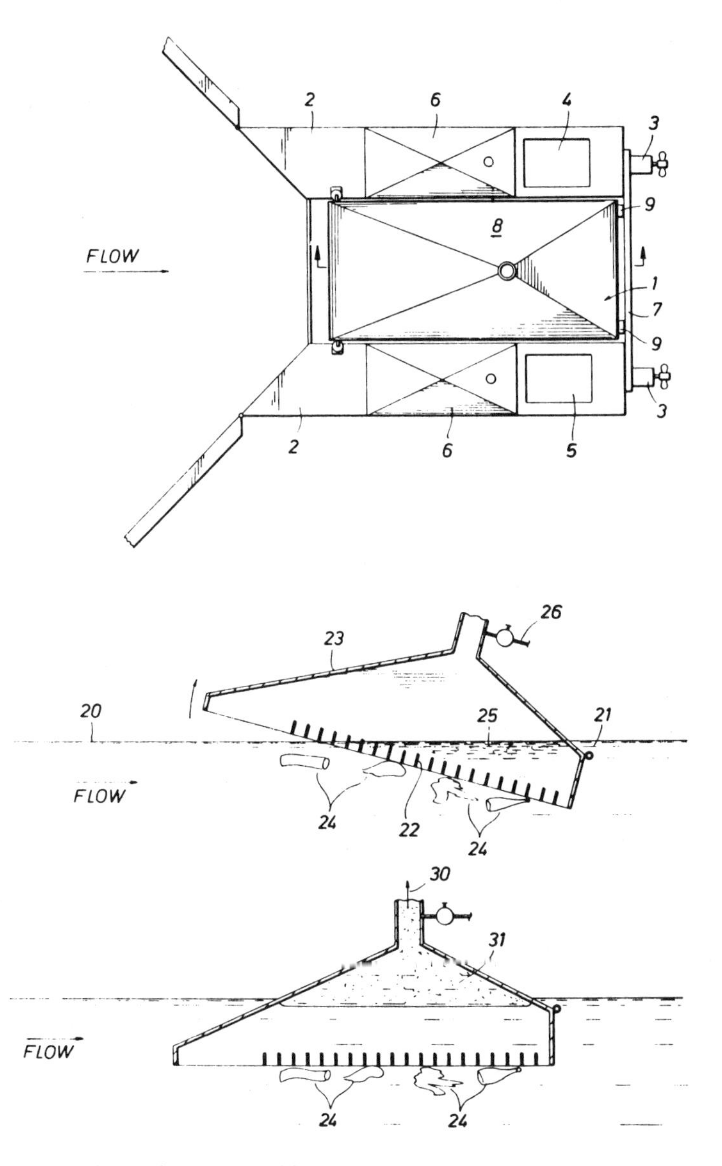

Source: U.S. Patent 3,865,730

The catamaran shown has a booming device **7** across the rear making the general plan shape to be that of a "U" although other plan shapes may be used with good results. Inasmuch as the rear portion of the funnel is the primary closure means between the two pontoons, device **7** has a primary function of joining the two pontoons so as to act as a structural unit. It can, however, be a frame covered by sheeting so as to act as a "backup" barrier. An inverted funnel **8** is suspended by hinges **9** or equivalent means within the confines of the "U" such that the lower portion of the funnel, preferably generally rectangular in shape, fits the inside contour of the vessel. The height of the funnel is of the order of the vessel draft, and the upper extremity of the funnel is typically 2 to 4 inches in diameter although such dimensions vary with the overall size of the skimmer. Short diverters shown in phantom, may be hinged to the forward ends of the pontoons with a means (not shown) for changing the angles thereof.

Generally, any conventional booms, hydraulic boom arms or other types of diverters may be employed in front of the skimmer to make skimming more efficient. Diverters are more practical for use with a large oil spill in an unconfined area than a small spill in a confined or open area.

In order for the device to collect oil, there should be a current or the device must be pulled or propelled through the water, or both, to provide relative movement between the device and the oil **20** ahead of the skimmer. The skimmer is pivoted about a hinge point **21** so that a baffle system **22** is exposed to the oil which passes through the baffle grid and inside the collector walls **23** of the skimmer which contains air. Large debris **24** does not pass into the skimmer inasmuch as the baffles are sized to permit substantially the entry of only oil into the skimmer. The oil collects in a pool **25** inside the skimmer which is provided with an air vent **26** to permit displaced air to pass out of the skimmer. When current or skimmer speed is low so that the baffles are submerged and the open forward end of the funnel is at the water level, the baffle incline can be decreased, thereby allowing oil to enter more readily. The baffles **22** can be merely screens near the front of the funnel becoming deeper at the rear or a checkerboard arrangement or transverse louvers. Such baffles are particularly essential in oil collection in waves and currents.

The inclined baffle system of this process has a number of advantages. Thus, oil can be concentrated at velocities greater than 2 feet per second which is not feasible with other known forms of skimmers. The baffles act as cells in damping the effects of waves and reducing turbulence inside the funnel during oil removal, which is a particularly valuable feature. Additionally, baffle sizing restricts the size of debris which can enter, so that the oversized materials are simply swept past. Finally, the baffles dissipate wave energy and minimize wave reflection and oil entrainment during collection.

Debris **24** is suspended beneath the baffles and does not enter into the skimmer. The skimmer during oil removal is positioned horizontally in the water so that the oil is trapped into the upper part of the funnel. If desired, the funnel can be further inclined so that the oil is pushed out of the top of the inverted funnel. Suction is provided at **30** by a positive displacement pump or other suitable means known in the art for removing oil **31** out of the skimmer. A primary advantage of the funnel is that the oil concentrates to a thick layer by buoyant forces so that there is a high oil-to-water ratio, whereby oil is easily removed without the removal of much water. As mentioned above, a serious disadvantage of other prior art skimmers is that a large quantity of water is removed along with oil, which creates difficult storage problems.

A very similar device is described by *R.R. Ayers and D.P. Hemphill; U.S. Patent 3,959,136; May 25, 1976; assigned to Shell Oil Company.*

A device developed by *K. Aramaki, H. Kawakami and M. Suzuki; U.S. Patent 3,907,685; Sept. 23, 1975; assigned to Bridgestone Tire Company Ltd., Japan* is a belt device for collecting floating matter from a water surface comprising a pair of pulleys, one of which being arranged above the water surface and the other pulley being arranged below the water surface, and an endless belt loosely engaged with the pulleys and having a lower loosening portion inclined from a relative direction of water flow by an acute angle.

Figure 102 illustrates the application of such a device in an oil recovery vessel. The ship **1** is provided at its central part with a water passage **2** so as to cause a water flow into the ship as the latter cruises on the water surface. In other words, the ship is made open in a substantial U shape ahead as viewed from above.

FIGURE 102: BRIDGESTONE TIRE CO. SKIMMER DESIGN

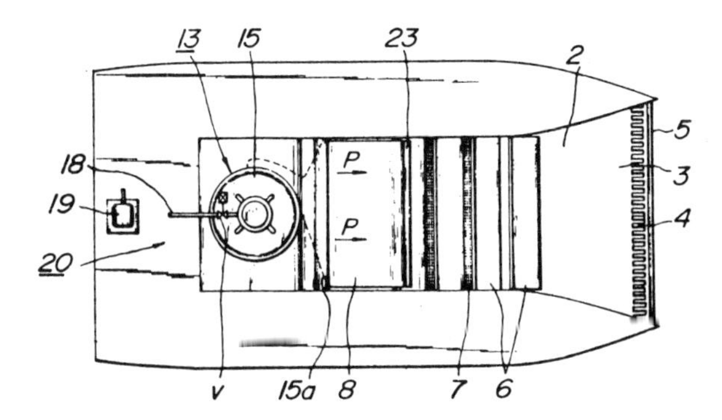

Source: U.S. Patent 3,907,685

The water passage is provided at the leading end of the ship with an inlet opening **3** so as to guide the water flow into the passage **2** when the ship cruises. The inlet opening is provided with a screen **4** for preventing foreign matter from entering the passage **2** together with the floating matter **A** such as drifted petroleum and the like, which is accidentally leaked from a tanker onto the water surface.

The screen **4** is of comb type and is pivotably mounted at its upper end on a horizontal shaft **5**. Thus, when the ship is not operating for recovery of the floating matter **A**, for example, in a normal cruising, the screen **4** can be held above the water surface. In the passage **2** behind the opening **3** are arranged a plurality of anti-reverse-flow boards **6** each inclined from a bow direction by an acute angle, whereby the water and floating matter **A** in the passage **2** are so controlled that they are not reversed in direction and prevented from becoming influenced by waves when the ship is moving forward. As a result, the water and floating matter **A** can be introduced into the belt device.

At the bottom of the passage **2** below the anti-reverse-flow boards **6** is disposed wave-suppressing filter **7** which serves not only to prevent the floating matter **A** from leaking to the outside of the ship through the bottom thereof due to water turbulence not eliminated by the anti-reverse-flow boards, but also to discharge an excess amount of water outside the ship.

That is, the wave-suppressing filter together with the anti-reverse-flow boards play a role of increasing a thickness of the layer of the floating matter **A** near the front of the belt device as will be described below.

The belt device for collecting the floating matter **A** comprises an endless belt **8** having the same width as that of the passage **2**, and a pair of pulleys **9**. The endless belt is driven in a direction shown by an arrow **P** by means of a suitable driving means (not shown). One of the pulleys **9** is arranged above the water surface and the other pulley is arranged below the water suface, an axis extending through the centers of both pulleys being inclined from a relative direction of water flow in the passage **2** by an acute angle.

The endless belt has such a circumferential length that the belt loosely engages with the pulleys **9**. By making the circumferential speed of the pulley arranged above the water surface large as compared with that of the pulley arranged below the water surface, a loosening portion **8a** can be formed in the lower travelling side of the belt which is guided into the water, and as a result, that part of the belt which is positioned near the water surface becomes inclined from the water surface by a smaller angle. This tendency can be promoted by using a belt material having an apparent specific gravity smaller than the specific gravity of the water, preferably having an apparent specific gravity of not more than about 0.7 which is smaller than a specific gravity of light oils.

As seen, the loosening portion collects a large amount of the floating matter **A** over a wide area on the water surface during the moving of the belt **8**, so that the contact area of the belt with the floating matter **A** becomes larger.

Moreover, in order to ensure the contacting of the belt with the pulleys, it is preferable to auxiliarily use a pair of touch rollers **23** which serve to adjust the length of the loosening portion **8a** of the belt.

The endless belt is driven by the pulleys, through the touch rollers **23**, which serve to make the contact area of the belt with the pulley large and to reduce a slipping of the belt.

When floating matter **A** such as petroleum or the like arrives at the loosening portion **8a** of the endless belt **8** it is sandwiched between the loosening portion and the water surface and then is pulled downwardly into the water along a portion **8b** of the belt **8**. After the belt **8** is turned upwardly around the pulley **9** arranged below the water surface, the floating matter **A** is separated from the belt by means of a wiper **22** and fed into a separating device **13** located at the rear portion of the water passage **2** together with the water accompanied by the floating matter **A** through a socket **15a** while floating upwardly. In the separating device **13**, the floating matter **A** is subjected to a gravity separation to reduce the water content thereof and finally recovered.

The separating device **13** shown is a downwardly directed whirl generating bucket **15** having a propeller **14**, a socket **15a** and a suction nozzle **16**. In the bucket **11**, whirl is produced by the rotation of the propeller, whereby the floating matter **A** is separated from the water due to the difference of the specific gravity between both the substances. The water is discharged through a bottom opening **15b** of the bucket **15**, while the floating matter **A** collected in the upper portion of the bucket is successively transported into a storage tank **17** through a conduit **18** by means of the suction nozzle **16**. Reference numeral **19** represents a vacuum pump and reference numeral **20** represents a suction means consisting of the storage tank, the conduit **18** and the vacuum pump **19**. Moreover, on the conduit **18** may be mounted a solenoid valve **v** controlled by a detecting means **21** so as not to operate the suction means **20** when the bucket **15** is filled with seawater or the like.

A vessel designed by *C. in 't Veld; U.S. Patent 3,909,416; Sept. 30, 1975; assigned to Hydrovac Systems International, Inc.* is a barge for separating oil-water mixtures which comprises a sealed tank into which the mixture is pumped for gravity separation. Pumping is effected by withdrawing the separated water from the bottom of the tank by means of a pump that produces a negative pressure in the tank, and this negative pressure sucks up the oil-water mixture from the surface of the water, without subjecting the mixture to mechanical pumping.

Figure 103 shows such a vessel in plan and elevation. Barge **1** floating on a water surface **2** comprises a separating tank **3** having a mixture supply pipe **4** connected at one end to the tank. The other end of the mixture supply pipe is submerged in the oily water mixture and has a guard means **5**. A discharge pipe **6** for water is connected to tank **3** near its bottom. The other end of discharge pipe **6** is connected to suction system **7**. The tank **3** is a hermetically closed reservoir extending throughout the greater part of the length of the barge.

Before operation tank **3** is completely filled or filled to just below its top wall **8** with water. The suction system **7** is started and the suction generated in tank **3** sucks up oily water mixture to tank **13** through guard means **5** and mixture supply pipe **4**. Water is discharged outboard through pipe **6**. It will be evident that in operation, the water oil separation is accomplished in the tank **3** near the surface; that is, an oil layer collects on top of the liquid in tank **3**. During operation the thickness of the layer increases continuously until the tank is filled with oil to an extent which prevents a further suitable separation after which the operation is stopped and the barge is moved to a place for draining away the oil by any conventional means.

FIGURE 103: HYDROVAC SYSTEMS SKIMMER VEHICLE DESIGN

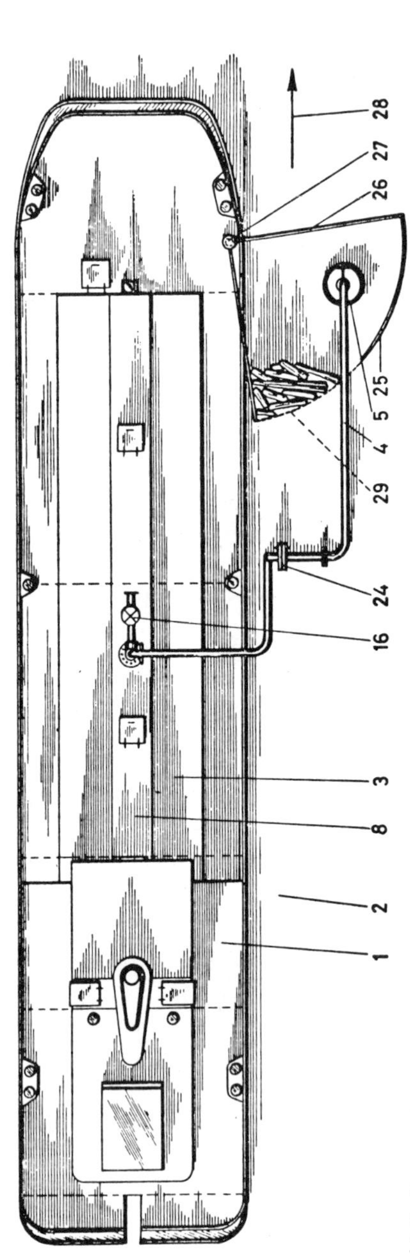

Source: U.S. Patent 3,909,416

By sucking the oily water mixture into tank **3**, the mixture entering the tank is less agitated than by pumping up the oily water mixture to tank **3**. Reduced agitation promotes the desired separation.

It will, of course, be apparent to those skilled in the art that when discharging water through pipe **6** smaller particles of oil having insufficient rising capacity are carried along by the water in the direction to the discharge pipe connection to tank **3**. In order to reduce as much as possible the amount of oil particles thus entrained, the distance between the oil layer in the tank and the connection of discharge pipe **6** to tank **3** should be as great as possible. For this purpose tank **3** has a cross section which increases upwardly.

Reduction of the oil-water mixture supply causes a fall in the normally substantially constant liquid level in the tank, which is observed by a float **15** in the tank coupled to a control system for reversing the operation of the suction system **7** dependent on the position of float **15** in the tank.

Oil-water mixture supply pipe **4** has a branch **16** controlled by a valve. In decreasing the oil-water mixture supply it is possible to introduce air or another suitable medium through the branch.

Preferably mixture supply pipe **4** consists of a plurality of pipe sections pivotally connected to each other. In this way the guard means can be positioned at a desired location in the area to be treated. When mounting the apparatus according to the process in a barge, these pipe sections are preferably of a nonflexible material. Further the hinge joint to which the pipe section having the guard means has been connected (hinge joint **24**) can be secured in a selected position.

In using pipe sections of solid nonflexible material and a hinge joint **24** which can be secured in a selected position, it is possible to keep guard means **5** at a predetermined depth beneath water level.

It will be evident that keeping the barge draft substantially constant and keeping the guard means at a predetermined distance beneath the surface of the water are beneficial to the stability of the barge. Accordingly both measures are benefical to a quiescent liquid condition in the tank which is necessary for a suitable separation of the two liquids of the mixture. If waves or swell are present the guard means **5** can be kept at a predetermined submersion by controlling positively the guard means **5**.

The plan view shows a catching screen **25**, in the form of a long plate extending over a circular arc of about 90°. The screen is movably suspended by means of a leg **26** of a stanchion **27** on the barge. In a preferred embodiment the stanchion **27** consists of two tubes one of which is mounted on the barge whereas the other one supporting the leg **26** is slidable and rotatable with respect to the first-mentioned tube.

In operation the catching screen is lowered outboard the barge so that guard means **5** is present in the space bounded by the barge hull and catching screen **25**. When the barge moves in the direction of the arrow **28**, for example with a speed of three miles per hour, the effect is obtained that a greater part of the oil-water mixture of the area is concentrated near guard means **5**. In this connection it is pointed out that the dotted line portion **29** of catching screen **25** is perforated. In operation, large solid objects such as pieces of wood and the like which pass guard means **5** or are freed when cleaning the guard means, will collect in the perforated catching screen portion.

A vessel designed by *O. Massei; U.S. Patent 3,915,864; Oct. 28, 1975; assigned to Costruzioni Battelli Disinquinanti SpA, Italy* for use in removing a floating contaminant liquid such as oil from the surface of water has a hull forming an immersed inverted channel into which surface layers flow as the vessel advances. The hull is shaped so as to guide the liquid into accumulation zones from which the liquid is drawn by a pump into settling tanks disposed in pontoons on each side of the hull, the water being discharged from the tanks and the contaminant being collected in the tanks.

A scheme developed by *P.J. Strain; U.S. Patent 3,922,225; Nov. 25, 1975* comprises cleaning up oil spills on water by effecting a first separation externally of the ship at the bow, and conveying the oil into the ship for a second separation through an elongated baffled conduit. Oil in excess of the ship's capacity may be discharged at the stern into waiting tankers. Intake means which may heat the oil and include when necessary surrounding wave-damping means connected to the ship's bow communicate flexibly with the ship's interior. One embodiment of the intake means is a wide shallow funnel. Novel for shallow coastal water oil skimming is a floating box having a water-level weir through which oil enters and a discharge port connected to a long flexible conduit communicating with a ship of substantial draft standing off in deep water. Mobile means such as shore tractors and powered boats tied to each box on the land side hold it in position and move it as desired.

A vessel designed by *F.A.O. Waren; U.S. Patent 3,928,206; Dec. 23, 1975* includes a collecting tank supported to float in water with its upper edge above water level, the tank having side walls, a rear wall and an open bottom. Water and any oil or debris present is caused to flow into the tank by means of a vaned impeller extending between the side walls of the tank and arranged for rotation about a horizontal axis positioned somewhat above water level. The oil and debris collected in the tank are retained therein by a nonreturn valve consisting of a plate which extends between the side walls of the tank rearwardly of the impeller and is pivotally mounted at its lower edge for rearward movement about an axis positioned below water level. The upper edge of the plate is provided with a float so as normally to maintain this edge above water level.

A side elevation of such a vessel is shown in Figure 104. As shown, the vessel comprises a tank **10** having a pair of opposed side walls **11** (only one of which is visible), and a rear wall **12**, which is shaped to improve the navigation performance of the apparatus. The bottom of the tank is open. The walls are made of any suitable material, such as metal, or resin-bonded glass fiber. Part or all of the top of the tank is closed by a deck **14** which is provided with a cabin **15** for the operator. This cabin also houses an engine (not shown) for driving a pump. The buoyancy of the apparatus is arranged to be such that the tank floats with its upper edge above water level.

The pump comprises a vaned impeller **16** and a cooperating floor member **17** both extending between the side walls **11**. The floor member which is below water level has a rear edge **18** below and adjacent the periphery of the impeller and a forward edge **19** of part circular section which is provided with a movable lip portion **20** that can be adjusted to a desired position below the water level.

A plate **21** which extends between the side walls is hinged at its lower edge to the rear edge of the floor member and is provided with a float **22** along its upper edge which ensures that this upper edge is normally maintained above water level. The arrangement acts as a nonreturn valve to prevent back flow from the tank past the impeller when this is not being rotated, but is so arranged that the plate can turn rearwardly and downwardly when the impeller is rotated to permit a free flow of water into the tank.

FIGURE 104: WAREN SKIMMER VEHICLE DESIGN

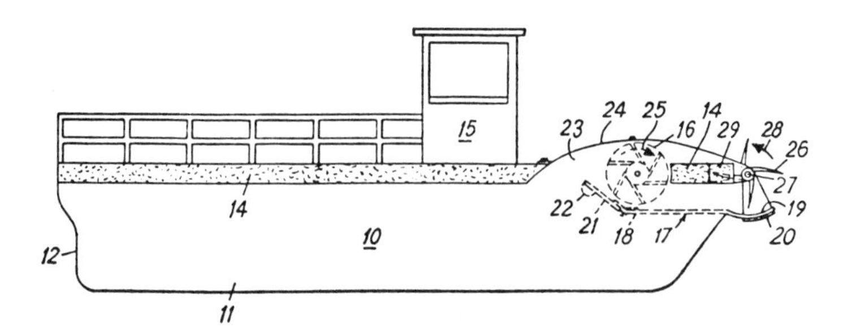

Source: U.S. Patent 3,928,206

The impeller is arranged in a transverse slot **23** in the deck which is covered by a housing **24** that can be opened to provide access to the impeller. Suitable drive means (not shown) connect the impeller with the engine in the cabin so that the impeller may be rotated in the direction of the arrow **25**.

In order to prevent floating debris from being drawn into the impeller a plurality of prongs **26** are provided to pick up such debris. These prongs are spaced in groups of four about the periphery of a shaft **27** at a series of locations spaced along the shaft. The shaft **27** is arranged to be rotated in the direction of the arrow **28** (i.e., in the opposite direction to the impeller) by the engine driving the impeller through suitable gearing (not shown). At least the ends of each group of four prongs are arranged to pass through slots **29** in the deck and housing which perform a wiping action on the prongs and ensure that debris picked up by the prongs is deposited on the deck, where it can be allowed to pile up, or removed manually or otherwise.

The engine used to drive the impeller and shaft may be a conventional internal-combustion engine. This engine may also be used to drive a pump (not shown) mounted on the deck and serving to discharge oil collected within the tank.

The tank may be provided with an outboard motor or other means for driving it and means may also be provided for steering the tank. Alternatively, the tank may be arranged to be towed by one or more craft.

In operation, the tank is taken to the area in which an oil slick or patch has been located and is then driven or towed into contact with the edge of the oil slick or patch. The engine for driving the impeller and the shaft is then started and a large volume of water and oil mixture is thus drawn over the lip portion and into the tank.

Since the oil being lighter floats on the surface of the water in the tank the water is displaced by the oil and consequently leaves the tank through the open bottom thereof. Thus the oil is collected in the tank and, when the tank is full, it may be taken back to harbor for the oil to be pumped out. Alternatively, the oil may be pumped out of the tank while the apparatus is in operation so that the oil may be collected in drums or other suitable containers, or in another vessel, and be taken away without stopping the collection of oil.

A vessel developed by *G.M. Fletcher; U.S. Patent 3,929,644; Dec. 30, 1975* is a self-propelled watercraft for scavenging oil spillage and other floating debris from the surface of a body of water as, for example, oil accumulations and other debris that floats about harbor areas. The watercraft has a catamaran-type hull providing spaced apart hull sections each of which is equipped with a receiving tank having a substantial volume thereof extending downwardly below the surface of the water body. Each tank has an inlet located along and above a generally horizontal deck extending between the hull sections at the water level, and each tank also has an exit opening adjacent the bottom thereof which places it below the deck. The craft is open at its bow to define a mouth which permits a surface layer of the body of water to wash rearwardly along the deck toward the inlet opening of each tank as the craft is propelled through the water.

Since the oil spillage and other debris are lighter than water and therefore float, the volume of each receiving tank is effectively increased by removing the water accumulations that settle toward the bottom of the tank; and such removal of water is effected by reducing the pressure at the exit openings via a venturi flow passage system associated with the exit openings and which reduces the pressure there automatically as the craft moves through the water.

An apparatus developed by *S.G. Fast; U.S. Patent 3,947,360; March 30, 1976; assigned to Sandco Limited, Canada* is a boat having a holding tank and at least one belt conveyor having its lower end submerged and its upper end in communication with the holding tank so that oil or other floating substances are removed from the water and conveyed to the holding tank as the boat progresses through the water.

A vessel developed by *W.P. Kirk and D.W. Reynolds; U.S. Patent 3,966,613; June 29, 1976* is a catamaran with pivotal fore and aft gates defining a well. When moved through a floating liquid spilling with the fore gate open the well collects spilling. The collected spilling may be pumped directly into a tank. Also, the catamaran is equipped with means to deploy a collapsible, floatable retrieval sheet over and into the well to confine the collected spilling, the aft gate being then opened to set the sheet adrift. A peripheral barrier of air bubbles concentrates the oil toward the center of the well.

The collapsible floatable retrieval sheet concept was discussed earlier in this volume in connection with patents by Kirk and Reynolds cited at the beginning of the section on oil spill containment.

A vessel designed by *R.R. Ayers; U.S. Patent 3,966,614; June 29, 1976;*

assigned to Shell Oil Company is a skimmer for removing oil from the surface of a body of water which is articulated from front to rear to be wave comformable and/or has a quiescent collection zone formed by bottom and/or forward baffles.

Figure 105 shows the basic essentials of this device as compared to prior art devices. The top view shows the advantages of articulation. It is apparent that the mouth of the articulated skimmer does not rise above the surface of the water. On the other hand, the mouth of the prior art skimmer does periodically rise above the level of the water so that oil is missed.

FIGURE 105: DIAGRAM SHOWING ESSENTIALS OF SHELL OIL CO. SKIMMER VEHICLE DESIGN

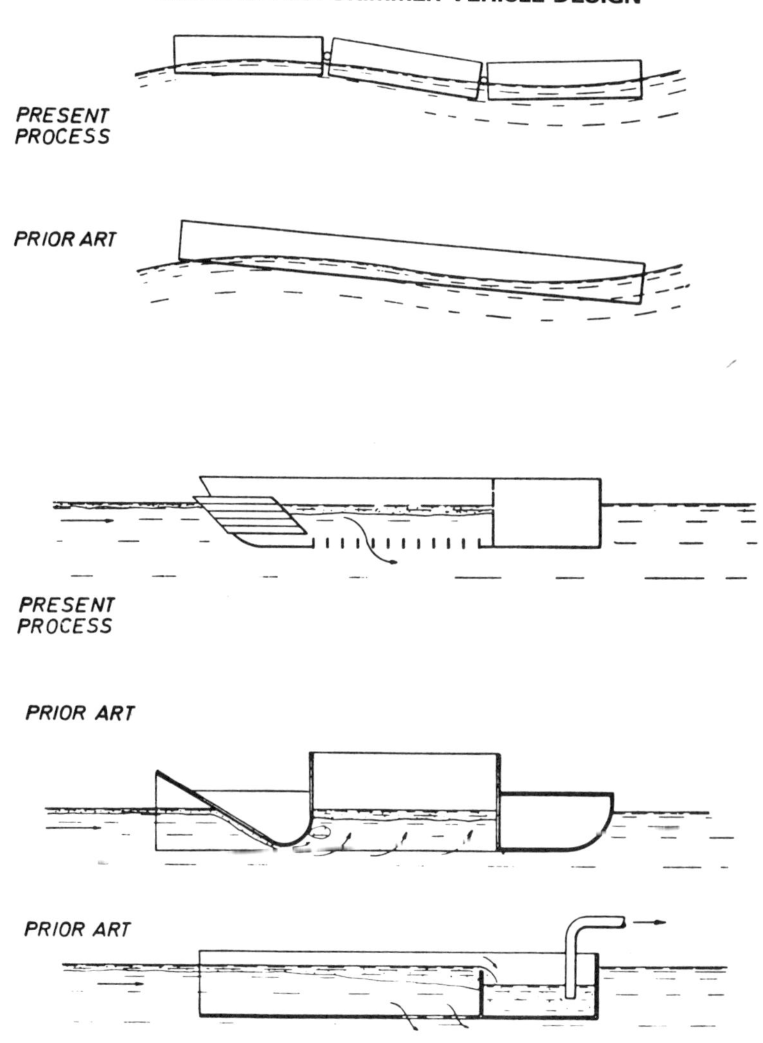

Source: U.S. Patent 3,966,614

The lower views show the advantages of providing a forward baffle which permits horizontal entry of the oil into the skimmer while still functioning to dampen wave action. Among the prior art devices, one device pushes the oil down before it can rise into the skimmer so that, especially with heavy oils, much does not rise in time to be collected, and the other device, a conventional floating box skimmer, allows waves to enter the skimmer with a resultant loss in efficiency.

A somewhat similar design is covered by *R.R. Ayers; U.S. Patent 4,049,554; Sept. 20, 1977; assigned to Shell Oil Company.*

An oil collection craft developed by *S.L. Petchul and R.K. Petchul; U.S. Patent 3,966,615; June 29, 1976* collects and then contains and concentrates oil spills on the surface of water. This action is effected by the flow of water and oil relative to a channel with an adjustable water and oil intake funnel at the forward end, an oil blockage wall at the rear, an adjustable flow splitter at the rear, a water vent at the rear bottom, and an adjustable vent plate at the rear bottom. These devices enable operation over large areas in a minimum of time and do so even given an adverse sea state.

A skimmer vehicle developed by *D.E. Wilson; U.S. Patent 3,983,034; Sept. 28, 1976; assigned to Chevron Research Company* consists of three pontoons, one located in each of the corners of a triangle made up of interconnecting structural members floating on a body of water. The corner pontoons are adjustably buoyant permitting the skimmer to move vertically up or down as a unit in the water so as to adapt to the wave height of the body of water.

At the base of the triangle is the skimmer mouth sloping upward towards a sump. The two pontoons on each side of the sloped mouth have a clamp easily attachable to an oil boom which guides an oil slick into the skimmer. The function of the mouth is to skim off approximately two inches of the water surface. The sloped mouth terminates at an impregnable deflector centrally located between two screens that serve as a wave quieting assembly as well as a separator of debris floating on the water. The skimmed water is then directed through the debris screens into a quieting area.

Once in this quieting area, the oil slick flows over a self-adjustable weir into a sump. The weir is made adjustable by a float that modifies the weir elevation with changes of liquid level in the sump. The skimmed liquid is then pumped to a storage tank from which it is recycled by letting the water at the tank bottom flow under gravity back into the oil boom.

This device is shown in Figure 106. In its usual application, the skimmer **50** is towed abreast of a floating vessel **63** equipped with storage tanks **65**, a pump **64**, interconnecting pipes or hoses **81** and a davit **71** for keeping the flexible hose from the aft end of the vessel **63**. An outrigger **68** extending above the water surface **56** and outward from the side of the vessel **63** supports the leading edge of an oil boom **66**.

Another section of oil boom **66** is similarly secured between the inboard front flotation pontoon of the skimmer **50** and the vessel **63** thus forming a V-shaped trap that directs the oil and water into the mouth of the skimmer shown.

The oil slick **69** enters the skimmer mouth then flows across an upwardly sloped entrance plate along the vertical sides of the pontoon connecting members through debris screens to a quieting area immediately behind the debris screens and a deflector plate. The screens, however, do not pass debris like oil absorbing material, wood, straw, or other nonliquid floating material.

FIGURE 106: CHEVRON RESEARCH CO. SKIMMER SYSTEM

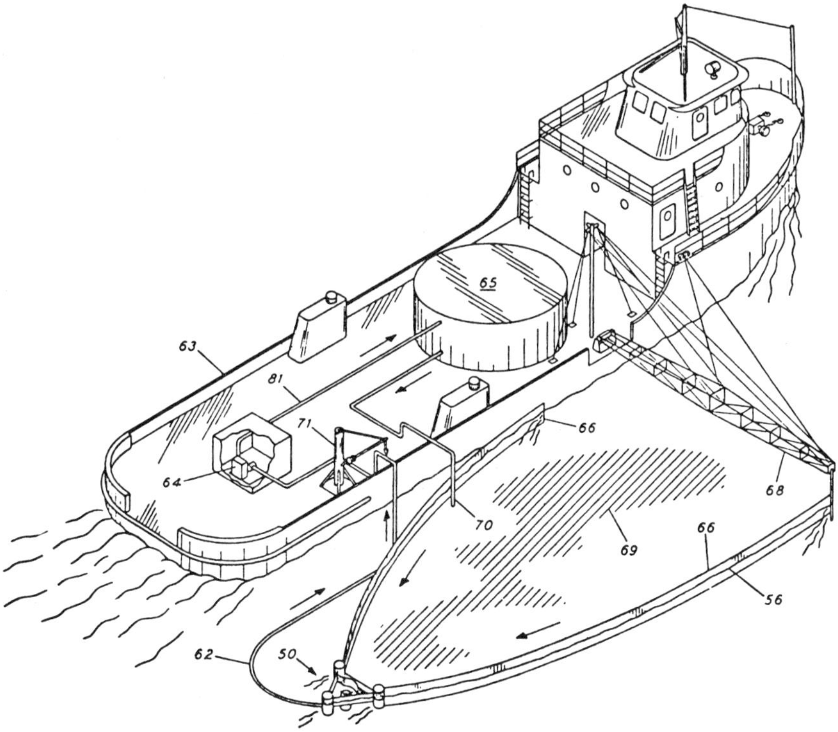

Source: U.S. Patent 3,983,034

Oil and water are pumped from the sump through a hose outlet located at the bottom of the sump through a flexible hose **62**, the pump **64** and a rigid pipe **81** to an oil/water separation tank **65** on board the floating vessel **63**. The water which settles to the bottom in the tank is bled through a hose or hoses **70** from the tank **65** back into the area enclosed by the oil booms **66**.

A field test of the skimmer indicates skimming capabilities in 4 to 5 foot waves. In such conditions, skimming efficiency is low; however, the oil readily separates from the water thus permitting bleed back to the sea. In effect, the efficiency is increased by this recycling method. In calm seas, on the other hand, skimming efficiency in the first cycle is up to 60%. Needless to say, this efficiency is greatly increased by the recycling procedure. During the skimming operations, the forward speed of the floating vessel **63** is approximately ±¾ knot. Experiments have indicated the assembly shown requires two hours for placing the apparatus into operation. In essence, this is a fast response skimmer designed for immediate reply to offshore spills.

A boom design developed by *C. in 't Veld; U.S. Patent 4,014,795; March 29, 1977; assigned to National Marine Service, Inc.* which is a device for sweeping and collecting oil from the surface of a body of water is constructed of a framework of structural members arranged in the form of a box beam, the framework supporting two parallel rows of floating barrier screen panels and being connected to end floats. Flow diverter vanes beneath the barrier screen panels cause a surface transport current to flow between the two rows of the panels towards a surface skimmer located adjacent one end of the boom when the boom is towed in a direction transversely of its length across a body of water to be swept.

The forwardmost row of barrier screen panels sweeps floating oil on the water surface towards the skimmer, and the transport current carries oil that is caught in the underflow beneath the forwardmost row of panels towards the skimmer. Extension arms including floating vertical barrier screen panels are pivotally attached to the boom to extend the sweep area, the arms being foldable inwardly against the central boom structure for stowage of the boom.

A skimmer vehicle developed by *J.L. McGrew; U.S. Patent 4,033,869; July 5, 1977; assigned to Marine Construction & Design Co.* utilizes spray booms creating fine water spray curtains angled to the water's surface from an elevation above water level. Mounted divergently at the bow of a skimmer vessel, these spray curtains are effective to funnel the oil into the vessel's pick-up device ahead of the vessel without dispersing or emulsifying the oil. The fine spray particles making up the curtains have carrying effect sufficient to project entrained air with the spray to the water's surface independently of variations in distance between the booms and the water's surface attending passage of waves and swells.

A skimmer vessel developed by *D.L. Cocjin and A.M. Masongsong; U.S. Patent 4,033,876; July 5, 1977* is a water craft having a pointed bow which incorporates curved vertically pivotable gates to partially form the pointed bow along with an underwater scoop having a horizontal surface beneath the water with the gates opening to permit water and oil to move onto that surface and upward over an inclined surface leading to a horizontal trough amidships above the watercraft line. The trough opens for gravity deposit of the oil and water into a water and oil separating tank within the rear of the watercraft hull.

REMOVAL FROM HARBOR AREAS

Harbor areas can be regarded as being of two types, either confined or open. Confined areas, the first type of harbor, are those around nested ships, adjacent to and under piers and pilings. Conditions are usually represented by little or no waves and low current. Some activities do have high current situations next to docks and piers. Procedures for these situations are discussed in the paragraph on river and tidal current areas which follows.

Figure 107 is an illustration of an oil spill in a confined harbor situation. The adjacent ship and quay wall provide excellent barriers to confine the oil to the immediate area. It is important that containment boom be deployed not only to enclose the spill but also to prevent oil, which is escaping the enclosure, from reaching those areas which would be very difficult and costly to clean up such as the dock piling structures shown. Where possible, booms should be deployed in such a way that skimmers can be located downwind of the contained slick.

It has been demonstrated that a reduction in total cleanup time of 300 to 400% can be obtained if the skimmer is tied off in a stationary position and the oil herded toward the skimmer by creating surface currents using water or air jets or boat propwash. A line of sorbents placed along the upwind edge of the slick provides a greatly increased sail area to allow the wind or artificially generated surface currents to rapidly push the slick toward the skimmer. Water streams, from a shoreside fire hose connection or other pump, must be carefully applied well away from the slick edge in a zigzag manner at a low angle of incidence to the water surface. If the water stream is played directly on the oil slick, oil is emulsified and driven into the water column, greatly complicating cleanup operations. Water streams from shoreside fire hoses have a range of approximately 100 feet.

Figures 108 through 110 illustrate possible uses in flushing oil toward a skimmer. Where a source of compressed air is available, an air jet can be used to push oil toward the skimmer. An air jet causes less turbulence at the oil-water interface; but its range is low, usually on the order of 10 to 30 feet. Propwash from available craft can be used in a variety of ways to push oil toward a skimmer. Propwash is particularly useful in flushing oil from beneath open piers. The craft should be kept at least 100 feet away from any boomed area to avoid excessive turbulence which may drive oil under the containment boom.

In all cases, these surface current generating techniques should be cautiously applied and sufficient time allowed for them to act on a slick. It should be kept in mind that even a low surface current of one knot, induced by these methods, is equal to a speed of about 1.7 feet per second or about 102 feet in one minute. Accordingly, even a low current can move an oil slick a great distance if given sufficient time.

Another technique, which is most effective, is to move the booms surrounding the spill area to reduce the boom area thereby increasing the oil slick thickness allowing for faster and more efficient pickup by the skimmer. After removal of the major part of a slick by these methods, a sheen almost always remains. The most effective method of handling sheen is to deploy piston film dropwise along its upwind edge, add a line of sorbents along the edge if necessary, and, using natural wind forces or water or air jets, move the sheen into the skimmer mouth, thus sweeping the water surface clear of oil.

Open harbor areas present a more difficult cleanup situation than confined areas. Open areas enable the wind to build up waves. Also, the oil slick may be over such a large area that it cannot be completely encircled with boom. Figure 111 illustrates the situation. The oil slick may either be over a large area or may be broken up by the wave system into windrows following the trough between waves. For either case, the first response should be to dis-

FIGURE 107: CONFINED HARBOR SPILL PROCEDURES

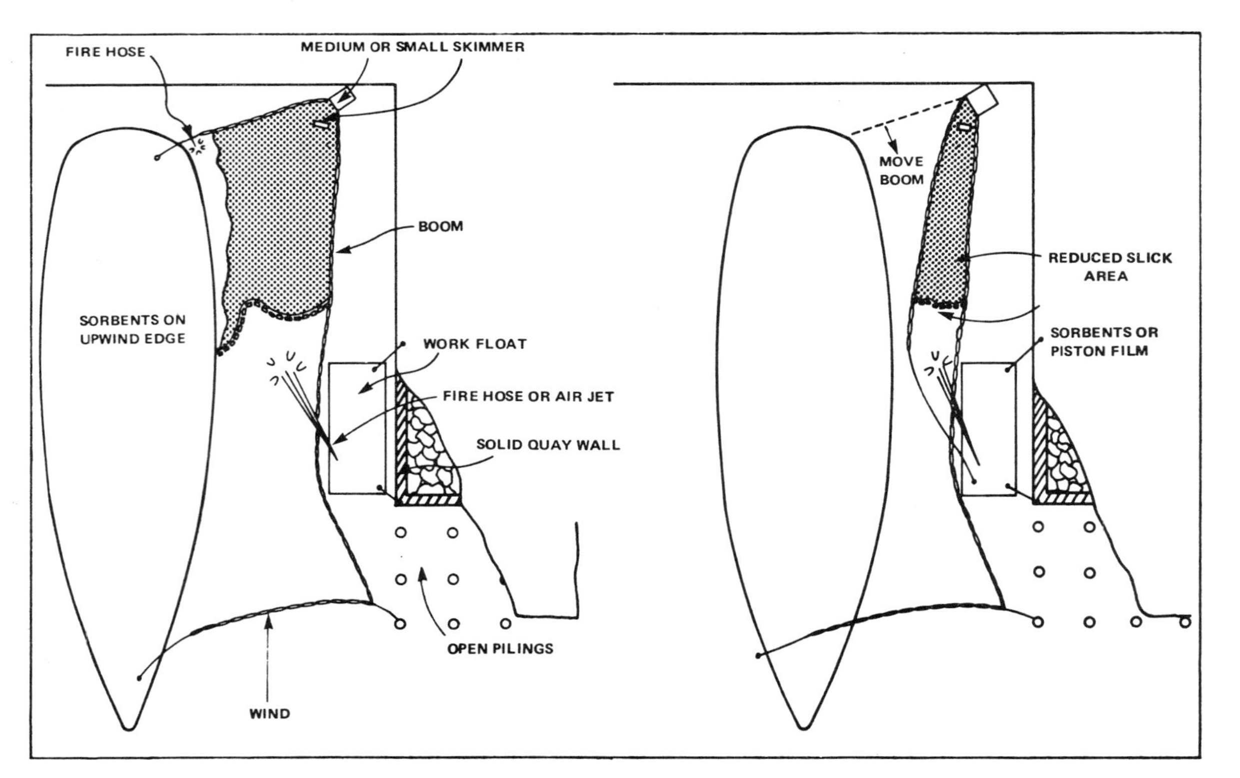

Source: Reference (9)

FIGURE 108: CLEARING OIL FROM UNDER PIERS WITH BOOM AND WATER JETS

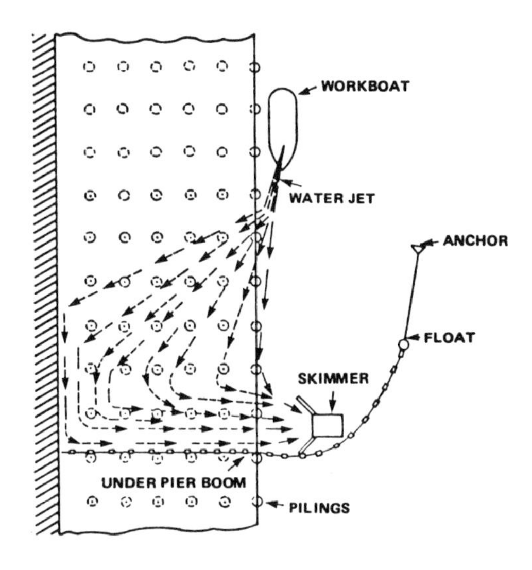

Source: Reference (9)

FIGURE 109: CLEARING OIL FROM UNDER BLIND PIERS

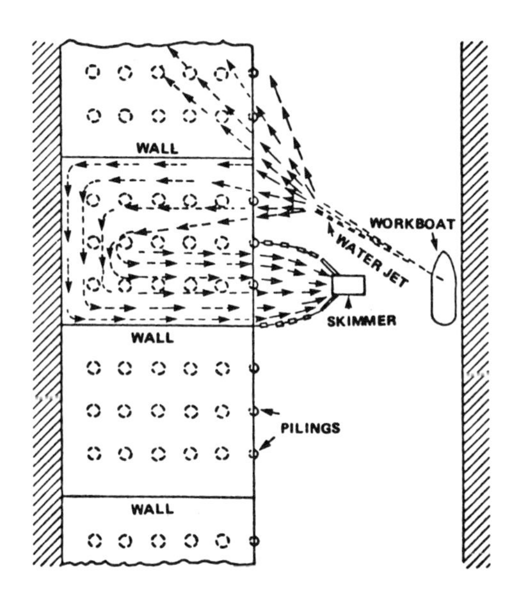

Source: Reference (9)

FIGURE 110: DRAWING OIL OUT OF A SLIP BETWEEN SOLID PIERS

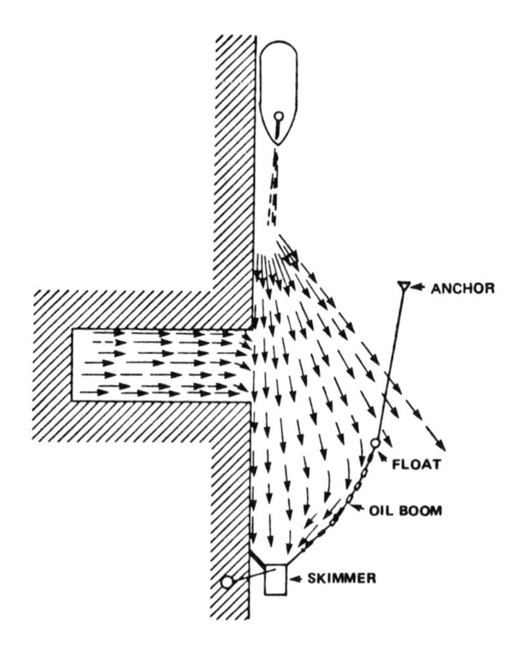

Source: Reference (9)

FIGURE 111: HARBOR OPEN AREA PROCEDURES WITH LARGE SKIMMER

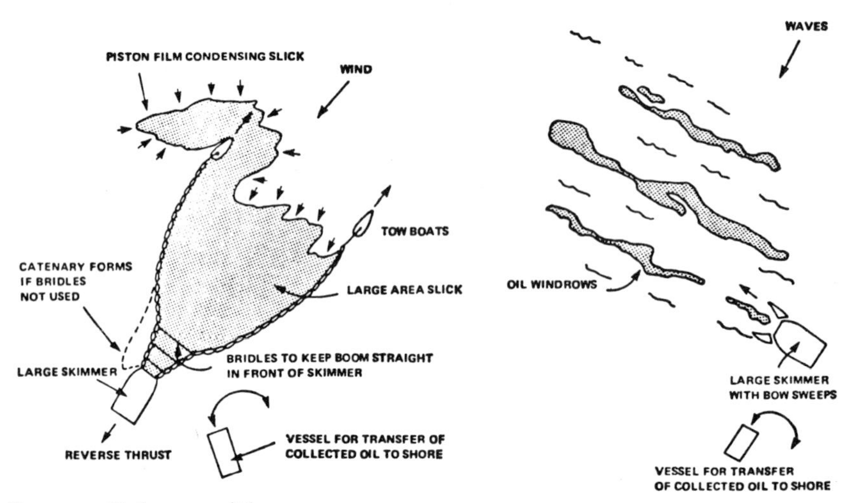

Source: Reference (9)

patch a high speed suface craft to the scene to estimate the extent of the slick, drop sorbents or floats around the slick perimeter to provide visibility for subsequent skimming operations, and surround the slick with piston film to thicken the oil and retard its spreading rate.

During open water field tests, it has been determined that piston film can reduce slick size by as much as 68%. A large, vessel-type skimmer is most effective in these situations and should be deployed with boom in a "vee" configuration if a large slick is encountered or alone with its bow sweeps down if the slick is concentrated in windrows. It is very difficult for crews to see the edge of an oil slick from the waterline. These sweeping operations are much more effective if they are conducted with radio contact to an airborne observer. To maintain the "vee" configuration with large skimmer and boom, it is necessary to operate the skimmer with a reverse thrust and to attach cross bridles to the boom in front of the skimmer to insure that all the oil flowing down the boom goes into the skimmer mouth and is not trapped in a catenary formed in the boom in front of the skimmer.

When a vessel-type skimmer is not available, the best procedure, as shown in Figure 112, is to deploy a boom in a catenary between two tow boats downwind and downcurrent of the slick, and maintain a low forward way to allow the slick to drift into the catenary. Then, using a craft such as an LCM, maneuver behind the boom at the bottom of the catenary and deploy a skimmer (small or medium) over the boom. Regardless of the type of skimmer or sweeping operation employed it is imperative that a support vessel be provided to transport the oil-water mixture to a shoreside storage facility.

In a study to develop methods of determining slick thickness from a distance, personnel from the Naval Research Laboratory have discovered that slicks exposed to wind in open waters are not of uniform thickness. The results of one field test, as shown in Figure 113, indicated that over 90% of the oil sick volume is contained in the shaded area. This thick oil area spreads very slowly and is always on the downwind edge of the slick. To the naked eye, the entire slick illustrated in Figure 113 appeared the same.

While these results are for oil released from a concentrated source, it is reasonable to expect that the shearing force of the water interface has a "bleeding" effect on slicks of initially near-uniform thickness. Therefore, while all parts of a slick visible to the naked eye must be eventually removed, it is more effective to first skim the downwind portion of a slick floating in open water exposed to wind. In all cases, it is more effective to rely upon an airborne observer to direct surface skimming operations to the areas containing the heaviest concentration of oil.

REMOVAL FROM ICY WATERS

Oil spill removal from icy waters may involve operations in arctic zones or simply cold weather operations in more temperate zones.

Low temperatures during oil recovery operations affect mechanical operation of equipment and cause an increase in oil viscosity. Start-up and operation of equipment in low temperatures are best handled by following

FIGURE 112: HARBOR OPEN WATER PROCEDURE WITH MEDIUM SKIMMER

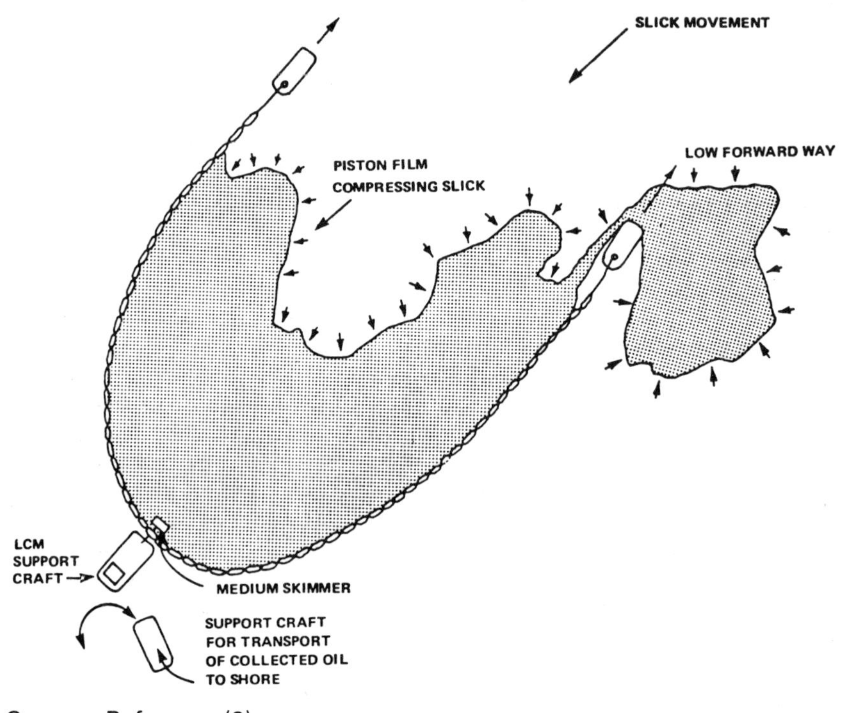

Source: Reference (9)

FIGURE 113: NRL TEST OF OIL THICKNESS OF SLICK IN OPEN WATER

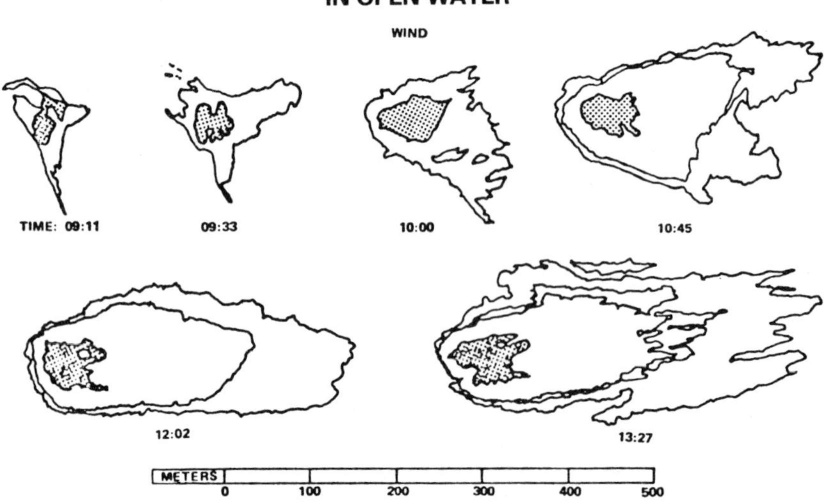

Source: Reference (9)

the manufacturer's instructions for cold weather. Hoses and pumps should be drained immediately after use to avoid freeze-up. The great increase in oil viscosity at low temperatures is both a help and a hindrance to oil pickup. The high viscosity results in a lower spreading rate than at warmer temperatures, causing oil to remain congealed in thicker slicks for easier pickup. High viscosity reduces the flow-rate delivery of some pumps. The positive displacement pumps used inthe NAVFAC central procurement program are able to handle high viscosity oils and oil-sorbent mixtures with little loss in output rate. However, even with positive displacement pumps, a reduced flow rate or excess discharge pump pressure can be experienced if oil is being pumped through long hose lengths. In these cases, one technique, as a last resort, is to inject water into the pump inlet to create an oil-water slurry having a lower equivalent viscosity than oil alone.

The rotating belt operating principle of the medium and large skimmers procured by NAVFAC is not sensitive to oil viscosity. However, some loss in discharge rate may be experienced when pumping oil from the collection wells behind the belts in these skimmers. The small skimmer system, with its overflow weir principle, is sensitive to oil viscosity. If the viscosity is high enough, oil may "dam up" at the mouth of the skimmer, while the much lower viscosity water is drawn under the thick oil and into the skimmer.

Tests have confirmed that piston film can still be effective on lighter oils in water temperatures as low as 40°F. Where spills occur on water free of ice, the procedures described in the previous paragraphs can be employed. For spills on snow, various USCG tests in arctic environments have shown that snow is a good sorbent. Therefore, oil does not permeate far into snow layers. The best cleanup procedure is to scrape off the oil-saturated snow and dispose of it in an approved landfill.

Once oil enters the water and is trapped under ice layers, location and removal of the oil are very difficult. In an actual field situation where this occurred, the technique illustrated in Figure 114 was successfully used.

FIGURE 114: PROCEDURE FOR OIL UNDER ICE

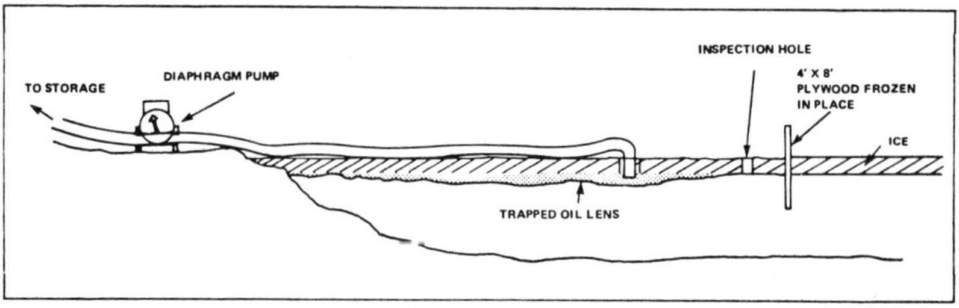

Source: Reference (9)

A portable auger drill, of the type used for ice fishing, was used to bore through the ice to define the extent of the trapped oil. Once this was established, a high-pressure, hot-water jet was used to cut a slot in the ice. Plywood sheets, four feet by eight feet, were then slid down the slot, down into the water below the ice and allowed to freeze in place. These plywood "booms" prevented the oil from spreading under the ice. Within this "boomed" area the major portion of the oil was pumped out through bored holes in the ice. To remove the oil remaining, the ice inside the "boomed" area was broken up and removed to an approved landfill.

A technique developed by *P.C. Deslauriers, R.P. Voelker, E.J. Lecourt and L.A. Schultz; U.S. Patent 4,053,406; Oct. 11, 1977; assigned to Artec, Inc.* utilizes an aquatic vessel, preferably a barge which is either self- or non-self-propelled equipped with ice cutting, oil recovery and processing apparatus. The ice covered area that is polluted with oil is approached by the vessel and the oil-contaminated ice is cut into slabs which are conveyed to a cleaning station where high pressure air and/or water jets and/or chippers remove the oil and some of the exterior of the slab. The oil and ice mixture that is removed from the slab is collected and heated. The oil is separated and purified and used as fuel and/or stored. The cleaned ice slabs are dropped back in the channel behind the vessel. Oil that is located beneath the ice, not adhering to the ice slabs, travels down an inclined plane under the vessel, rises in a well, and is separated out and used as fuel or stored.

The vessel can also be used to clear an area, such as a harbor, of ice and to transport the collected ice to a designated area, since it has the capabilities for loading, storing, and off-loading ice. The propulsion means for the barge may include Archimedes' screws for propelling the barge at slow speeds.

REMOVAL FROM STREAMS

The primary procedure for fast current rivers and tidal areas is to divert the oil out of the fast current into areas of low current. Figures 115 and 116 show the recommended deployment of booms and skimmers for a river and fast current tidal situation. A key item to note is that a portion of the diverting boom is directed upstream along the bank to minimize the amount of oil contaminating the shore.

FIGURE 115: RIVER PROCEDURE

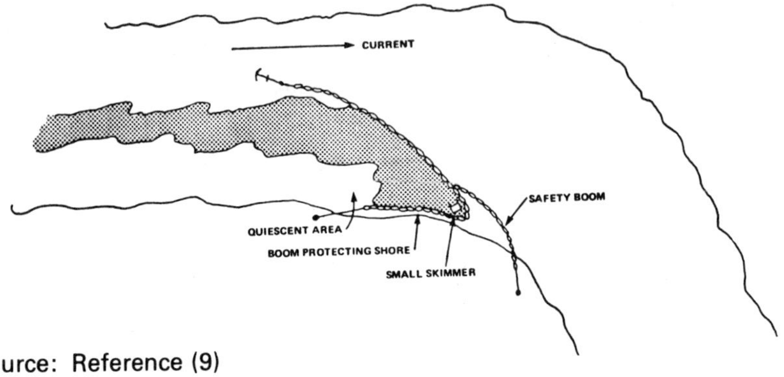

Source: Reference (9)

FIGURE 116: FAST CURRENT TIDAL SITUATION

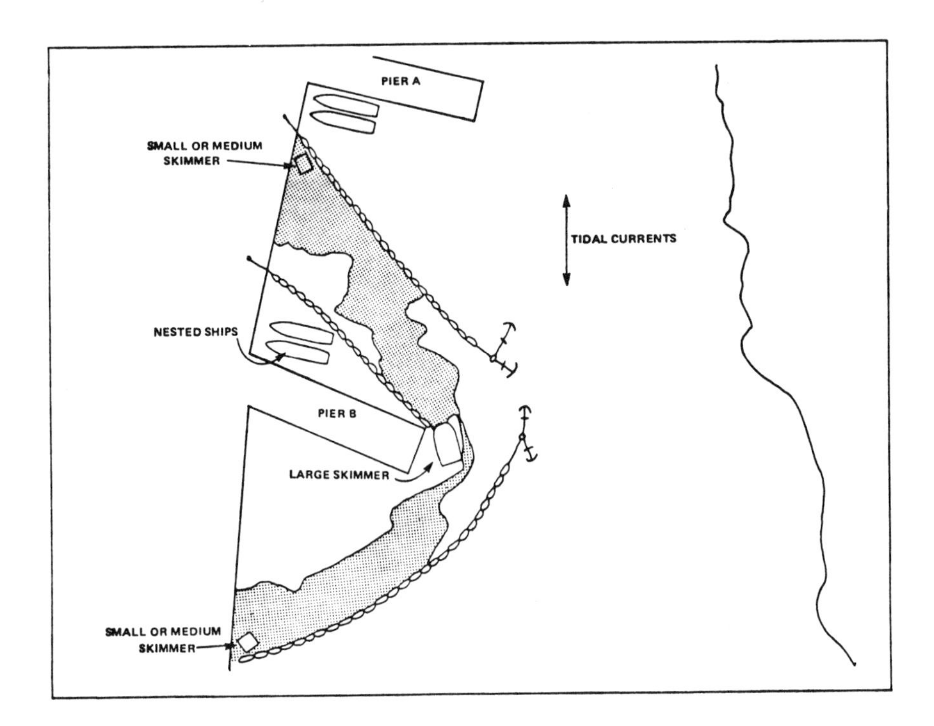

Source: Reference (9)

Also, a second boom attached behind the primary deflecting boom is a good practice should the oil escape from the thick layer concentrated in the bottom of the diverting boom. The small and medium skimmers are well suited to this application. The small skimmer has a draft of only 10 inches and can be particularly useful in shallow water areas adjacent to some shorelines.

Figure 116 illustrates a situation where tidal variations require booming of a spill on the upstream and downstream sides. As the tidal currents change from ebb to flood, oil will be driven to the top or bottom of the spill area to be collected by waiting skimmers. Spill situations involving a small stream are best handled by constructing an open mesh support across the stream and placing sorbent material on the upstream side as shown in Figure 117.

Success has been exprienced in actual field cases using chicken wire or chain link fencing material. The sorbent should be in the form of chunks larger than the mesh of the supporting screen. This results in a large sorbent surface area and allows the sorbent to be gradually removed by hand as oil is sorbed. Depending upon stream depth and water current speed, a pool of oil may form on the upstream side of the sorbent and can be removed by a small skimmer.

FIGURE 117: SMALL STREAM PROCEDURE

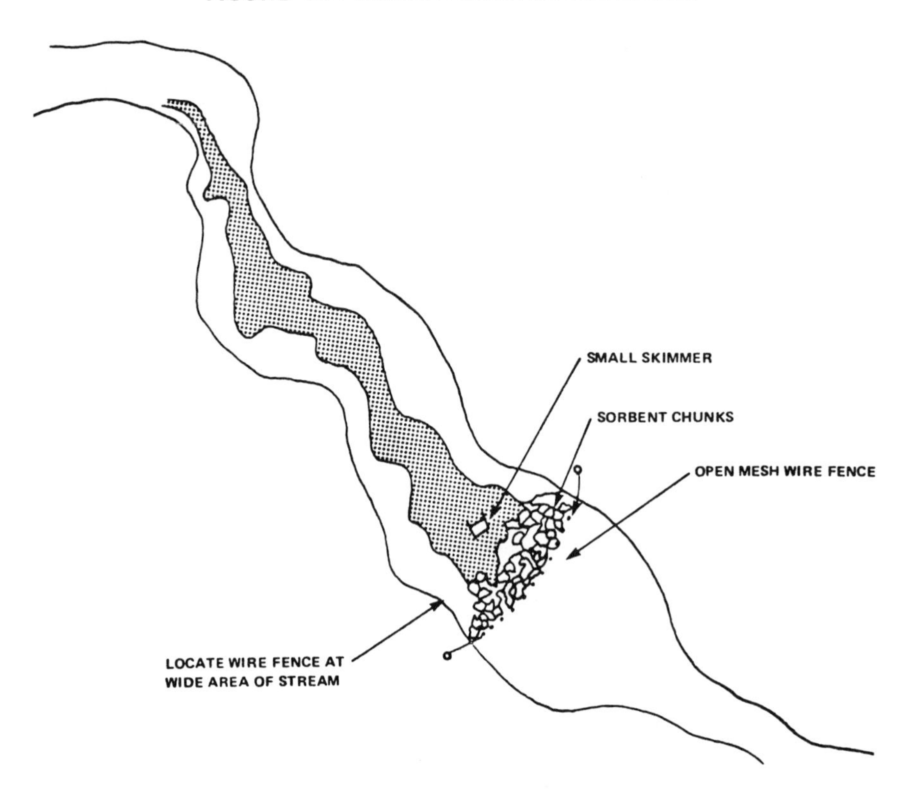

Source: Reference (9)

A device developed by *J.L. Dubouchet; U.S. Patent 3,815,742; June 11, 1974; assigned to Societe Generale de Constructions Electriques et Mecaniques (Alsthom), France* comprises a separator which is located alongside a liquid stream bed and in communication with the latter so that there is diverted thereunto at least a selected depth of the top of the stream extending below the maximum expected depth of pollutant material in the liquid.

An apparatus developed by *J.R. Laman; U.S. Patent 3,834,538; Sept. 10, 1974; assigned to The Firestone Tire & Rubber Company* is a floating apparatus to be anchored across a polluted stream to skim the water and collect the debris. The apparatus includes a floating boom anchored within abutments by means which also provide barrier seals. The upstream side of the boom presents a baffle surface. Adjacent the baffle surface a series of paddle means skim the water and collect the debris. At one end of the boom the paddle means traverse a ramp. The coaction of paddle means against the ramp moves skimmed liquids and collected debris to a sump.

A system developed by *L.W. Jones; U.S. Patent 3,850,807; Nov. 26, 1974; assigned to Amoco Production Company* is a system for removing dispersed oil from a flowing inland stream. A bed of granular sulfur is provided within a unique container buoyantly supported by the stream. In most oily streams the oil, due to gravity, accumulates on the surface in a very thin film, sometimes referred to as an oil slick or sheen. This top portion of the stream is directed through the bed of granular sulfur where the oil is coalesced and removed, leaving clean water flowing from the container.

A fence design developed by *L.F. Stovall; U.S. Patent 4,000,618; Jan. 4, 1977; assigned to Exxon Production Research Company* includes a series of vertically "stacked" baffles. When in operating position the barrier extends across the breadth of the stream and to any desired distance above and below the surface of the liquid. The baffles are spaced apart with each baffle overlapping a portion of the baffle or baffles adjacent to it to form a passageway between each baffle for flow of the liquid stream through the barrier.

Such a fence design is applicable to fluids flowing in a stream or channel. One installation of such a device is shown in plan and elevation in Figure 118.

FIGURE 118: SKIMMER FENCE FOR USE ACROSS STREAMS

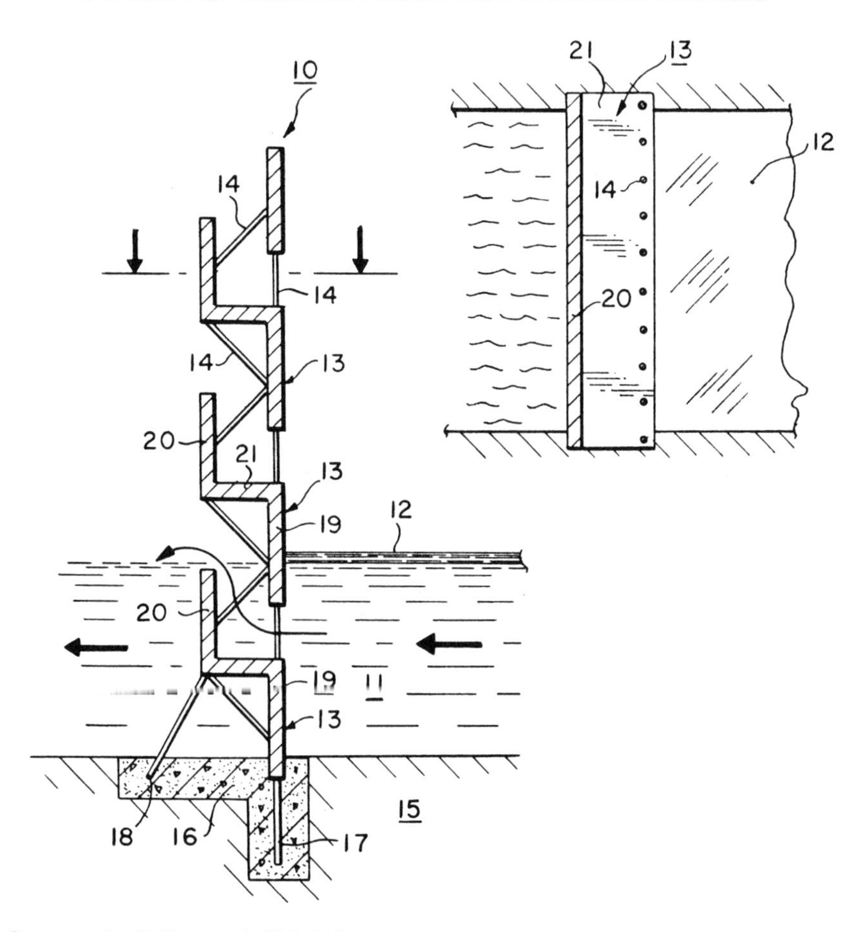

Source: U.S. Patent 4,000,618

The barrier, generally designated **10**, is positioned in a stream of liquid **11** flowing in the direction of the arrowed lines. The stream contains a floating layer **12** of oil or other petroleum product or any contaminant it is desired to remove from stream **11**. Barrier **10** includes a series of vertically arranged or stacked baffles, each designated **13**, connected together by structural support members **14**, two of which, **17** and **18**, are secured to the land **15** under the water **11** in cement **16**. As also shown each baffle **13** forms a "step" extending across the stream having two parallel spaced apart vertical members **19** and **20** connected together by a horizontal member **21**. Each member **19** of one baffle overlaps the member **20** of an adjacent baffle (except for the lowermost member **19**).

SPILL REMOVAL EQUIPMENT—OPERATION AND MAINTENANCE

Additional equipment required for the effective cleanup of an oil spill includes field radios, small boats for positioning of boom, and water pumps and firehoses to herd oil slicks toward skimmers. Where a source of air is available near the spill, compressed air hose and fittings can be used to herd oil toward a skimmer. Personnel morale and effectiveness can be greatly increased if sufficient rags, gloves, work life vests, low cost disposable coveralls, and foul-weather gear are made available during major spills, which can and do last for days or weeks. This aspect is most efficiently handled if a central station is established for daily issuing and collection of these items and for the check-in and check-out of work crews. Providing hot coffee in cold weather and drinking water in hot weather also greatly increases personnel effectiveness.

Pumps

One of the primary considerations during an oil spill operation is to minimize the volume of oil-water emulsions which must be handled and ultimately disposed. Gravity separation of collected oil-water mixtures at various points during the process of skimmer pickup to disposal is the best method of accomplishing this. Separation of collected oil and water occurs both at the various skimmers themselves and during the temporary storage of the collected mixture prior to ultimate disposal.

Proper selection of transfer pumps is essential to minimize the amount of mechanical oil-water emulsion generated during the operation. Positive displacement pumps of the diaphragm or progressing cavity (screw) type are best for this type of service. Documented tests have verified that oil-foam sorbent mixtures of up to 30% sorbents by volume can be pumped by these types of pumps with no loss in flow rate. Mixtures containing 70% foam sorbents result in a 20% reduction of flow rate.

The use of centrifugal or vane type pumps should normally be avoided if possible. The NAVFAC-procured skimmer systems are equipped with positive displacement pump units to transfer collected oil and water to temporary storage tanks. Both the small and medium skimmer systems use diaphragm pumps, while the large system is provided with a screw type pump. These pumps have the advantages of placing minimum turbulence into the oil-water mixture being pumped and can pass most sorbents and some debris without clogging. Should these pumps become clogged with rags, seaweed or other debris, they can quickly and easily be cleared by partial disassembly, as in the flapper valve assemblies of the small and medium skimmer diaphragm units; or operated in reverse, as in the screw type pump of the large skimmer. Should ancillary transfer pumps be required in an oil spill operation, diesel or air-operated diaphragm pumps are recommended.

Temporary Oil/Water Storage

After oil and water are skimmed from a spill area, the mixture is usually transferred to temporary holding tanks prior to disposal. This practice provides a retention time, which can be effectively controlled to allow oil-water separation by gravity. By carefully draining the settled water, the volume of product which must be disposed of can be substantially reduced. Equipment which can be used for temporary storage includes waste oil rafts (donuts), shoreside tanks of various types, and Ships' Waste Offload Barges (SWOBs).

(a) Open Bottom Donuts: Waste oil rafts are depicted in Figure 119. They are also called oil rings, oil disposal rings, and ODR's, but are most commonly called donuts. Open bottom donuts have been used in the Navy since the 1940s to collect oily bilge and ballast waste from ships.

FIGURE 119: CROSS SECTION OF CLOSED BOTTOM (LEFT) AND OPEN BOTTOM (RIGHT) DONUT

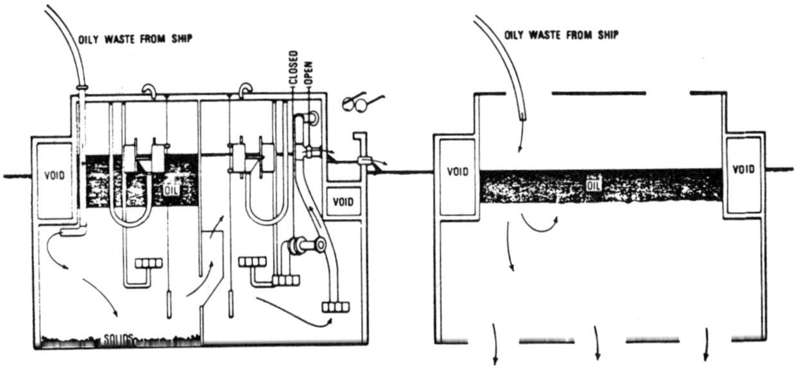

Source: Reference (9)

This donut is an oval shaped floating tank 25 feet long, 15 feet wide, and 18 feet high with an open bottom and an open top. When it is floating in its normal operating condition, it holds approximately 26,000 gallons and draws about 13 feet. Buoyancy is provided by a circumferential flotation collar around the upper portion of the tank. Basically, the donut is a floating gravity-operated oil/water separator. The fact that the bottom of the donut is open, allowing free flow in and out of the donut, means that the donut maintains a constant volume and a constant draft. Oily ballast and bilge wastes are pumped directly from the ship into the donut through its open top. Inside the donut, the oil and water separate by gravity. The oil rises to the surface, and an equal volume of water is displaced out the open bottom.

There are about 100 old open bottom donuts still in use throughout the Navy. As concern over environmental protection increases, the open bottom donut becomes more and more unaccepable. Although, if used carefully, it does an adequate job of separating oil from water, the open bottom does not provide sufficient protection from oil spills. Oil spills are caused by pumping oily waste into the donut at too high a rate and towing the donut too fast. Additionally, oil-coated solid particles settle down through the donut, pass through the open bottom, and contaminate the bottom of the harbor.

(b) New Closed Bottom Donuts: Both open bottom and closed bottom donuts are depicted in Figure 119. Although the outward shape and size of the new donut have not changed, the design of the inside has been modified to make it a better oil/water separator and to reduce the possibility of causing an oil spill. The bottom of the donut has been sealed off, and a series of pipes and valves has been added to maintain a constant volume and draft. Oily waste from the ship enters the donut through the two fill pipes. The hose used to pump from the ship to the donut must be supplied by the ship. No special connection is required since the hose is merely inserted into the six-inch diameter fill pipe and secured with a line. Maximum total flow into the donut should be 200 gpm. Oil rises to the surface inside the donut, and solids gravitate downward to collect on the bottom of the donut.

Water is displaced through the donut and passes up and out a discharge pipe at the same flow rate at which oily waste enters the donut. The water is discharged from the donut at a point above the waterline where it can be visually checked for oily sheen before it is returned to the harbor. Other improvements include the addition of a center bulkhead to give a two-stage gravity separation capability, a closed top with lockable access hatches to keep out trash, two floating oil skimmers to facilitate the removal of separated oil, and an indicator to measure the amount of oil collected.

(c) Donut Servicing Subsystem: In order to service the donut and maintain it so that the water discharged from the donut creates no visible sheen, NAVFAC has developed an Oil/Water Separation and Removal (OWS&R) subsystem. The five modules that make up the OWS&R subsystem include diesel-driven pumps, three 2,000-gallon oil storage tanks, and a filter/coalescer type oil/water separator. The modules are skid-mounted and have fork-lift slots for operational flexibility and ease of mobility.

The subsystem is self-contained and may be operated independently of outside power sources ashore, on a pier, or on a small craft such as an LCM-6. The OWS&R subsystem services the donut by removing and storing separated oil and solids, cycling harbor water through the donut to prevent stagnation, and mechanically separating oil/water emulsions that may inadvertently be discharged into the donut. Each subsystem comes equipped with eighteen 25-foot sections of three-inch diameter hose. The hose is fitted with three-inch, Kamlock couplings to connect the donut to the OWS&R subsystem. Generally, one OWS&R subsystem can service five to seven donuts.

(d) Ships' Waste Offload Barges (SWOBs): Need for SWOBs—The Navy (9) is procuring barges to collect ship-generated, oily waste and transfer it to shore facilities for treatment and disposal. Reasons for using the SWOB include the following:

a. A need to collect oily waste from ships berthed at remote locations.
b. Strict environmental regulations that preclude water from donuts being discharged to surrounding waters.
c. A need to collect oily waste from ships that is mechanically or chemically emulsified so that gravity separation will not occur in a donut.
d. A need to collect oily waste that is discharged from ship at too high a rate to be handled by donut.
e. Not enough donuts available to handle the oily waste generated.

General Description of the SWOB:

Hull Characteristics:

Hull Characteristics

Length, overall, molded	106 ft
Breadth, molded	26 ft
Draft, light, without margin	1 ft, 11 in
Draft, full load, without margin	5 ft, 11 in
Depth (molded) to deck at side	8 ft, 3 in
Depth, (molded) to deck at center line	8 ft, 3 in
Displacement, full load (oily waste) without margin	394.4 long tons (2,240 lb each)
Displacement, light, without margin	105.3 long tons
Capacity	75,000 gal

Waste Collection and Transfer

Load oily waste	400 gpm
	800 gpm
Offload oil waste	75,000 gal*

*In 4 hr using two pumps.

It should be noted that these rates are the design flow rates for the piping system aboard the SWOB. If, for example, oily waste is loaded into the barge through the manhole on top of the expansion trunk, the loading rate could be increased.

(c) Power and Pump Systems: One generator is provided to supply the required power for off-loading oily waste. The generator is rated for 75 kW at 450 volts, 60 hertz, three-phase, and 0.8 power factor for continuous operation. The generator has a maximum capacity of 85 kW. The generator is driven by a diesel engine designed for marine operation. Two electric, motor-driven oily waste pumps are installed for transferring the oily waste cargo. The pumps are designed only to offload oily waste from the barge. Pumps selected are the Blackmer rotary vane type, each with a capacity of 160 gpm and 175-foot water total discharge head. The barge is non-self-propelled, and requires a tug for movement around the harbor.

(d) Hoses and Hose Handling System: Four 50-foot lengths of collapsible two-and-one-half-inch oily waste cargo hose are provided with each barge. Hose storage racks are also provided. The hose is fitted with two-and-one-half-inch Kamlock couplings to connect to both the bilge riser on board ship and the riser on the pier side oily waste collection pipeline. Two booms are provided on king-posts for handling cargo hose. Each boom is designed for a lifting capability of not less than 1,270 pounds. The boom reach is sufficient to deposit cargo hoses at a point 25 feet from the full load waterline and three feet inboard of the shell of the ship being serviced.

(e) Vacuum Trucks or Shoreside Tanks: When vacuum trucks or shoreside tanks are used for temporary storage, a 55-gallon, open-top drum filled with sorbent (preferably polymeric foam as in Figure 120) can be used to insure that no oil re-enters the harbor as water is drained from the bottom of these units. Figure 121 shows the use of a small, simple filter cartridge on the tank or vacuum truck drains which, although untested by the Navy, has been used successfully by industry in preventing traces of oil from being flushed out with the drain water.

FIGURE 120: DRAINING SETTLED WATER FROM TEMPORARY STORAGE TANK

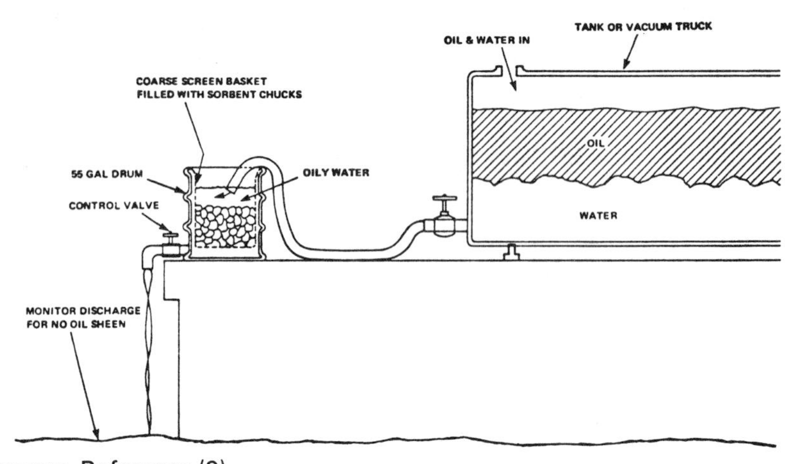

Source: Reference (9)

FIGURE 121: TREATING DRAINWATER WITH IMBIBER BEAD CARTRIDGE

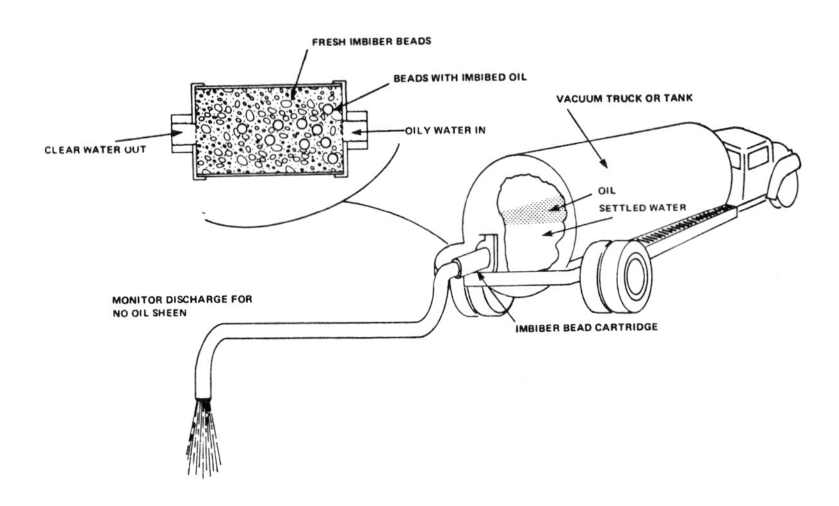

Source: Reference (9)

The cartridges contain Imbiber Beads, small-diameter pellets which allow water to pass but trap oil as it is carried out with the water. The beads "imbibe" or capture the oil by permeation through their walls and in so doing swell to 27 times their original volume. When a majority of the beads have captured oil in this way, the packed bed of beads swells to such an extent that the flow from the tank is automatically blocked off until a fresh supply of beads is placed in the cartridge.

The beads are a proprietary product of Dow Chemical, and while untested by the Navy, they appear to be a cost-effective way of reducing the volume of collected oil and water which must be subjected to disposal. When available, oil-water separators of the types tested by the Civil Engineering Laboratory, Naval Construction Battalion Center, Port Hueneme, California, and the Naval Ship Research and Development Center, Annapolis, Maryland, can be used to treat the settled water drained from vacuum trucks, shoreside tanks, WOR's or SWOB's.

A combination of commercially available units, which both of these groups have found to be effective, is a parallel plate gravity separator, as a first stage, followed by a filter-coalescer separator, as a second stage, to insure that the oil content of drained oily water is below the EPA discharge restrictions of no visible oil sheen (approximately 10 parts of oil per million parts of water).

Shore Support Facilities

In order to provide adequate protection and maintenance of oil spill control equipment, NAVFAC has identified shore support equipment requirements in the form of boom support, skimmer support, utility boat support, and general storage. Examples of shore support facilities for oil spill control equipment are:

a. Boom Support. This includes reel storage (Figure 122), boom containers, chafing gear, and boom cleaning facilities (Figure 123).
b. Boat Support. This includes launching ramps, storage cradles, and boat slip preparation.
c. Skimmer Support. This includes skimmer slip preparation.
d. General Storage. This includes open and covered storage, security fencing, security posts, and concrete pads.

FIGURE 122: OIL SPILL BOOM REEL

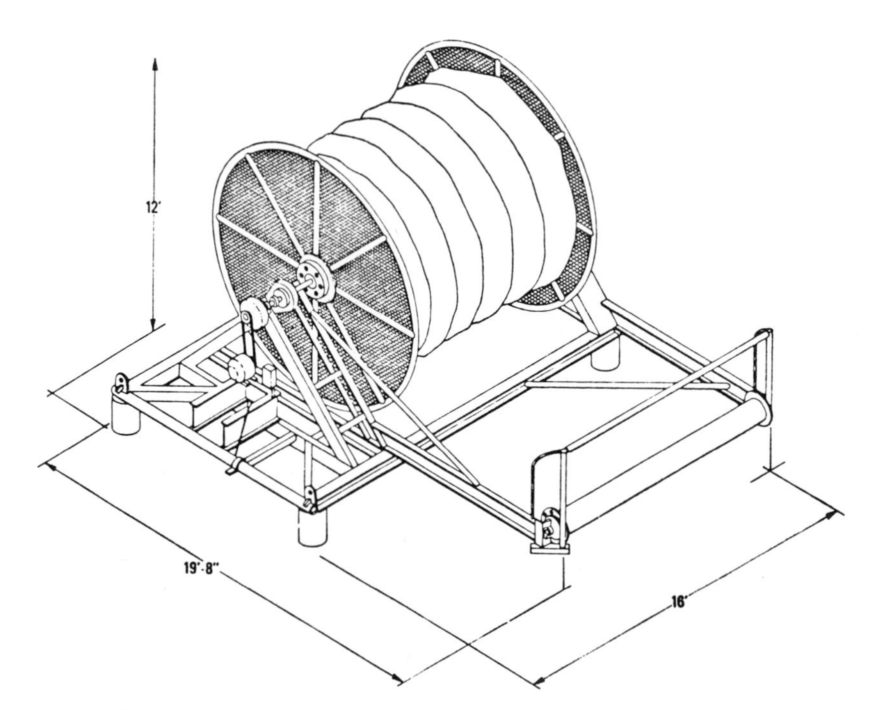

Source: Reference (9)

Cleaning and Restowage of Equipment

A major part of fast and effective response to an oil spill is having equipment cleaned and properly restowed for future use. Aside from a desire for general "good housekeeping," oily equipment can create additional expense during an oil spill. For example, equipment oily from a previous spill, especially containment booms, is difficult to handle and can bleed oil sheens into the water and result in further effort and expense when it is deployed on the next spill. Because of its large surface area, containment boom occupies the major part of the equipment cleanup effort.

FIGURE 123: BOOM CLEANING FACILITY

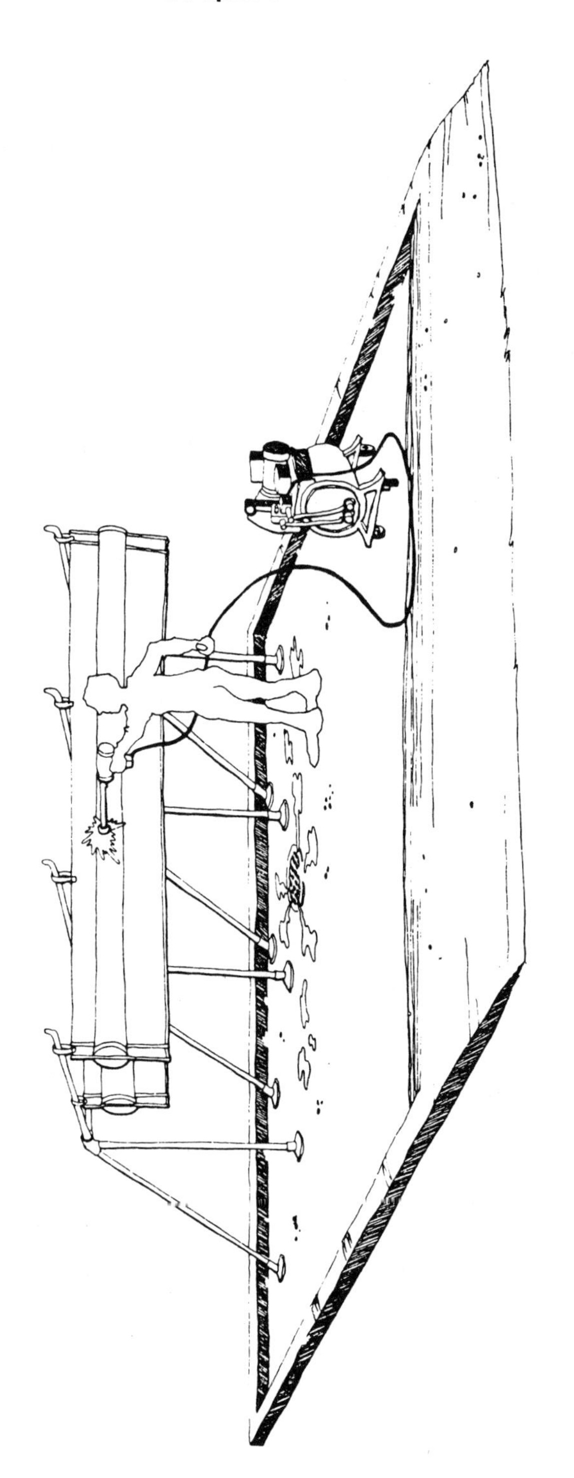

Source: Reference (g)

Recommendations of the boom manufacturer for cleaning solvents and maximum wash water emperature should be followed in all cases. High-pressure, heated water (120°F, 1,200 psi) from small, portable, washing units is very effective in loosening weathered oil from boom surfaces. Detergents, kerosene or other petroleum or water-based dispersant can also be used.

Figures 124, 125 and 126 show possible facilities for cleaning boom and miscellaneous small equipment following a spill. In Figure 124 the boom is rolled across the top of the cleaning trough as brushes and a high-pressure, washing unit are used to remove accumulated oil residues.

Figure 125 shows a much larger boom cleaning facility which has been successfully used on past oil spill operations. In this design the construction materials (two-by-fours and 4′ x 8′ sheets of marine plywood and polyethylene sheeting) are procured locally when needed. A sandy beach is ideal for this unit since long lengths of boom can be pulled from the water directly into the cleaning rack.

Brushes and high-pressure, water jets with biodegradable detergents are used to clean the boom. The wastewater from this operation flows down the graded centerline trench into a sump area at one end of the rack, where it is pumped to a suitable disposal site, e.g., municipal sewer (if allowed) or to an approved landfill via vacuum or tank truck. This type of cleanup facility could be constructed on site for a large oil spill at a remote location.

FIGURE 124: BOOM CLEANING TROUGH

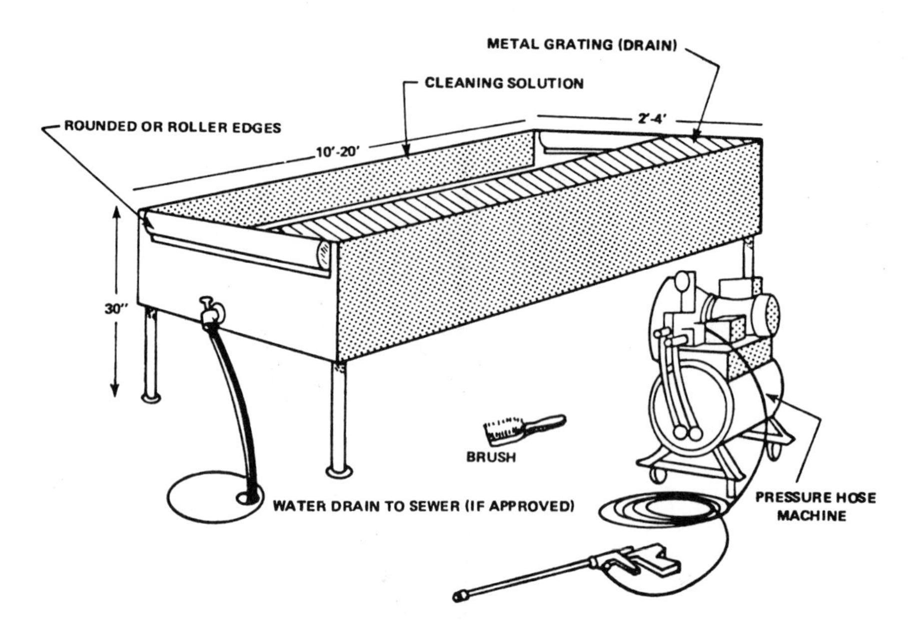

Source: Reference (9)

FIGURE 125: BOOM AND SMALL EQUIPMENT CLEANING STATION

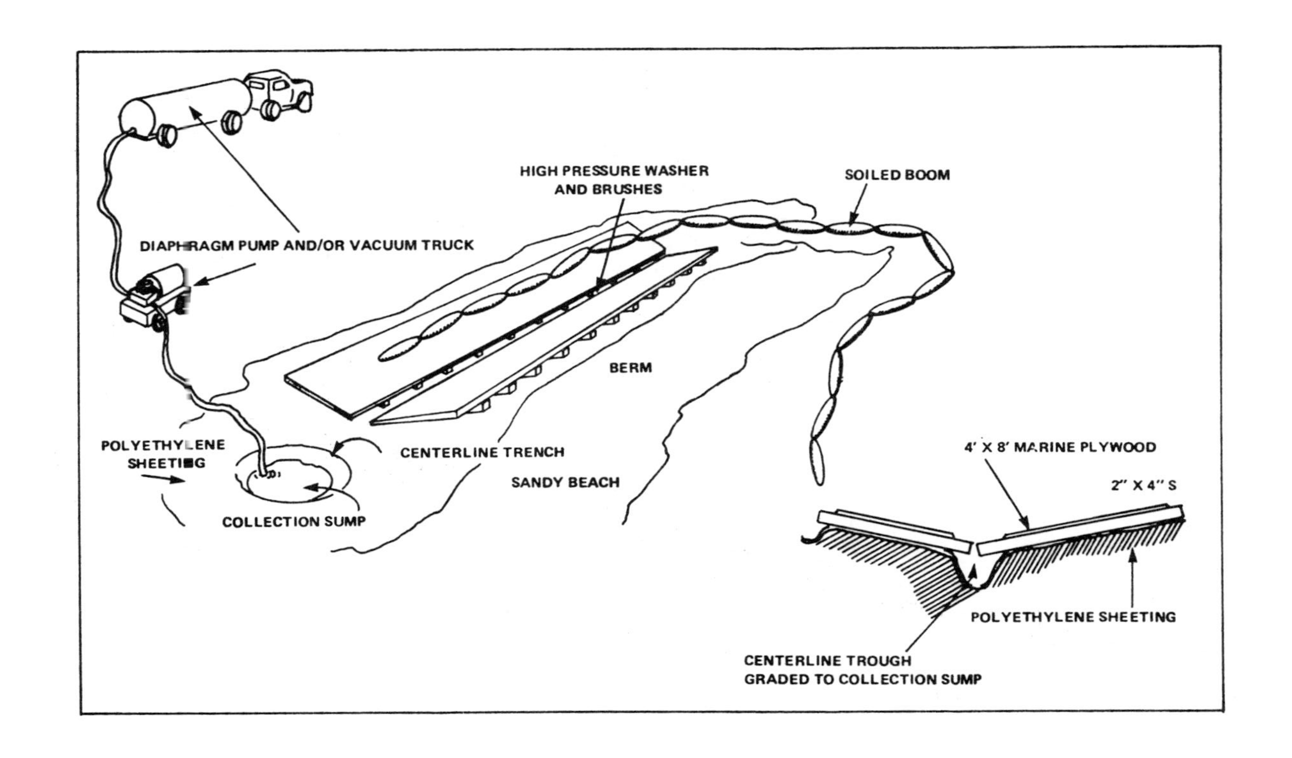

Source: Reference 9)

FIGURE 126: POSSIBLE BOOM CLEANING FACILITY

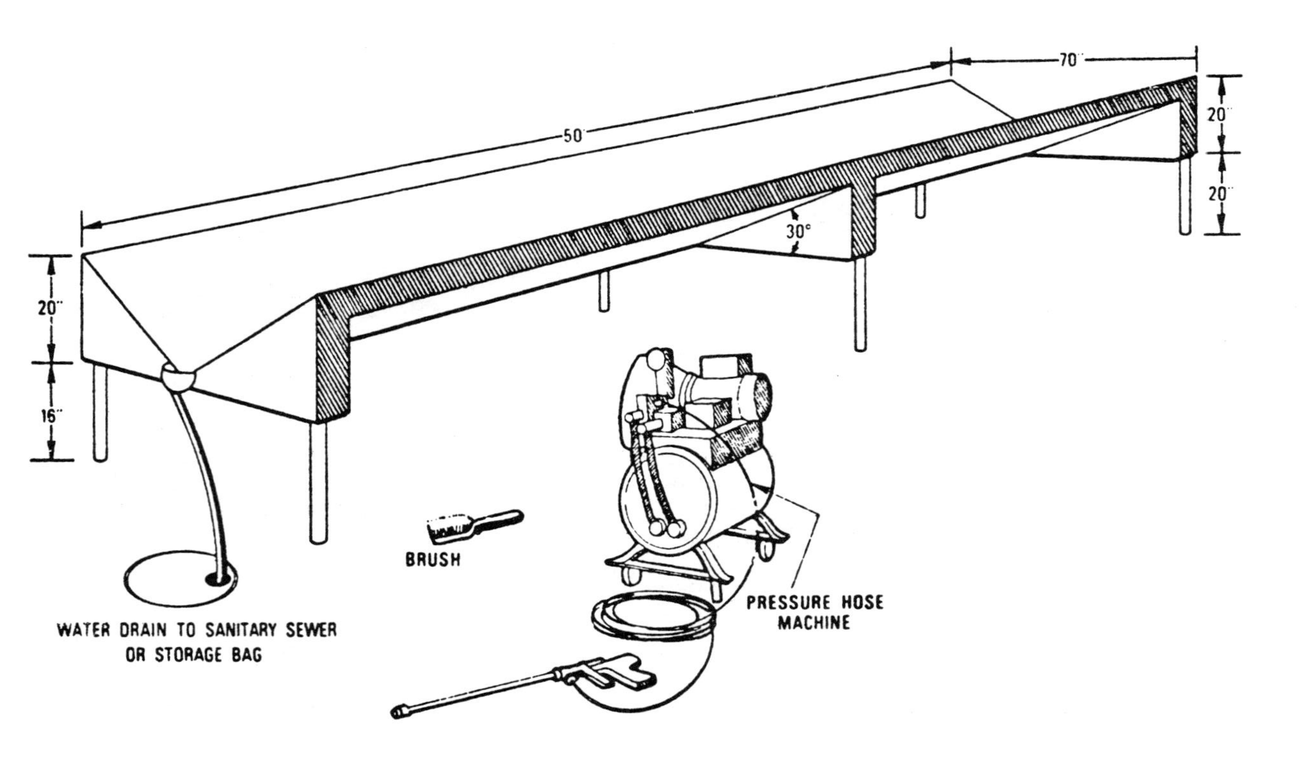

(continued)

FIGURE 126: (continued)

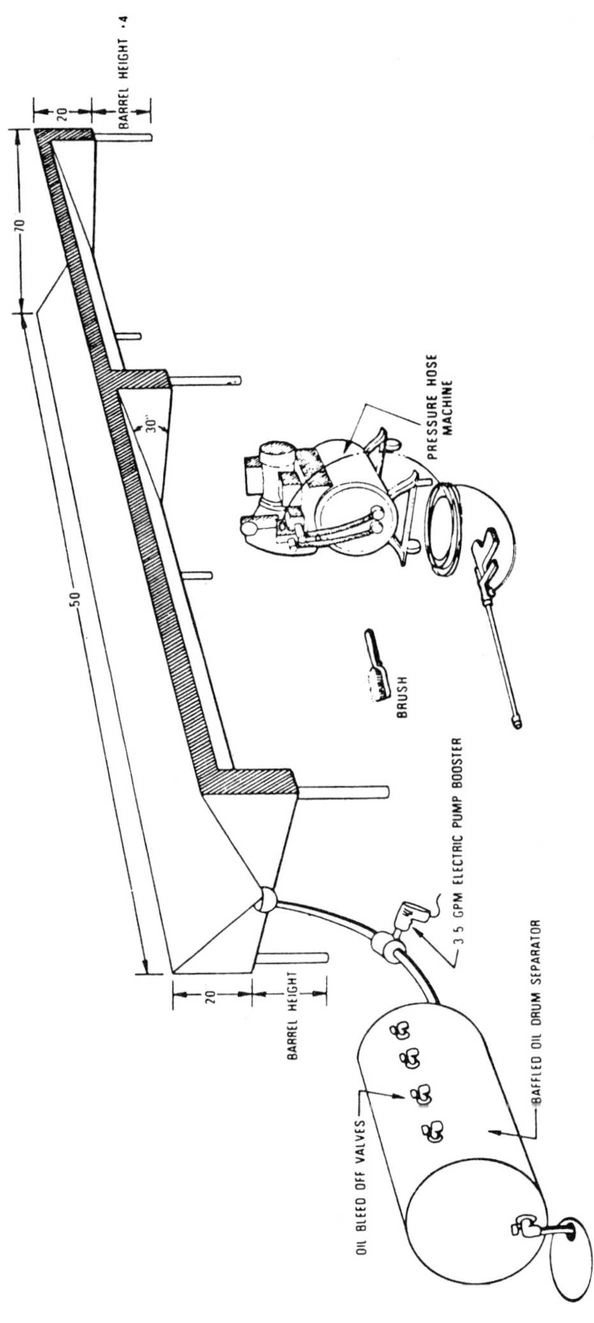

(continued)

FIGURE 126: (continued)

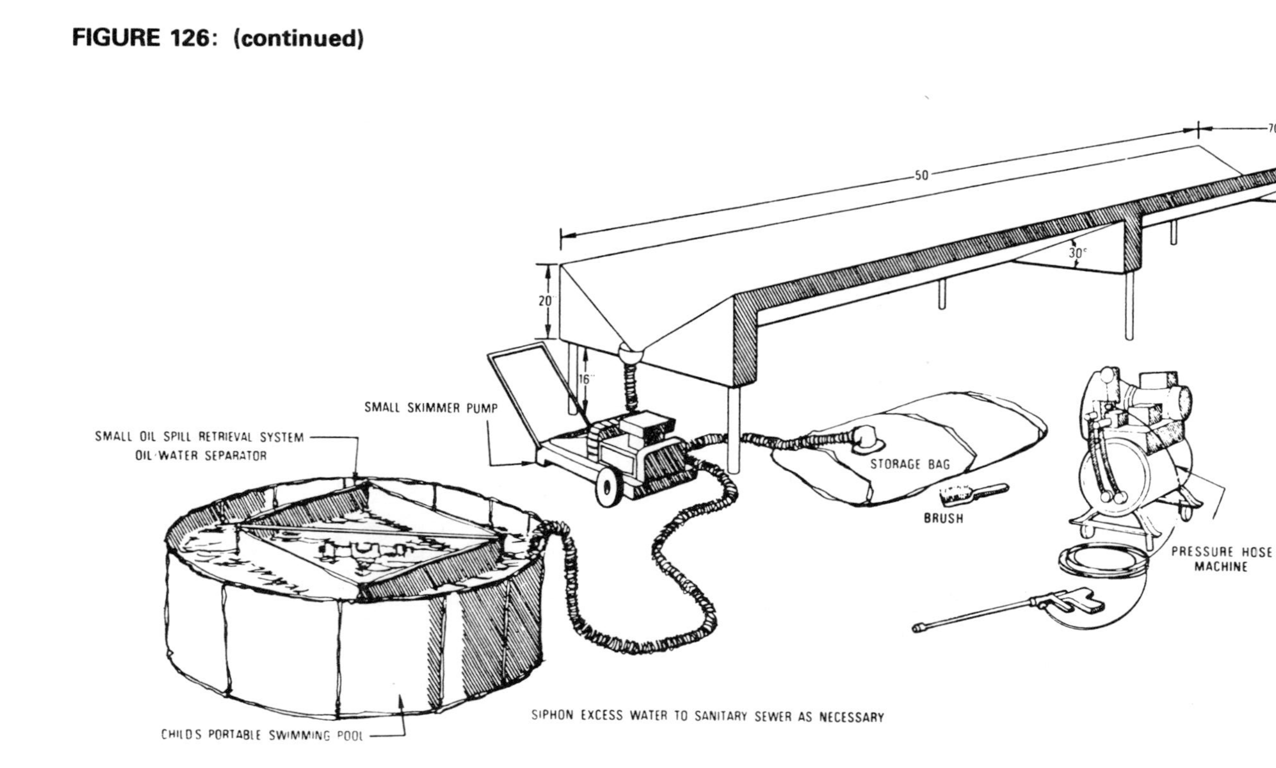

Source: Reference (9)

Manpower Requirements

It is important that adequate pools of manpower, either military or contractor-supplied, be identified before a spill incident occurs. The number and skills of personnel required will vary with the location, size of the spill and even during the course of a single spill operation. For a given spill situation, the waterfront conditions, spill size, and available equipment will be the main factors in determining manpower requirements.

As a planning guide, manpower estimates are provided in Tables 23 and 24. Table 23 lists the manpower and skills required for operation and maintenance of the major equipment components. Table 24 presents operational manpower estimates for these major equipment components when used in various type spill situations. Manpower and equipment required for a given spill may exceed that shown in Table 24, especially if the spill is spread over a large area.

TABLE 23: MANPOWER ESTIMATES FOR MAJOR EQUIPMENT COMPONENTS

Component	Operation (men)	Maintenance
Utility small boat	2	1 engine mechanic 1 small boat hull repairman
Small skimmer system	2	1 small diesel engine mechanic 1 laborer
Medium skimmer system	2	1 small diesel engine mechanic 1 pneumatic mechanic 1 laborer
Large skimmer system	2-3*	1 diesel engine mechanic 1 hydraulic mechanic 1-2 laborers
Containment boom	3**	4-5 laborers for cleaning and re-stowing

*Crane and riggers required for initial deployment.
**Additional men may be required for retrieval of boom at end of operation.

Source: Reference (9)

TABLE 24: MAJOR EQUIPMENT/MANPOWER ESTIMATES FOR VARIOUS SPILL SITUATIONS

Spill Location	Spill Size	Major Equipment	Manpower (men)
At pier side	100 gal or less	1 each: utility boat, small skimmer, boom, or	4
		1 each: utility boat, medium skimmer, boom, or	4
		1 each: utility boat, large skimmer, boom	5

(continued)

TABLE 24: (continued)

Spill Location	Spill Size	Major Equipment	Manpower (men)
At pier side	More than 100 gal, less than 1,000 gal	1 each: utility boat, small skimmer, boom, or	4
		1 each: utility boat, medium skimmer, boom, or	4
		1 each: utility boat, large skimmer, boom	5
Away from pier	100 gal or less	2 utility boats, 1 small skimmer, boom, or	7-8
		2 utility boats, 1 medium skimmer, boom, support LCM, or	8-10
		2 utility boats, 1 large skimmer, boom	6-7
Away from pier	More than 100 gal, less than 1,000 gal	2 utility boats, 1 small skimmer, boom, or	7
		2 utility boats, 1 medium skimmer, boom, support LCM, or	8-10
		2 utility boats, 1 large skimmer, boom	7
Away from pier	In excess of 1,000 gal	Minimum manpower requirements regardless of equipment.	15

Source: Reference (9)

OIL/WATER SEPARATORS

Oil/water separators may be used in various stages of spill and pollution control. Some proprietary designs for such separators are discussed in the paragraphs which follow.

An apparatus developed by *B. Valibouse and J. Pichon; U.S. Patent 3,789,988; Feb. 5, 1974; assigned to Societe Grenobloise d'Etudes et d'Applications Hydrauliques (Sogreah), France* is an apparatus for removing layers of pollutants, such as oil, floating on a heavier liquid, especially water which depends for its operability on the effect of the relative displacement speed of the heavier liquid and the overlying pollutant. Such relative speed is used to direct the removed layer of heavier liquid and pollutant, by tangential introduction, into at least one cyclone chamber wherein the induced rotation of the removed materials is caused as to create a whirlpool area in which the pollutant is concentrated.

In the central portion of such concentration, the pollutant is extracted through a pipe fitting in the axis of and crossing the ceiling of the cyclone. The centrifuged water, free of pollutant, is discharged at the end of the cyclone.

A device developed by *H.M. Rhodes; U.S. Patent 3,794,583; Feb. 26, 1974; assigned to Oil Mop, Inc.* for separating oil from an oil and water mixture involves first subjecting the mixture to passage through a chamber containing an oil mop structure made from fibrillated strips of polypropylene secured to a polypropylene line so that the oil will be attracted by the polypropylene strips and the water will pass through the mop structure.

The water is thereafter subjected to passage through a fabric membrane having a 2% fluorocarbon solution impregnating the fabric which will permit passage of the water through the membrane rejecting the oil and permitting the water to pass on and be drawn off separately from the oil which may likewise be drawn separately from the chamber.

Another oil/water separation apparatus developed by *H.M. Rhodes; U.S. Patent 3,810,832; May 14, 1974; assigned to Oil Mop, Inc.* functions by directing the oil/water mixture through a barrier of filaments of polypropylene arranged across the path of mixture flow which barrier is anchored at its base at the bottom of the fluid confining means such as an API oil separator or a ditch or canal. The free ends of the strips of polypropylene are directed upwardly forming an inclined plane up which the oil droplets amalgamate assisted by the buoyancy of the oil and the force flow vector of the mixture passing through the fluid confining means.

An apparatus developed by *L. Mercuri; U.S. Patent 3,844,944; Oct. 29, 1974* is one in which a plurality of containers is provided each of which acts as a separator and each of which receives processed fluid by gravity flow from the container immediately preceding it. In effecting separations, the lighter of the two liquids is allowed to go to the top and the heavier of the liquids, successively containing a lesser amount of the lighter in successive separators, is passed to the next separation container. In the separation chamber, there is a wall that affords an opening near the bottom of the container to allow flow to the next container, each container thus, in effect, having two chambers.

The first chamber receives the mixture containing a greater amount of the lighter material and as separation occurs, the second chamber receives a mixture which contains lesser amounts of the lighter material. The second chamber contains an outlet to the separation container next to it which is shorter in height. Thus, the separations and the flows are effected by gravity. The lighter of the immiscible liquids is either skimmed off the top or, when a given separation chamber contains practically all of the lighter material, the lighter material is passed to a separate storage means.

A device developed by *P. Preus and J.J. Gallagher; U.S. Patent 3,862,040; Jan. 21, 1975* is an oil/water separator which is adapted to function at constant high efficiency regardless of fluctuation of liquid flow.

Such a separator is shown in cross section in Figure 127. The separator **10** comprises a skimmer tank **12**, communicating with a separating tank **14** and a concentrating tank **16**.

An effluent discharge line **18** feeds one end of the skimmer tank **12** with liquid containing water and a less dense hydrocarbon component. The skimmer tank is provided with a series of baffles **20** extending upwardly from the bottom to a point below the surface **22** of the liquid in the tank. The hydrocarbon component separates out, due to its lower density, as a layer **24** on top of the water **26** in the tank. A series of 90° elbow discharge pipes **28** are disposed through a wall of the skimmer tank opposite the effluent discharge line with the lower ends of the vertical segment thereof disposed in spaced relationship to the bottom of the tank at a point below the level of the upper edge of the baffle. The vertical disposition of the horizontal segment of the elbow determines the liquid level and should therefore be disposed above the level of the upper edge of the baffle. The size and number of discharge pipes are determined by the maximum quantity of effluent anticipated from the discharge line, the multiple pipes being required to provide a maximum variation in liquid processed with a minimum variation in the liquid level.

FIGURE 127: PREUS AND GALLAGHER OIL/WATER SEPARATOR DESIGN

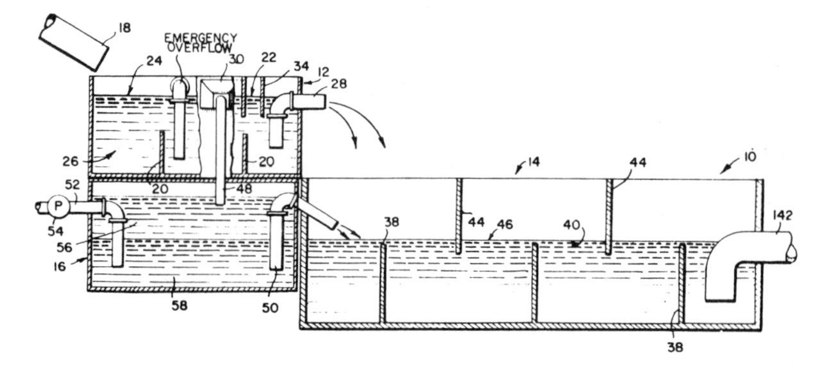

Source: U.S. Patent 3,862,040

The plural pipes could obviously be replaced with a single discharge device such, for example, as a rectangular cross-sectional duct conforming to the general discharge configuration provided by the plural pipes if so desired. A weir skimmer **30** is provided in the side of the skimmer tank at the level of the liquid surface. The weir lip should be disposed, with relation to the lower surface of the horizontal component of the discharge pipes such that the lip is between ¼ and ½ inches below the liquid surface during optimum flow conditions. A vertically adjustable lip (not shown) may be provided for the purpose of obtaining this disposition.

Other types of skimmers such as mechanical pump skimmers, belt or roller skimmers, or other kinds of weir skimmers, such, for example, as slotted tube type skimmers, can be substituted for the weir skimmer specifically disclosed without altering the basic operation of the process.

An inverted weir is disposed in the skimmer tank to intersect the surface of the liquid, and oriented to channel the hydrocarbon layer into the skimmer weir. A second inverted weir **34** is disposed to similarly intersect the liquid surface upstream of the discharge pipes. The number and disposition of the inverted weir and baffles may be changed as the particular installation requires. The primary purpose of these components is to provide a maximum travel path for liquid to flow through the skimmer tank with a suitable number of changes in direction to preclude entrainment of hydrocarbons in the liquid flow and permit separation out of the hydrocarbons to the liquid surface. The particular configuration of each installation depends upon the rate of discharge effluent and the amount of mechanical mixing and emulsion which exists in the effluent.

Discharge of the major portion of the water from the skimmer tank through the discharge pipes is directed into one end of the separating tank **14**. The separating tank is provided with a series of baffles **38** extending similarly to those of the skimmer tank from the bottom of the tank to a point below the suface of the water **40**. Discharge conduits **142**, having 90° elbows and of suitable size and number to handle the anticipated maximum effluent discharge volume, are placed in the same manner and for a similar purpose as the discharge pipes **28** of the skimmer tank. A series of inverted weirs **44** are disposed to intersect the water surface. The inverted weirs and baffles serve the same purpose and function in the same manner as those of the skimmer tank; and, as described above, may similarly be varied and reoriented to accomodate particular requirements of specific installations.

The primary function of the separation tank is to serve as a backup for the skimmer tank and associated systems in the event of malfunction or overloading thereof, and to remove traces of hydrocarbons which may occasionally escape from the skimmer tank. In this respect, a buoyant oleophilic-hydrophobic material **46** such, for example, as the particulate, fiberous material, Sorbent C, can be maintained on the water surface to absorb traces of hydrocarbons in the separator tank, and to act as a tell-tale to indicate malfunction or overloading of the skimmer tank or associated systems. Saturated sorbent material may be removed periodically by vacuum device, nets or the like.

The skimmer weir **30** discharges through a conduit **48** into the concentrating tank **16**. The tank **16** is provided with a decanting line **50** which extends from the point proximate the bottom thereof vertically to a point proximate the top thereof, and then horizontally out through one end to discharge into the separating tank. The horizontal component of the decanting line is preferably sloped downward for at least a portion thereof to facilitate discharge therethrough into tank **14**. The spacing of the lower end of the vertical component of the line **50** from the bottom of the tank should be minimum consistent with keeping the inlet clear of debris and silting in the tank **16**. A discharge line **52**, preferably having means, such as a pump **54** for facilitating discharge, is disposed in the other end of the tank **16**. The discharge line could possibly be valved and sloped, in a manner similar to the decanting line **50**, to provide for gravity discharge into a storage tank or the like if so desired.

In operation, hydrocarbons and water, skimmed from the liquid in the skimmer tank, enter the condensing tank **16** and stratify under the influence of gravity into a hydrocarbon layer **56** on a water base **58**. As indicated above, the ration may be on the order of 58% water or higher so that, as the tank **16** fills, a considerable base of water is present therein. As the liquid level in the tank reaches the horizontal component of the decanting line, liquid will start to flow through the decanting line into the separating tank **14** under the influence of gravity.

Due to the placement of the intake of the decanting line, this liquid will be water until the water base is almost entirely decanted at which point the tank will contain almost pure concentrated hydrocarbons. By periodic checks of the hydrocarbon level in the tank, or by observation of the discharge from the decanting line, the tank can be then pumped or discharged through the discharge line for final removal of the hydrocarbons.

In the event the separator is to be installed in effluent discharges subject to appreciable quantities of surface run off, skimmer bypass provision should be made for unusually heavy rainstorms since skimmers can only be economically designated for effective operation for reasonable variations from normal flow.

A device developed by *J.D. Conley, D.E. Belden and R.D. Terhune; U.S. Patent 3,878,094; April 15, 1975; assigned to Fram Corporation* consists of a mechanical emulsion breaker for removing emulsified hydrocarbon from a water stream, and, upstream of the emulsion breaker, a separator for removing from the stream free and entrained hydrocarbon. A preconditioner is preferably located upstream of the emulsion breaker for removing solids and initiating the separation of hydrocarbon.

Controls are provided for maintaining the hydrocarbon-water interface levels in the separator and the emulsion breaker within predetermined limits despite variation of the hydrocarbon concentration in the incoming stream, to thus prevent remixing of water and hydrocarbon, the controls operating in a closed, pressurized system by sensing the interfaces and adjusting the hydrocarbon and water discharge rates; a monitor is provided for continuously measuring hydrocarbon concentration in the treated water discharge, providing a signal to recycle that discharge in the event effluent quality is too low.

A process developed by *V.L. Traylor; U.S. Patent 3,884,803; May 20, 1975; assigned to Union Oil Company of California* is a process for separating low API gravity oil from water in which a small amount of a high API gravity oil is added to the oil-contaminated water and the mixture thereafter contacted with finely divided gas bubbles to float the oil and included particulate solid matter to the surface for mechanical separation. A low oil-content product water suitable for discharge into public waters or sewer sytems, or for use in industrial and agricultural applications is produced.

Figure 128 illustrates the element of this process.The oil-contaminated water flowing through conduit **10** and a controlled amount of high API gravity oil flowing through conduit **12** are admixed at the suction of pump **14**. Gas is introduced through conduit **16** into pump discharge conduit **18**, and this mixture introduced into pressurization tank **20**, which is maintained at a pressure of about 20 to 100 psig, and usually from about 30 to 50 psig, by means of pressure control valve **22**. Tank **20** provides sufficient retention time to permit the gas to dissolve in the liquid. Excess gas is vented through valve **24** in vent pipe **26**. The mixture of oil-contaminated water and high API gravity oil having the gas dissolved therein is introduced into flotation chamber **30** through the conduit **32**. Flotation chamber **30** is maintained at a pressure lower than the gas pressurization pressure, causing the gas to effervesce and form minute bubbles which float the particulate solids and immiscible liquids to the surface. The oily scum or froth containing both the oil and the solid matter is removed from the surface by mechanical skimming and withdrawn from flotation chamber **30** through conduit **32**. Clarified effluent water is withdrawn through conduit **34**.

In another modification of this process, the high API gravity oil and gas are admixed with only a portion of the oil-contaminated water, with the remainder of the oil-contaminated water being introduced directly into flotation chamber **30** through conduit **36** having valve **38** therein. This modification has the advantage that pump **14** and the attendant equipment piping and gas pressurization equipment can be of smaller size.

FIGURE 128: PROCESS FOR SEPARATING LOW API GRAVITY OIL FROM WATER

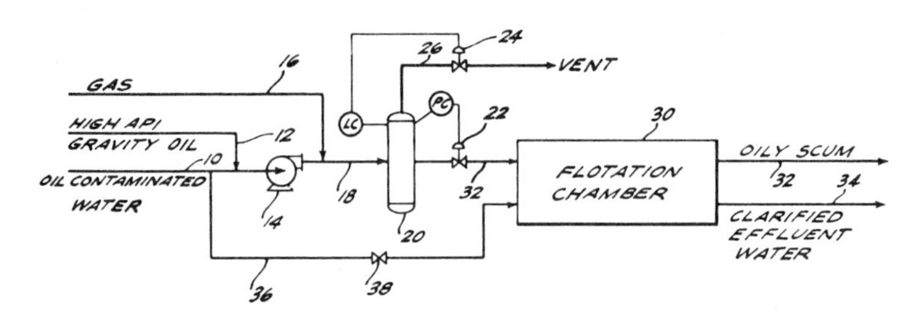

Source: U.S. Patent 3,884,803

An apparatus developed by *G.H. Sundin, F.H. Riedel, W.R. Niemi and R.C. Slocumb; U.S. Patent 3,915,859; Oct. 28, 1975; assigned to Conwed Corp.* is an apparatus for separating two immiscible liquids and preferably for removing oil from an oil-water mixture. When the term "oil-water mixture" is used it will be understood that the term includes emulsions, dispersions, etc.

The problem of separating oil from water is one which is difficult but yet very necessary in these days of environmental awareness. There are, of course, many known processes and apparatuses for the removal of oil from water. Typical of these are the centrifugal separators in which the oil and water are separately positioned by centrifugal force because of their varying densities and are then separated from each other by decantation. Other known processes include mechanical coalescers, those which absorb oil from water with a selective material such as a plastic foam (U.S. Patent 3,520,806) or with a cellulose fiber mat (U.S. Patent 3,630,891).

The prior art processes which are known are reasonably good for removing large quantities of oil from water. Their failing is that they do not get the water as pure as it should be. For example, with centrifugal separators it is difficult to get below 100 ppm oil even using the most sophisticated equipment. With the mat type products it is difficult to get down even to this level.

In this device there is provided an apparatus for separating oil from water and for getting the oil level down to 10 ppm or even less.

The essentials of the apparatus are shown in Figure 129. As shown, an oil-water mixture **10** is enclosed within a spool **12**. The spool preferably comprises two end discs and a central arbor but it may be perforated drum or the like if desired. Wrapped about the spool is a sheet of selectively absorbent material **14**.

The selectively absorbent material may be a cellulose fiber mat, plastic foam mat or the like, depending on application. The essential features of this material are that it must be capable of passing one of the liquids, it must be capable of retaining the other liquid, it must be flexible enough to pass around the spool and yet it must have sufficient structural integrity so as not to tear or hydraulically erode in operation. It will of course be understood that the structural integrity referred to could be in part supplied by a reinforcing member or by a reinforcing support along which the absorbent material travels. Suitable for this purpose are the cellulose fiber mats disclosed in U.S. Patent 3,630,891. These mats comprise a wood fiber held together with a water resistant binder such as a phenolic and are usually reinforced with a plastic net to give additional tensile strength. In most cases, a wax coating is employed so that the mat will repel water and float which is generally its intended application.

It is preferable to reduce the amount of wax or other sizing used with the cellulose fiber mats since this will permit the water to permeate more freely through the material. It will be understood that while specific reference is made to the fiber mat of U.S. Patent 3,360,891 there are many other materials which are suitable for use as for example, polyurethane foam, ethylene propylene diene foam and the like.

FIGURE 129: CONWED CORP. APPARATUS FOR ABSORPTION OF OIL FROM WATER

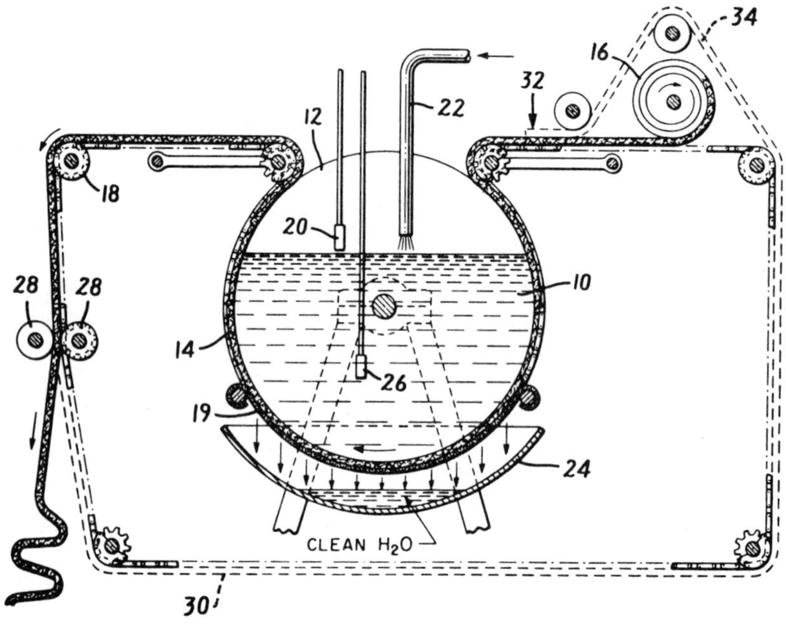

Source: U.S. Patent 3,915,859

Oil absorbent material is contained on a supply roll **16** and is advanced over the spool by driven shaft **18**. The spool is preferably rotated at the same rate that the selectively absorbent material is advanced so that there is no undue friction between the two. The oil absorbent material as it passes around the spool may be supported as for example by a wire screen **19** on its bottom side if desired. This wire screen can either be fixed in place or can be continuous and move with the sheet of absorbent material. It will also be understood that while only a single supply roll is shown, a plurality of supply rolls could be employed to supply a plurality of layers of absobent material, either of the same or of different composition.

Supply pipe **22** introduces the oil-water mixture to the spool, and a sensor **20** is provided for determining the level of the oil-water mixture.

In operation, the oil-water mixture will be introduced to the spool and water will pass through the absorbent material whereafter it will be collected in collecting trough **24**. The oil which is in the water will be absorbed by the oil absorbent material. As the oil abosrbent material becomes more saturated with oil, it will become less easily penetrable by the water (because its air spaces will tend to fill up) and the oil-water level in the spool will increase. When the oil-water level reaches the high level sensor, the sensor will be activated and will send a signal to the driven shaft telling it to advance the oil absorbing sheet.

As the sheet of oil absorbent material is advanced, the water will start to flow more rapidly through the fresh oil absorbent material and this will reduce the oil-water level in the spool. When the oil-water level again falls below the level of the sensor, the driven shaft will no longer be signaled to operate until such time as the oil absorbent material shall again become oil laden and the oil-water mixture shall again rise to the level of the sensor.

If desired, a float type sensor could be employed. This would preferably be employed in a column where it would trip an "on" switch as it rose to a certain point and then it would trip a second lower located switch to the "off" position when it dropped to that point. Alternatively, a solid state control having upper and lower "on" and "off" positions respectively could be employed.

While the automatic sensor switch is preferred, there are other ways of automatically advancing the sheet of oil absorbent material. For example, the driven roll could be driven at a constant speed thus advancing the oil absorbent material at a constant rate. This is especially applicable where the quantity of oil in the oil-water mixture remains substantially constant and thus the rate of advance of the oil absorbent material can be constant. The constant driven roll should preferably be rheostat or similarly controlled so that the constant speed can be set based on the amount of oil in the water, the amount of oil removal desired, etc.

No matter what type of advance is employed, the strength of the absorbent material must be kept in mind in determining the height to which the oil-water level will be permitted to rise. It is, of course, not desirable to permit the liquid level to rise to such a height that it will put undue strain on the absorbent material which might make it break or at least deleteriously space from the discs of the spool.

In addition to the automatic means for advancing the driven shaft, there could also be employed a lower level sensor **26** which could increase the flow of oil-water solution through the pipe **22** should the oil-water level begin to fall that low.

After the oil laden absorbent material passes over the driven shaft, it may suitably be passed through squeeze rolls **28** whereby the absorbed oil may be squeezed out and recovered. Thereafter, the used absorbent material may either be discarded or may be recycled along line **30** as shown back to starting point **32** where it can be employed as a second (or third, etc.) layer of absorbent material. It will be appreciated by those skilled in the art that if the recycled layer is made the top layer by following the path **34** this facilitates use of the absorbent material in two cycles whereafter it can conveniently be discarded.

A device developed by *L.L. Fowler; U.S. Patent 3,931,019; Jan. 6, 1976; assigned to Products and Pollution Controls Co.* is a cell for coalescing oil droplets dispersed in a water emulsion including a perforated core into which the emulsion is injected, a layer of emulsion breaking fibrous material wound about the core through which the emulsion passes, and a helical wire wrapping on the outside of the fibrous material. The wire wrapping is of selected tension and spacing.

A process developed by *G. Hirs; U.S. Patent 3,992,291; Nov. 16, 1976; assigned to Hydromation Filter Company* is a process for filtering suspended oil contaminants from an aqueous-based liquid. The method includes flowing the contaminated liquid through a bed of granulated black walnut shells which have a relatively weak affinity for oil. The suspended oil is coalesced during flow through the filter bed to form globular oil particles that are large enough to become entrapped in the interstices of the filter bed.

After a substantial quantity of oil has been accumulated in the filter bed, the bed is rejuvenated by flowing backwash liquid therethrough to provide a high velocity scrubbing to free at least a portion of the accumulated oil from the bed, the shells' weak affinity for oil facilitating such rejuvenation. Next, the bed is reformed and a subsequent filtration cycle is initiated.

A device developed by *J.E. Morgan, W.M. Lynch and P.M. Pelton; U.S. Patent 4,010,103; March 1, 1977* is an automatic oil-water separating device used within an oil-water separating container for skimming the oil off the top of the water and directing it through a collecting conduit to an associated container. The device includes a skimming trough supported by first flotation means mounted on opposite sides of the trough, and a second flotation means transversely disposed to the trough. The specific shape of both the first and second flotation means is designed to immediately restore the trough to the skimming position in the event the unit is accidentally submerged below the surface of the water and to automatically pivot the unit to accelerate oil separation.

A separator designed by *D.H. Fruman; U.S. Patent 4,022,694; May 10, 1977; assigned to Hydronautics, Incorporated* is a cartridge-type filtering apparatus containing a block of squeezable foam that is encased in an envelope of flexible impervious material to avoid channeling and provide a more efficient and effective oil-water separation process.

The device is shown schematically in Figure 130. The oil-water separation apparatus includes a housing **10** having an oil-contaminated water inlet **12** and outlets **14** and **16** for the separated oil and water, respectively. Housing **10** is divided into three interconnected chambers, preferably arranged horizontally as shown in the drawings. These chambers consist of a separation chamber **20** essentially in the middle of the housing, a feed chamber **22** communicating with the influent side of separation chamber **20** for distributing the flow of feed uniformly through that chamber, and a flotation chamber **24** on the effluent side of the separation chamber for receiving the effluent from the separation chamber and permitting gravity separation of the effluent.

FIGURE 130: HYDRONAUTICS INC. OIL/WATER SEPARATOR DESIGN

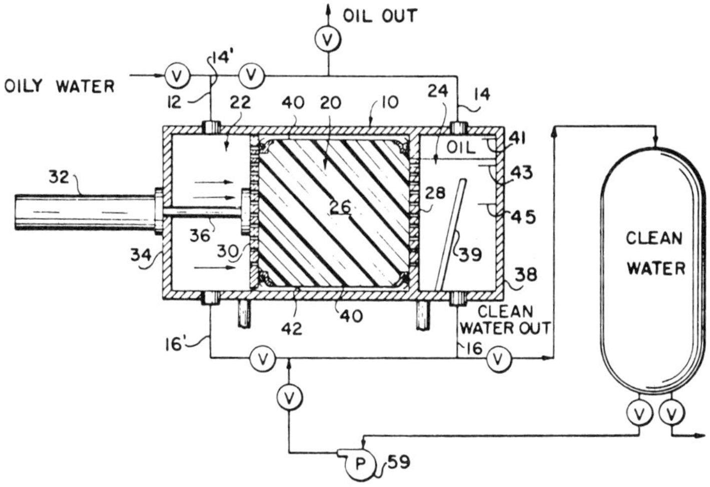

Source: U.S. Patent 4,022,694

The separation chamber is filled with a cartridge or block of porous material **26**, having filtering characteristics with respect to the contaminated feed and, more particularly, oil-absorbing and oil-coalescing characteristics, that is regenerable by compression to remove absorbed contaminant from the material. Block **26** is held in position in housing **10** between two perforated plates **28** and **30**, the rear perforated plate **28** being fixed in the housing and the front plate **30** being mounted for axial movement in response to applied pressure with respect to fixed plate **28**.

As shown, the means for applying pressure to plate **30** to compress porous block **26** and regenerate it for further use comprises a hydraulic cylinder **32** mounted on the front wall **34** of housing **10** and having a piston **36** that extends through feed chamber **22** and is connected to movable perforated plate **30**. Thus, it can be seen that, as hydraulic cylinder **32** is actuated, perforated plate **30** pushes towards fixed rear plate **28** compressing material block **26** and expelling absorbed contaminant out of the block and into chambers **22** and **24**.

Alternatively, if it is desired to express the absorbed contaminant during compression in only one direction, for example, only into flotation chamber **24**, suitable one-way valves would be incorporated in plate **30** that would be open during the flow of feed through the system, but that would close during compression and regeneration.

Feed chamber **22** of housing **10** is defined by the space between the inside of front wall **34** and the front face of perforated plate **30** in its retracted position. While the provision of a feed chamber is not essential, it is desirable to permit the feed to be uniformly distributed over perforated plate **30** for uniform flow through porous block **26**.

A flexible envelope **40** of impervious material completely surrounds the outer surface of foam block **26** and is located between the block and the inner wall surface **42** of separation chamber **20**. Envelope **40** is of less diameter than wall surface **42** and is connected at either end to perforated plates **28** and **30**, thus essentially providing a cartridge of filter material between the end plates that is separated from the inner walls of the housing by an annular space.

The selection of the material comprising porous block **26** is not critical provided the material has an open, reticulated structure and when used as an oil-water separator desirably has both oil-absorbing and oil-coalescing properties. A material found to be ideally suited for such use is a block of polyurethane foam having a pore size of between 30 and 100 pores per linear inch (ppi) and preferably a pore size of around 60 ppi. It has been found that such polyurethane foams demonstrate excellent oil-scavenging properties with good flow characteristics over a large range of oil properties and concentrations.

Material **25** may consist of a plurality of discrete chips or a plurality of segmented pads lined up in series within the envelope, but preferably is a single, integral block of the porous foam. The block, in general, is found to be more efficient than filling up the envelope with a plurality of small chips of a compressible material.

Envelope **40** is preferably a sheet of impervious fabric having a rubberized coating, such as a Hypalon coated sheet of a polyester fabric or nylon.

Preferably, housing **10** and foam block **26** are circular and the block is oversized with respect to envelope **40** to precompress the foam against the envelope and avoid channeling between the envelope and the foam and the envelope is of less diameter than the inside wall of the chamber to prevent the envelope from contacting with the walls of the separation chamber.

Preferably, flotation chamber **24** is contiguous with separation chamber **20**, forming an integral part of housing **10** and defined by the section between fixed rear perforated plate **28** and the rear wall of the housing. This provides a more compact and efficient separator, but it will readily be apparent to those skilled in the art that chamber **24** could be remote from the housing with suitable piping to transfer the effluent to the chamber.

Chamber **24** is preferably divided into two parts, as shown, by a baffle **39** to avoid contamination of the clean water outlet **16** and to increase gravity separation by creating an upward flow of the effluent in the chamber. When the oil is a light oil, the coalesced droplets accumulate at the top of chamber **24** and the upward flow of effluent along the baffle **39** assists in purging the droplets in this direction and out through oil outlet **14**.

To control the level of accumulated oil in chamber **24**, a plurality of oil-water interface detectors **41**, **43**, and **45** are provided in rear wall **38** of housing **10**. These detectors are conventional and operate on the differences in electrical conductivity of water and oil. Thus, for example, when the oil level in chamber **24** falls below middle detector **43**, thus changing the conductivity between detectors **43** and **45**, an oil pump (not shown) would be activated to pull the oil out of outlet **14** and thereby prevent the oil from falling to a level where it could contaminate the clean water outlet. Oil withdrawal is continued until the oil-water interface reaches detector **41** and then the pump is deactivated. This, of course, would be the arrangement for light oil-contaminated water, suitable modifications being made when heavy oil-contaminated water was being passed through the system.

In operation, the contaminated feed is fed through inlet **12** into feed chamber **22** of separator **10** where it is evenly distributed throughout the cross-sectional dimensions of the housing. The porous regenerable material **26** meeting the above-defined criterion is typically a piece of polyurethane foam substantially filling up the space in separation chamber **20** bounded by envelope **40** and plates **28** and **30** and having 60 pores per linear inch, a thickness of approximately 16 inches, and a cross-sectional area of approximately 4 square feet. The contaminated feed flows through foam block **26** where the oil is absorbed and/or coalesced depending upon the degree of saturation of the foam.

In the beginning, practically all of the oil is absorbed by the foam, but after the foam begins to become saturated, large droplets of coalesced oil start to appear in the effluent as a result of the oily water passing through the foam. Discharging this effluent in chamber **24**, however, permits the large coalesced droplets of the oil to be readily separated by gravity, so that it is not necessary to shut off the feed at this point and regenerate the foam just because oil appears in the effluent. The separated droplets are then withdrawn through oil outlet **14** and the heavier, clean water through outlet **16**.

The separation step cannot be sustained indefinitely. At a certain level of oil saturation of foam pad **26**, the pressure drop across the foam increases to such an extent that it tends to block the flow of contaminated water. The flow of water is then stopped and regeneration of foam block **26** is accomplished by moving perforated plate **30** toward fixed plate **28** by hydraulic cylinder **32** to compress the foam and expel the absorbed oil. The oil expelled from the foam flows into both feed chamber **22** and flotation chamber **24** where it rises to the top of the chambers and is drawn off through oil outlets **14** and **14'**.

To assist in purging the expelled oil, clean water can be pumped back into chambers **22** and **24** during regeneration by pump **59**, to push accumulated oil towards oil outlets **14** and **14'**. Following regeneration, hydraulic cylinder **32** is retracted, water inlet **16'** and oil outlet **14'** to feed chamber **22** are closed, and the oily water is again fed to the separator.

By permitting gravity flotation of the effluent from the polyurethane foam, both the absorbing and coalescing properties of the foam can be utilized resulting primarily in longer periods of operation between regeneration without sacrificing the quality of the effluent or the flow rate through the separator.

REMOVAL FROM COASTS AND BEACHES

Many of the beaches of the world have been despoiled in recent years by oil spills. Oil tankers wrecked at sea and discharge of bilge water by ships at sea have accounted for large quantities of oil which has eventually washed onto beaches. Not only have such beaches suffered a considerable aesthetic loss, but also marine life and water fowl in the vicinity of the beaches have been harmed. As a result, considerable efforts by many parties have been devoted to developing various means not only to prevent oil from contacting such beaches, but also to clean such beaches after they have been contacted by oil. Such efforts have to date been very expensive and only modestly successful.

One of the more usual methods of protecting beaches is through the use of straw. The straw is spread at the water's edge in order to catch incoming oil slicks. Once the straw has been soaked with oil it must be immediately retrieved or it will sink, making retrieval difficult. At best, the straw is highly inefficient and much oil passes through it and soaks into the beach sand. The oil soaked sand must then be picked up and taken to waste disposal and fresh clean sand deposited in its place. As is evident, the use of straw alone is not entirely satisfactory for several reasons.

The large fetch for wind and waves approaching coastal areas makes the use of the large skimmer system, with larger support craft for boom towing, such as LCM's or YTB tugs, and the large Class III boom a necessity. When deploying equipment in this situation, the primary concern should be to keep oil away from beaches, shoreline or harbor jetties. Figure 131 illustrates the situation. While operations proceed on the pickup of the oil slick, another effort should be the booming of sensitive beach areas and marinas where the slick may come ashore. Procedures for oil pickup are essentially the same as described previously for open water harbor areas with the exception that larger support craft may be required for personnel safety and operational effectiveness. Airborne monitoring of slick movement and direction of the skimming operation is essential.

Every effort should be made to contain and pick up oil when it is on the water. The cleanup and restoration of beaches or shoreline following an oil spill are major tasks and are typically labor-intensive, slow, costly processes. Oil slicks can be diverted away from beach areas by booming out past the surf zone. If it seems likely that oil slicks will strike a beach area, there are many procedures and various pieces of equipment which can be used for swift, efficient cleanup. In some situations the most effective procedure may be to "do nothing" because of one or more of the following reasons:

1) Less human/equipment disturbance will be created in the area, thus minimizing crushing or exposing marine life normally under rocks, etc.
2) Thin oil films will be washed off by wave action, sand scour, cobble rotation, etc., in several tide cycles.
3) Light oil fractions, once ashore, will evaporate rapidly.
4) Weathered oil in some instances will become "nontoxic," almost biologically inert, and may even serve as a substratum to some extent.

FIGURE 131: COASTAL AREA PROCEDURES

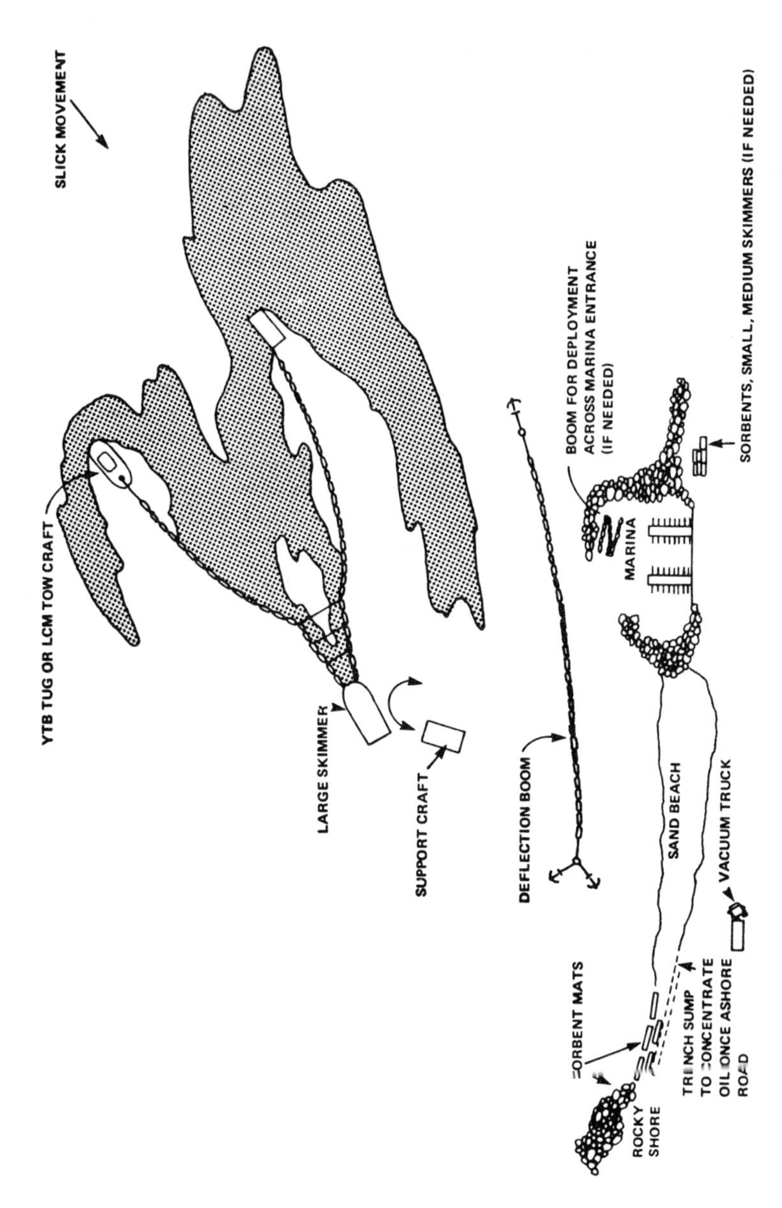

Source: Reference 9

5) Shoreline areas are relatively ecologically barren and isolated so that a spill has a minimum aesthetic impact.

Whether or not to launch a beach cleanup operation is a decision of the On-Scene Coordinator, after consultation with state wildlife and Environmental Protection Agency (EPA) personnel. In most cases involving large spills adjacent to recreational shorelines or ecologically sensitive areas, cleanup and restoration effort must be undertaken. Part of the immediate response to a spill should be a prediction of slick movement and the protection of any beaches or shoreline in its path by proper positioning of diversionary booms. If it is likely that oil will come ashore and must be physically removed, a beach cleanup operation plan should be initiated. A beach cleanup operation consists of three major phases:

1) Physical removal of contaminated material.
2) Disposal of contaminated material or its treatment to remove contained oil.
3) Replacement of removed material with uncontaminated material.

The type of cleanup equipment and techniques required for a given beach depends upon an on-site evaluation of the environmental sensitivity of the area and the type and extent of oil contamination. In any case, the best procedure is the physical removal of the oil together with as little beach material as necessary. The decision to use either manpower alone, or in combination with motorized earthmoving equipment, will depend upon the load-bearing qualities of the beach soil, the slope of the beach and access from paved roads. Regardless of the extent of equipment or manpower used, a most effective procedure to minimize the amount of contaminated beach material is to spread sorbents over rocky areas and on sandy or cobble beaches at and above the existing tide line before the arrival of the oil slick. Figure 131 shows suggested deployment of sorbents on various types of shoreline areas. Since the sorbents must be picked up after becoming soaked with oil, it is advisable to use sorbents in large flat strips or chunks surrounded by a large mesh nylon bag which can be easily lifted into disposal containers.

When using manpower alone to physically remove oil soaked sorbents and beach material, teams of no less than six men should be formed to work a specified length of shoreline. These work parties must be supported with sufficient containers, e.g., open top drums or wheelbarrows, to deposit contaminated material. Where possible, filled drums should be removed from the cleanup area by power equipment, such as front end loaders or cranes operating from nearby roadways. A primary Navy resource which should be considered for major cleanup operations is the Seabees, who are skilled in earthmoving projects requiring specialized field rigging on adverse terrain. Manual labor may be effectively used by digging trenches parallel to the surf line as shown in Figure 131.

Oil will be washed up on the beach and into the trench by wave action. Since the water has a much lower viscosity than oil, the water percolates into the sand leaving the oil behind to be manually scooped or pumped out into the open top drums or a waiting tank or vacuum truck. Oil residues will not stick to the wet sand, and these residues can be manually scrapped into the trench using a square-end shovel or other frabricated scraper.

Oil is very difficult to remove from large rocks or boulders. A procedure which has worked well on viscous weathered oil adhering to rocks is as follows. The hose from a vacuum truck is attached to a bell reducer. The small diameter pipe (one-and-one-fourth inches in diameter) results in a higher intake air velocity which aids in sucking oil from the rough and porous surface of rocks. Since oil does not adhere well to wet surfaces, a precautionary tactic which will aid in recovery by a vacuum pipe is to hose down rocky areas before the arrival of an oil slick. Large rocks and concrete surfaces can be cleaned with high-pressure, hot-water units (120°F, 1,200 psi) followed by sand blasting, if necessary, to remove oil stains.

The use of steam cleaners and blow torches to remove oil from rock or concrete surfaces should be avoided due to explosive spalling (cracking) of the surface which not only disfigures the surface but is also dangerous to operating personnel. Shoreline areas comprised of loosely packed cobbles can also be cleaned with high-pressure, hot-water units (120°F, 1,200 psi). The oil loosened by these units can be directed back into the water for pickup by skimmers; or, if a surf zone is present, the oil can be directed up on the beach into trenches dug parallel to the surf line, where the loosened oil can float to the top for pickup by a small skimmer (e.g., the skimmer of the Small Skimmer System) powdered by a pump or vacuum truck.

During a beach cleanup operation, which typically lasts over a number of days, attention should be paid to the effect of the wave and tidal action in removing oil from the shoreline. It has been observed in actual field operations that a contaminated section of shoreline rocks was cleaned just as effectively by wave and tidal action overnight as an adjacent section which had been worked by a crew of six men for seven hours.

Burning of oil on a contaminated beach is not effective since it is difficult to ignite the oil patches and maintain combustion. Complete combustion is usually not achieved and leaves a heavy, black residue of tar and charcoal which itself must be physically removed. From an effective operational point of view, burning of oil in place on beaches is not recommended. If judged necessary in a particular situation, burning operations are subject to the prior approvals prescribed by the National Contingency Plan.

The use of dispersants on beaches is expressly prohibited by the National Contingency Plan. Dispersants are not effective in beach cleanup operations because they loosen compaction of beach sand. Also, they cause oil to penetrate deeper into the beach, only to percolate up, over a long period of time, to the surface where the oil must be continuously picked up.

If the extent of contaminated shoreline is large and/or the beach area is firm and wide enough to support and allow maneuvering of motorized equipment, it is very cost-effective to use this equipment in place of manpower alone. The type of equipment to be used depends upon the type of beach soil, the oil penetration depth and access to paved roads. Fine sand and small gravel beaches are usually of sufficient compaction and gentle enough in slope to support motorized equipment.

Beaches composed of larger cobbles are usually of such loose compaction and steep slope that manpower, aided by dragline or front end loading machinery located off the cobble area, must be used. Removal of contaminated material from flat, sandy or gravel beaches can be accomplished by various types of mechanical equipment. The most effective type depends upon the depth of oil penetration.

1) For oil penetration up to one inch, the combined use of road graders and elevating scrapers is recommended.
2) For oil penetration from one to nine inches, the use of elevating scrapers only is recommended.
3) For oil penetration greater than nine inches, wheeled, front end loaders and bulldozers are effective, with the former being more efficient for firm ground and the latter more efficient for softer ground.

These procedures are summarized for various types of beaches in Table 25. Use of mechanized equipment in a given situation will depend upon beach access and the load-bearing properties and slope of the beach. Machinery may have to be modified with large, low-pressure tires. Operators of this type of equipment, either commercial or military (e.g., Seabees) should be brought to the site for technical advice before major equipment movements are undertaken.

The proper disposal of removed beach material is a complicated problem. Various techniques to treat the removed beach sand on site to remove oil and recover clean beach sand for replacement in the affected spill area have been evaluated in EPA-funded programs. Two of these, a fluidization and a froth flotation method, show promise but still require additional developmental work. The best current procedure is to minimize (as indicated in the preceding discussion) the amount of contaminated material and to dispose of this material in an approved sanitary landfill pre-selected and identified in the governing Navy Contingency Plan.

In an EPA-sponsored research study, selected earthmoving equipment was evaluated for use in the restoration of oil-contaminated beaches to determine the cost of the operations and the effectiveness in removing oil-contaminated sand and debris. Specifically, the objectives of the study included:

1) Determination of modifications and cost required to improve the capacity of the selected equipment.
2) Development of optimum operating procedures for each method.
3) Determination, through field testing, of the operating cost of each method evaluated.

The evaluation tests conducted indicated that several restoration procedures provided considerable savings in effort and cost over methods previously used. Full-scale demonstration tests of each restoration procedure were conducted to evaluate the operaing procedures and to determine the cost and effectiveness of each restoration procedure. As a result of the tests conducted in this study, the restoration procedures listed in Table 26 were recommended for use in the restoration of oil-contaminated beaches.

TABLE 25: METHOD/EQUIPMENT USED IN BEACH CLEANUP FOR ACCESSIBLE AREAS

Area of Contamination/ Type of Oil	Depth of Oil Penetration	. . . . Type of Beach . . . . Sand/Gravel	Cobbles	Boulders
Large area/ heavy oil		Mechanical removal:	Mechanical removal: Dragline or WFEL*	High pressure, hot water units (120°F, 1,200 psi) then sand blast
	Shallow (½" to 1")	Grader and ES* combination		
	Moderate 1" 9")	ES*		
	Deep (>9")	WFEL* for firm ground; bulldozer for soft ground		
Large area/ light oil		Mechanical removal: Harrow plow or beach cleaning machine	High pressure hot water units	High pressure, hot water units
Small area/ heavy oil		Manual removal and replacement of sand	Manual removal	High pressure, hot water units
Small area/ light oil		Manual removal: rake	High pressure, hot water units	High pressure, hot water units

*ES means elevating scraper and WFEL means wheeled front end loader.

Source: Reference (9)

TABLE 26: RECOMMENDED RESTORATION PROCEDURES

Restoration Procedure	Method of Operation
(A) Combination of motorized grader and motorized elevating scraper	Motorized graders cut and remove surface layer of beach material and form large windrows. Motorized scrapers pick up windrowed material and haul to disposal area for dumping or to unloading ramp-conveyor system for transfer to dump trucks. Screening system used to separate beach debris such as straw and kelp from sand when large amounts of debris are present.

(continued)

TABLE 26: (continued)

Restoration Procedure	Method of Operation
(B) Motorized elevating scraper	Motorized elevating scrapers, working singly, cut and pick up surface layer of beach material and haul to disposal area for dumping or to unloading ramp-conveyor system for transfer to dump trucks. Screening system used to separate beach debris such as straw and kelp, from sand when large amounts of debris are present.
*(C) Combination of motorized grader and front end loader	Motorized graders cut and remove surface layer of beach material and form large windrows. Front end loaders pick up windrowed material and load material into following trucks. Trucks remove material to disposal area or to conveyor-screening system for separation of large amounts of debris from sand.
*(D) Front end loader	Front end loaders, working singly, cut and pick up surface layer of beach material and load material into following trucks. Trucks remove material to disposal area or to conveyor-screening system for separation of large amounts of debris from sand.

*Use restoration procedures (C) and (D) only in instances where motorized elevating scrapers are not available. Operations of front end loaders on oil-contaminated beach areas should be kept to a minimum.

Source: Reference (9)

The surface conditions and topography of an oil-contaminated beach and the manner in which the oil has been deposited onto the beach will dictate the choice of equipment to be used and the operating procedures to be followed. The procedures described are those recommended for the restoration of relatively flat, sandy beaches contaminated under one or both of the following situations:

1) Beach material uniformly contaminated with a layer of oil up to the high-tide mark and/or deposits of oil dispersed randomly over the beach surface. Oil-deposit penetration is limited to approximately one inch.
2) Agglomerated pellets of oil-sand mixture or oil-soaked material, such as straw and beach debris, distributed randomly over the surface and/or mixed into the sand.

The procedures use the following equipment, singly or in combination: motorized graders, motorized elevating scrapers, front end loaders, conveyor-screening systems, and mulch spreaders.

Restoration activities can range from shoveling up asphaltic or tarry residues of the spill to applying hot water washes on rocky shorelines, to extensive manual or mechanized efforts to collect, reclaim and reestablish affected beach sand or trenching of estuaries to remove as much oil as possible. Restoration efforts have focused on beach areas, where procedures selected vary with the type, age and amount of spilled oil and the type of beach affected. Generally the lighter oils (less viscous) will penetrate the sand more readily, requiring uses of techniques that might range from harrowing in sorbent material to foster degradation; to sand pickup, reclamation and/or replacement. Treatment of beach sand to remove oil can only be justified where replacement beach sand is scarce and costly, because current methods for beach sand reclamation are either very expensive or not proven. Table 27 summarizes some of the techniques used to remove oil from shorelines.

TABLE 27: SUMMARY OF TECHNIQUES USED TO REMOVE OIL FROM SHORELINES

Type of Oil	Condition of Oil	Possible Removal Technique
Oil remaining at sea for long periods of time	Pebbles or streaks	Raking. Mechanical scraper.
Crude oil remaining at sea for long period of time	Water-oil emulsion	Dry screening. Mechanical scraper.
Fresh petroleum products (fuel oil and crude oil)	Less viscous oils penetrate the sand (most difficult to remove)	Apply straw. Disc harrow to expose surface to wind and sunlight. Dig collecting pools.
	More viscous–surface soiling oils	Manual (shovels). Mechanical grader to form rows. Scraper to collect the soil/oil.
	Either type of oil on rocky coastline	Wipe with sorbent. Wash with hot water.

Source: Reference (9)

A technique developed by *P.R. Scott; U.S. Patent 3,941,694; March 2, 1976; assigned to Shell Oil Co.* is one in which the adherence of oil to siliceous material, such as beach sand, is reduced by contacting the material with a blend of a primary long chain alcohol and an aliphatic solvent either before or after the oil contacts the siliceous material.

A technique developed by *M. Goldman; U.S. Patent 3,962,083; June 8, 1976; assigned to RRC International, Inc.* involves a combination of devices for treating a shoreline against contamination from an oil spill on adjacent waters.

It consists of a first vehicle for laying down a web of oil absorbing material on the shoreline and taking up the web after it has absorbed oil, a second vehicle for transporting the first vehicle and the webs to the area of the shoreline, and a third vehicle for storing oil extracted from the web by extraction mechanism on the first vehicle. An all-terrain vehicle for laying a web of oil absorbing material on a shoreline consits of a vehicle body supported by a plurality of wheels, means on the body for supporting a coiled web of the oil absorbing material and for guiding the web onto the shoreline, and means for coiling the web back onto the vehicle and extracting oil therefrom.

The web for use in absorbing oil deposited on a shoreline is made up of spaced layers of plastic netting confining shredded polyolefin fibers therebetween.

A device developed by *R.F. Wendt, J.R. Acker and N.R. Braton; U.S Patent 4,043,140; August 23, 1977* is a cryogenic beach cleaner.

The machine travels on the beach and sprays liquid nitrogen onto the contaminated area, thereby solidifying the oil and sand mixture so that the mixture can be separated from the underlying uncontaminated sand and be efficiently removed from the beach and transported to a remote site for disposal or further treatment.

DISPOSAL OF RECOVERED SPILL MATERIAL

As oil is recovered from the spill area, it may be pumped to a storage area or container where oil/water separation is initiated or continued. Gravity separation, centrifugation, and other separation techniques are available in commercial equipment. The concentrated oil is then removed to transport facilities and conveyed to recycle or disposal sites. This procedure is schematically represented in Figure 132.

FIGURE 132: DISPOSAL PROCESS

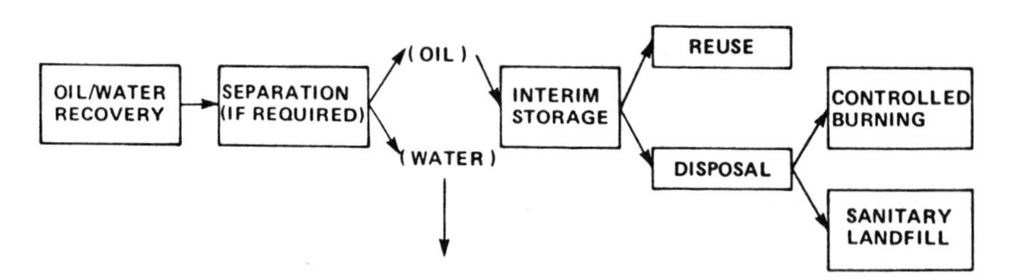

Source: Reference (9)

Once oil has been removed from the spill site, the major battle may have been won; but the conflict goes on, because unless the oil waste or oiled debris is properly disposed of by the Navy activity or contract operator, it can and will become a problem at the disposal site. The conventional disposal methods listed in Table 28, for example, may allow the oil to recontaminate surface or ground waters, degrade air quality, or present a fire hazard. Damages from any unauthorized disposal of oil by the Navy or its contractor may lead to litigation.

TABLE 28: SUMMARY OF SPILLED OIL DISPOSAL TECHNIQUES

Disposal Technique	Equipment Required	Advantages	Disadvantages	Comments
Controlled burning	Incinerator and appropriate feed-storage device.	Good volume reduction with a small amount of inert ash for disposal.	Could be the cause of smoke, odor and particulate emissions.	May need air pollution control system. May be used beneficially in fire-fighting training.
Open burning	A method of land application (pumps, nozzles, hoses, etc.).	Very economical and simple in its concept and operation.	Will create air pollution, especially smoke. May be a safety problem.	Should not be considered except under favorable meteorological conditions and when there is no other choice.
Sanitary landfil	Approved land area. Earth moving equipment.	Satisfactory disposal method.	May create safety hazard. May cause a leachate problem.	May not be acceptable, in many areas, to regulatory officials.
Burial	Acceptable land area. Earth moving equipment.	Simple technique. Economical disposal.	May create a leachate problem.	Subject to regulatory control.
Land spreading	Method of land application (pumps, nozzles, hoses, etc.).	Very inexpensive. Very simple.	Safety and fire hazard.	Subject to regulatory approval.

Source: Reference (9)

The disposal options are essentially limited to reuse; to disposal by soil cultivation techniques; to controlled burning; or placement in an "approved" sanitary land fill, which is not always readily available.

Costs associated with disposal are often the major economic impact of the spill.

Recent attempts to resolve some of the problems of oil and oily waste disposal have included the use of special land fill "cells" constructed in accordance with EPA guidelines. These cells have alternating layers of one foot of oiled debris and six inches of clean fill. The layers are entirely surrounded by about a two-foot earth seal. Monitoring wells operated on such a landfill cell in Rhode Island indicate no appreciable leaking after eight months.

Reuse of the oil collected from the spill is the preferred disposal mode where it is possible. The recovered oil may be "re-refined" and recycled for beneficial use. Re-refining facilities are not always readily accessible from spill sites but this possibility should always be considered. The Association of Petroleum Re-Refiners should be consulted to identify re-refining capability in a particular region.

A technique developed by *B.P. Martinez and M.D. Zeisberg; U.S. Patent 3,923,472; December 2, 1975; assigned to E.I. Du Pont de Nemours and Co.* involves the use of a filter containing melt-spun thermoplastic synthetic fibers for absorbing the oil from water. When the fibers become saturated with oil, they are heated at an elevated temperature until they become liquid. The resultant liquid is then drawn from the filter for use as fuel. The general elements of such a scheme are shown in Figure 133.

A mixture of oil and water from a setting basin **10** is directed through valves **2, 4** and **6** by pump **3** into two filter boxes **12, 14**. The filter boxes are arranged so that one is on line while the other is on a stand-by basis. For example, the oil-water mixture passes through the filter box **12** and the oil is absorbed by synthetic fibers (less than 100 denier per filament) in the filter box and retained, while essentially oil-free water is returned to the stream **16** through pipe **13** attached to one outlet of filter **12**. When the first filter **12** becomes saturated with oil, the oil-water stream is directed into the second filter box **14** by means of valves **4, 6,** i.e., by closing valve **4** and opening valve **6**.

After removal of the oil-saturated filter **12** from service, the entire filter is heated with steam by means of steam coil **18** in filter box **12** to convert the fiber-oil mixture into a liquid. Advantageously, the temperature may be maintained on the liquid to insure that it will remain pumpable. This liquid is then pumped by pump **20** through pipe **15** from the other outlet of filter **12** to a boiler where it is used to generate steam. The procedure is repeated using steam coil **18** to heat filter **14** when it becomes saturated with oil.

Synthetic fibers will absorb many times their own weight of oil, depending on type and composition. Typical numbers are 30-35 for polyester and 10-170 for polylefins. The oil-polymeric material mixture may be converted to a liquid by heating at atmospheric pressure or above, using temperatures in the range from 110°-300°C. The process may be used with a wide variety of oils, including crude oil, fuel oil, used motor oil and light oils such as textile finishes. The process also may be useful in the disposal of waste synthetic fibers.

FIGURE 133: Du PONT PROCESS FOR OIL SPILL DISPOSAL BY ABSORPTION IN THERMOPLASTIC FIBERS WHICH ARE THEN BURNED

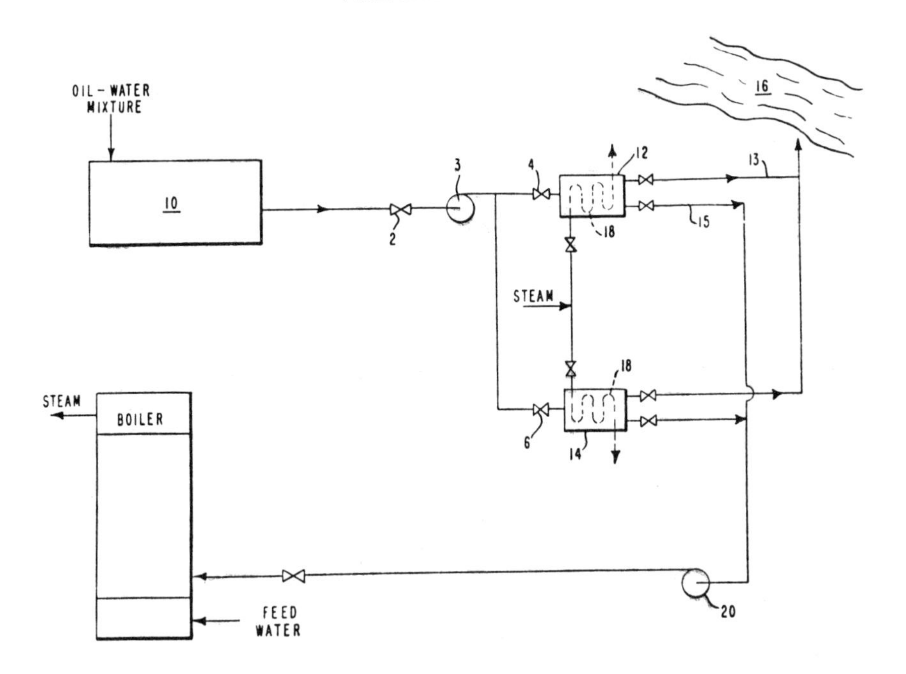

Source: U.S. Patent 3,923,472

A similar process can be envisioned for collection of oil from oil spills at sea. In this case, the fibers, after absorption of oil, would be transferred to a holding tank on board ship. Heating the oil-fiber mixture would convert the mass to a liquid where it could either be used as fuel on board ship or transported back to shore and pumped into a tank for subsequent use.

Bibliography

(1) Jones, H.R., *Pollution Control in the Petroleum Industry,* Park Ridge, N.J., Noyes Data Corp. (1973).

(2) Sittig, M., *Oil Spill Prevention and Removal Handbook,* Park Ridge, N.J., Noyes Data Corp. (1974).

(3) U.S. Environmental Protection Agency, *Development Document for Interim Final Effluent Limitations Guidelines and Proposed New Source Performance Standards for the Oil and Gas Extraction Point Source Category,* Report EPA 440/1-76/055a, Washington, D.C. (Sept. 1976).

(4) Travers, W.B. and Luney, P.R., "Drilling, Tankers and Oil Spills on the Atlantic Outer Continental Shelf," *Science* 194, 791-796 (Nov. 19, 1976).

(5) Wardley-Smith, J., *The Control of Oil Pollution,* London, Graham and Trotman, Ltd. (1976).

(6) Iammartino, N.R., "Oil Spill Control Nears for Two Pesky Problems," *Chemical Engineering,* 76,78,80 (May 10, 1976).

(7) D'Alessandro, P.L. and Cobb, C.B., "Oil Spill Control," *Hydrocarbon Processing,* 145-148 (March 1976).

(8) U.S. Environmental Protection Agency, *Development Document for Proposed Effluent Limitations Guidelines and New Source Performance Standards for the Petroleum Refining Point Source Category,* Report EPA 440/1-73/014, Washington, D.C. (Sept. 1973).

(9) Department of the Navy, *Oil Spill Control for Inland Waters and Harbors,* Report NAVFAC P-908, Alexandria, Va., Naval Facilities Engineering Command (Jan. 1977).

(10) Cavanaugh, E.C., Colley, J.D., Dzierlonga, P.S., Felix, V.M., Jones, D.C. and Nelson, T.P. (Radian Corp.), *Environmental Problem Definition for Petroleum Refineries, Synthetic Natural Gas Plants and Liquefied Natural Gas Plants,* Report PB-252,245; Springfield, Va., Nat. Tech. Information Service (Nov. 1975).

(11) Little, Arthur, D. Inc./Learning Systems, *Guide to Water Cleanup: Materials and Methods,* Cambridge, Mass. (1974).

(12) Milgram, J. (Sea Grant Program-MIT), "Being Prepared for Future Argo Merchants," Rockville, Md., National Oceanic and Atmospheric Administration (April 1977).

Company Index

Aer Corp. - 74
Amoco Production Co. - 56, 176, 312
Anheuser-Busch, Inc. - 240
Artec, Inc. - 310
Avco Everett Research Laboratory Inc. - 146
Bailey Meters & Controls Ltd. - 161
Ballast-Nedam Groep, NV - 284
Banque pour l'Expansion Industrielle "Banexi" SA - 255
Bennett Pollution Controls, Ltd. - 225
Bertin and Cie - 268
Bioteknika International, Inc. - 254
Bridgestone Tire Co., Ltd. - 192, 214, 226, 233, 291
British Petroleum Co., Ltd. - 235, 283
Cascade Industires, Inc. - 229
Chevron Research Co. - 176, 194, 300
Cities Service Oil Co. - 24, 249
Col-Mont Corp. - 238, 240
Commonwealth Scientific and Industrial Research Organisation - 252
Construzioni Battelli Disinquinanti SpA - 296
Continental Oil Co. - 148
Conwed Corp. - 333
Creusot-Loire - 244
Dresser Industries, Inc. - 21
John Dunn Agencies Ltd. - 254
E.I. Du Pont de Nemours and Co. - 350
Electrolysis Pollution Control Inc. - 246
Entreprise de Recherches et d'Activites Petrolieres-Elf - 36, 268
Exxon Production Research Co. - 313
Exxon Research and Engineering Co. - 76, 103, 114, 158, 177, 254
Firestone Tire & Rubber Co. - 312
Fischer & Porter Co. - 169
Fram Corp. - 332
GAF Corp. - 252
Gamlen Maintre SA - 199
B.F. Goodrich Co. - 134
Gotaverkin Oresundsvarvet AB - 153
W.R. Grace and Co. - 253
Gulf Research and Development Co. - 242
Hirt Combustion Engineers - 86
Hydromation Filter Co. - 336
Hydronautics, Inc. - 336
Hydrovac Systems International, Inc. - 293
Idemitsu Kosan Co., Ltd. - 243, 246
Imperial Chemical Industries, Ltd. - 235
Imperial Chemical Industries of Australia - 252

Institut Francais du Petrole, des Carburantes et Lubrifiants - 274
Kleber-Colombes - 204, 211
Kritbruksbolaget I Malmo AB - 241
Kyoei Senpaku Kogyo KK - 281
Lion Fat and Oil Co. - 243
Loctite (Ireland) Ltd. - 249
Marine Construction & Design Co. - 279, 302
Merritt Division of Murphy Pacific Marine Salvage Co. - 215
Metropolitan Petroleum Petrochemicals Co., Inc. - 208
Minnesota Mining and Manufacturing Co. - 284
Mitsubishi Denki KK - 226
Murphy Pacific Marine Salvage Co. - 279
National Marine Service, Inc. - 226, 302
New Zealand Ltd. - 252
NL Industries, Inc. - 62
Ocean Ecology Ltd. - 284
Offshore Devices, Inc. - 226
Oil Mop Inc. - 279, 329
Owens- Corning Fiberglas Corp. - 247
Pacific Pollution Control - 204, 224
Patents and Developments A/S - 274
Phillips Petroleum Co. - 89, 248
Pneumatiques Caoutchouc Manufacture - 204
Process and Pollution Controls Co. - 164
Products and Pollution Controls Co. - 336
RRC International, Inc. - 347
Salen & Wicander AB - 161
Salvage Oil Systems, Ltd. - 117
Sandco Ltd. - 298
Sanera Projecting AB - 197, 202
Seppic - 247
Shell Oil Co. - 78, 212, 216, 268, 276, 288, 291, 299, 300, 347
Snam Progetti SpA - 257
Societe Chimique des Charbonnages - 247
Societe Generale de Constructions Electriques et Mecaniques (Alsthom) - 312
Societe Grenobloise d'Etudes et d'Applications Hydrauliques (Sogreah) - 328
Societe Nationale Elf Aquitaine (Production) - 63
Sorbent Sciences Corp. - 245
Standard Oil Co. - 245
Submarine Engineering Associates, Inc. - 105, 212
Sun Oil Co. - 38
Sun Oil Co. of Pennsylvania - 82, 84, 272
Sun Research and Development Co. - 254
Sun Shipbuilding & Drydock Co. - 107, 136
Teijin, Ltd. - 246
Tenneco Oil Co. - 248
SA Texaco Belgium NV - 284
Texaco, Inc. - 58, 156, 167
Texas Instruments Inc. - 145
TRW Inc. - 271
Union Carbide Corp. - 248
Union Oil Company of California - 332
Uniroyal Inc. - 219, 220, 244
U.S. Secretary of Agriculture - 177
U.S. Secretary of the Navy - 228, 250, 256
U.S. Secretary of Transportation - 170
Westdeutsche Industrie-und Strassenbau-Maschinen GmbH - 240

U.S. Patent Number Index

3,783,129 - 238
3,783,284 - 144
3,783,621 - 188
3,783,622 - 189
3,785,972 - 248
3,786,637 - 192
3,786,773 - 132
3,788,079 - 175
3,788,481 - 285
3,788,984 - 240
3,789,988 - 328
3,792,589 - 194
3,794,175 - 268
3,794,583 - 329
3,795,315 - 197
3,796,656 - 268
3,798,158 - 240
3,798,911 - 197
3,798,913 - 199
3,800,950 - 240
3,800,951 - 268
3,802,201 - 199
3,802,456 - 138
3,803,848 - 202
3,804,661 - 240
3,806,727 - 146
3,807,177 - 202
3,807,178 - 204
3,807,617 - 204
3,810,546 - 268
3,810,832 - 329
3,810,835 - 176
3,811,285 - 204
3,812,973 - 241
3,815,682 - 32
3,815,742 - 312
3,816,359 - 248
3,817,687 - 74
3,818,708 - 205
3,819,514 - 242
3,821,109 - 248
3,822,789 - 271
3,823,828 - 286
3,831,387 - 117
3,831,756 - 271
3,832,966 - 117
3,834,538 - 312
3,835,049 - 249
3,836,004 - 272
3,838,775 - 272
3,839,869 - 208
3,839,870 - 33
3,842,270 - 148
3,843,306 - 252
3,843,517 - 253
3,844,743 - 56
3,844,941 - 176
3,844,944 - 329
3,844,950 - 272
3,845,196 - 24
3,846,290 - 254
3,846,335 - 240
3,847,815 - 273
3,847,816 - 286
3,848,417 - 208
3,849,989 - 211
3,850,206 - 76
3,850,807 - 312
3,852,964 - 211
3,852,965 - 211
3,852,978 - 211
3,853,767 - 274
3,853,768 - 274
3,855,152 - 243
3,856,667 - 254
3,859,796 - 211
3,859,797 - 212
3,860,019 - 58
3,860,519 - 287
3,862,040 - 329
3,862,963 - 243
3,863,694 - 138
3,865,722 - 249
3,865,730 - 288
3,867,817 - 214
3,868,824 - 215
3,869,385 - 177

3,870,599 - 254
3,871,956 - 254
3,871,957 - 254
3,875,998 - 36
3,878,094 - 332
3,879,951 - 38
3,880,758 - 274
3,882,682 - 216
3,883,397 - 254
3,884,803 - 332
3,885,418 - 150
3,886,067 - 244
3,886,070 - 247
3,886,750 - 216
3,888,086 - 219
3,888,766 - 244
3,890,224 - 252
3,896,312 - 170
3,899,213 - 170
3,900,421 - 255
3,901,254 - 62
3,901,818 - 244
3,902,998 - 245
3,903,701 - 220
3,904,528 - 245
3,905,902 - 274
3,906,732 - 134
3,907,684 - 274
3,907,685 - 291
3,908,443 - 153
3,909,416 - 293
3,912,635 - 274
3,915,859 - 333
3,915,864 - 296
3,916,674 - 156
3,917,528 - 245
3,919,083 - 249
3,919,112 - 255
3,921,407 - 222
3,922,225 - 296
3,922,860 - 224
3,923,472 - 350
3,923,661 - 271
3,924,412 - 225
3,924,449 - 158
3,928,205 - 276
3,928,206 - 296
3,929,631 - 249
3,929,644 - 298
3,931,019 - 336
3,933,632 - 246
3,939,663 - 226
3,941,692 - 102
3,941,694 - 347
3,943,720 - 226
3,947,360 - 298
3,948,770 - 103
3,957,009 - 104
3,958,521 - 98
3,959,127 - 256
3,959,134 - 177
3,959,136 - 291
3,960,722 - 246
3,962,083 - 347
3,962,875 - 226
3,963,617 - 175
3,964,295 - 161
3,965,004 - 107
3,965,920 - 161
3,966,597 - 246
3,966,603 - 162
3,966,613 - 298
3,966,614 - 298
3,966,615 - 300
3,971,220 - 226
3,973,430 - 164
3,977,969 - 249
3,979,175 - 78
3,979,291 - 226
3,980,566 - 246
3,983,034 - 300
3,984,987 - 136
3,985,020 - 114
3,986,959 - 277
3,990,975 - 279
3,992,291 - 336
3,992,292 - 279
3,995,440 - 81
3,996,134 - 235
3,996,975 - 84
3,998,060 - 228
3,998,733 - 235
4,000,618 - 313
4,004,453 - 167
4,006,079 - 247
4,006,082 - 279
4,006,086 - 281
4,009,985 - 86
4,010,012 - 21
4,010,103 - 336
4,010,779 - 89
4,011,159 - 247
4,011,175 - 243
4,014,795 - 302
4,016,726 - 228
4,021,344 - 283
4,022,694 - 336
4,030,304 - 229
4,031,707 - 250
4,032,438 - 283
4,032,449 - 284
4,033,137 - 232
4,033,869 - 302
4,033,876 - 302
4,035,289 - 63
4,038,182 - 284
4,042,495 - 257
4,043,131 - 233
4,043,140 - 348
4,045,671 - 167
4,046,691 - 284
4,048,854 - 169
4,049,170 - 233
4,049,554 - 300
4,052,306 - 284
4,052,313 - 284
4,053,406 - 310

Inventor Index

Acker, J.R. - 348
Ahmadjian, M. - 170
Alquist, H.E. - 248
Anusauckas, A.V. - 208
Appelblom, H.R. - 216
Aramaki, K. - 192, 214, 226, 291
Arceneaux, T.J. - 32
Atlas, R.M. - 256
Aulisa, G.D. - 272
Avey, R.L. - 268
Ayers, R.R. - 212, 216, 276, 288, 291, 298, 300
Azarowicz, E.N. - 254
Bagnulo, L. - 274
Bagot, H.E. - 277
Ballu, L. - 211, 204
Bartha, R. - 256
Battaerd, H.A.J. - 252
Belden, D.E. - 332
Bennett, J.A. - 225
Benson, R.A. - 205, 211
Bertram, L.E. - 245
Bhuta, P.G. - 271
Biechler, F.J. - 247
Blanchard, P.M. - 235
Bolger, B.J. - 249
Bourg, R.G. - 32
Braton, N.R. - 348
Brown, C.W. - 170
Brown, C.W. - 170
Bucheck, D.J. - 284
Bucker, E.R. - 254
Bunn, Co. - 238, 240
Byrd, G.H., Jr. - 254
Campbell, F.J. - 228
Canevari, G.P. - 76, 177
Case, C.E. - 245
Cavallero, L.T. - 74
Chang, C.H. - 146
Charpentier, P. - 36
Chastan-Bagnis, L. - 273
Chiasson, R.J. - 32
Cirulis, U. - 164
Clampitt, B.H. - 242
Cocjin, D.L. - 302
Cole, E.L. - 240
Conley, J.D. -332
Cooper, W.M., Jr. - 76
Corino, E.R. - 103
Cox, J.C. - 218
Creamer, C.E. - 248
Crisafulli, A.J. - 271
Dale, G.H. - 89
deAngelis, A.L. - 285
de Bourguignon, F.E. - 216
Degen, L. - 257
Degobert, P. - 274
Derzhavets, A.Y. - 286

Deslauriers, P.C. - 310
deVial, R.M. - 161
De Visser, E.V.M. - 284
DeYoung, W.J. - 244
Dille, R.M. - 167
Di Perna, J. - 104, 286
Dixon, A.L. - 253
Dubois, B.J. - 199
Dubois, E.M.R. - 268
Dubouchet, J.L. - 312
Durand, G. - 244
Elnicki, W.J. - 74
Fantasia, J.F. - 170
Fast, S.G. - 298
Favret, U. - 272
Ferm, R.L. - 176
Fisher, E.N. - 138
Fletcher, G.M. - 298
Fossberg, R.A. - 211
Fowler, L.L. - 336
Fruman, D.H. - 336
Funkhouser, S.P. - 277
Fusey, P. - 255
Galicia, F. - 274
Gallagher, J.J. - 188, 329
Gambel, C.L. - 189
Garber, D.C. - 107
Garcia, E.C. - 117
Gauch, G.J. - 220
Geist, J.J. - 232
Ghyselen, K.I. - 284
Gilchrist, R.E. - 248
Gill, J.A. - 62
Goldman, M. - 347
Goma, G. - 244
Goodrich, R.R. - 103
Graham, D.J. - 228, 271
Grant, M.G. - 162
Gratacos, J. - 63
Green, L.G. - 208
Gregory, M.D. - 148
Griffin, P.H., III - 21
Grimes, E.L. - 279
Gripshover, D.F. - 167
Guillerme, M. - 63
Gutnick, D. - 102
Gwyn, J.E. - 78
Hakansson, E. - 153
Halko, R.A. - 158
Hansel, W.B. - 84
Harwell, K.E. - 242
Hemphill, D.P. - 288, 291
Herzl, P.J. - 169
Hess, H.V. - 240
Hirs, G. - 336
Hirt, J.H. - 86
Hoegberg, R.G. - 274
Hoshi, H. - 243
Hoult, D.P. - 199
Imoto, T. - 246
Ingrao, H.C. - 170
in 't Veld, C. - 226, 293, 302
Irons, D.E. - 279
Irons, E. - 284
Jenkins, R.S. - 284
Johnson, R.L. - 271
Jones, J.W., Jr. - 242
Jones, L.W. - 56, 176, 312
Jordan, R.L. - 253
Kaneda, K. - 246
Katoh, T. - 246
Kattan, A. - 78
Kawaguchi, Y. - 214, 226
Kawakami, H. - 214, 226, 233, 291
Kermarrec, F. - 274
Kinase, T. - 226
King, J.A. - 249
Kirk, W.P. - 175, 298
Kitakoga, H. - 226
Koblanski, J.N. - 283
Kogan, P.G. - 286
Kondo, Y. - 192
Kriebel, A.R. - 150
Laman, J.R. - 312
Langlois, R.E. - 247
Larsson, A.A.R. - 272
Lecourt, E.J. - 310
Leonard, D.A. - 146
Lerch, D.W. - 279
Light, S.C., Jr. - 136
Lorentzen, A.P. - 284
Lucas, J.M. - 240
Lynch, P.F. - 170
Lynch, W.M. - 336
Maeda, I. - 243
Marbach, A. - 247

Marconi, W. - 257
Martineau, J. - 247
Martinez, B.P. - 350
Mason, C.M. - 38
Masongsong, A.M. - 302
Massei, O. - 296
McAllister, I.R. - 225
McCormack, K. - 144
McGrew, J.L. - 302
McKinney, R.W. - 253
McLellan, C. - 279
Meeks, D.G. - 235
Memoli, S.J. - 98
Mercuri, L. - 329
Meyer, J.E. - 252
Milgram, J.H. - 199, 226
Miller, G.H. - 156
Miranda, S.W. - 244
Miura, K. - 226
Mohan, R.R. - 254
Mohn, F. - 274
Monsan, P. - 244
Moreau, J.O. - 114, 158
Morgan, J.E. - 336
Morrison, C.R. - 247
Mourlon, J.-C.J. - 268
Mugino, Y. - 246
Muntzer, E. - 240
Muntzer, P. - 240
Murakami, K. - 243
Muramatsu, T. - 192
Nadaud, Y. - 274
Nagaoka, I. - 233
Neal, J.H. - 222
Nicks, P.F. - 235
Niemi, W.R. - 333
Nixon, J. - 254
Norton, M.G. - 235
Oberg, P.O. - 197, 202
Oddo, N. - 257
Ohkawa, H. - 246
Okamura, I. - 246
Okubo, K. - 226
Omori, A. - 246
Orban, J. - 245
Osborn, P.G. - 235
O'Sullivan, D.J. - 249
Oxenham, J.P. - 268
Pareilleux, A. - 244
Pelton, P.M. - 336
Petchul, R.K. - 300
Petchul, S.L. - 300
Peterson, E.C. - 246
Pichon, J. - 328
Pitchford, A.C. - 248
Pittman, A.G. - 177
Pollock, L.W. - 89
Preus, P. - 132, 188, 191, 211, 228, 243, 329
Pyler, R.E. - 240
Raymond, R.L. - 254
Rehm, W.A. - 21
Renfro, W.E. - 21
Renick, E.O., Jr. - 156
Reynolds, D.W. - 175, 298
Rhoades, V.W. - 24
Rhodes, H.M. - 329
Riedel, F.H. - 333
Robertson, G.W. - 219
Rolleman, J. - 117
Rolls, G.H. - 284
Rosenberg, E. - 102
Ross, S.L. - 250
Rudd, C.H. - 211
Ryan, M.M. - 33
Sayles, J.A. - 194
Schultz, L.A. - 310
Schwartz, M.G. - 284
Scott, P.R. - 347
Semenov, V.N. - 286
Sessions, B.J. - 233
Seymour, E.V. - 216, 276
Sharki, M.J. - 21
Shuffman, O. - 250
Shull, D.L. - 167
Sirvins, A. - 63
Slocumb, R.C. - 333
Smith, M.F. - 208
Stanley, W.L. - 177
Stearns, M.O. - 62
Stein, C. - 247
Stenstrom, B.H. - 161
Stern, L.E. - 241
Stewart, J.K. - 268
Stoddard, P.C. - 249
Stolhand, J.E. - 148

Stovall, L.F. - 313
Strain, P.J. - 296
Sturgeon, T. - 219
Sundin, G.H. - 333
Susuki, R. - 243
Suzuki, M. - 291
Tabachnikov, V.I. - 286
Talley, M.L. - 21
Tanksley, N.D. - 204, 224
Tayama, H. - 226
Teague, L.P. - 58
Tedeschi, E.T., Jr. - 134
Teng, J. - 240
Terhune, R.D. - 332
Tezuka, T. - 226
Thurman, R.K. - 215
Thyrum, P.T. - 167
Tingle, G.D. - 252
Tomikawa, M. - 246
Townsley, P.M. - 254
Tramier, B. - 63
Traylor, V.L. - 332
Tsukagawa, Y. - 233
Tsunoda, A. - 246
Tsunoi, I. - 281
Tyler, W.S. - 274
Valibouse, B. - 328
Van Stavern, M.H. - 167
Van't Hof, G. - 202
Voelker, R.P. - 310
Waren, F.A.O. - 296
Weatherford, D.J. - 287
Webb, M.G. - 283
Weiss, D.E. - 252
Wendt, R.F. - 348
Wengen, G.E. - 81
West, R.E. - 229
Whittington, J.M.C. - 252
Wilson, D.E. - 300
Winkler, A. - 249
Wittgenstein, G.F. - 138
Yano, I. - 226
Yocum, C.H. - 245
Yost, M.E. - 148
Zacharias, E.M., Jr. - 164
Zall, D.M. - 249
Zeisberg, M.D. - 350

Notice